普通高等教育“十四五”规划教材

低碳废水生物脱氮理论与技术

Theory and Technology of Biological Nitrogen Removal from Low Carbon Wastewater

陈重军　李大鹏　主　编
孙法迁　毕　贞　丁　静　副主编

中国石化出版社
·北京·

内容提要

本书是作者根据长期从事低碳废水生物脱氮相关科研和教学的经验编写而成，主要介绍了常见低碳废水生物脱氮工艺的基本原理、影响因素及工艺应用，并着重介绍了亚硝酸盐型厌氧氨氧化工艺的快速启动及影响因素、匹配型亚硝化的调控、颗粒污泥的形成及控制、工艺调控及应用等内容。书中既重视对低碳废水生物脱氮理论的介绍，也注意融合作者在低碳废水强化生物脱氮研究方面的经验和最新研究成果，注重知识的系统性，并力求做到重点突出、内容精练。

本书可作为环境工程专业学生的教材，亦可供从事低碳废水生物脱氮研究、生产、销售和应用的技术人员和管理人员参考。

图书在版编目（CIP）数据

低碳废水生物脱氮理论与技术：英文 / 陈重军，李大鹏主编 . -- 北京：中国石化出版社，2025.1.
（普通高等教育“十四五”规划教材）. -- ISBN 978-7-5114-7749-1

Ⅰ. ①X703.1

中国国家版本馆 CIP 数据核字第 202537T643 号

中国石化出版社出版发行

地址：北京市东城区安定门外大街 58 号

邮编：100011　电话：(010) 57512500

发行部电话：(010) 57512575

http://www.sinopec-press.com

E-mail：press@sinopec.com

北京富泰印刷有限责任公司印刷

全国各地新华书店经销

*

787 毫米 ×1092 毫米 16 开本 19.75 印张 474 千字

2025 年 1 月第 1 版　2025 年 1 月第 1 次印刷

定价：58.00 元

前言
PREFACE

“水污染控制工程”是环境类专业必修的一门专业基础课，其中废水生物脱氮是核心内容。为了响应国家“一带一路”倡议，为国外留学生提供适宜教材，并为国内学生对外输出奠定基础能力，特编写此教材。

本教材由2021年苏州科技大学研究生教材建设项目资助。本教材由苏州科技大学与浙江师范大学共同编写，编写组的专业教师具有废水处理、环境微生物、生态学等背景，均具有多年的教学经验，从工程视角入手，围绕低碳废水生物脱氮的技术原理、常见工艺、参数调控及微生物群落分析等基础知识和前沿热点，结合不同应用场景，提出技术的应用前景。本书可作为以工程应用创新能力培养为主的高等院校环境类专业学生的教材，可强化环境类专业学生废水生物脱氮理论的知识，并提升专业英语水平。

全书由陈重军和李大鹏主编并负责统稿，参加教材编写工作的有苏州科技大学陈重军（4~10章）、李大鹏（第1章）、毕贞（第11章、第12章）、丁静（第13章），以及浙江师范大学孙法迁（第2章）、董峰（第3章）、肖霄（第14章）。在编写过程中，听取了有关教师和学生的意见，得到了大家的热情关心和支持。同时本书从立项到顺利出版离不开中国石化出版社的帮助和支持，在此表示感谢！

本书编写过程中参考了大量相关教材和书籍，在此向有关作者表示感谢。由于编者水平和实践有限，疏漏之处在所难免，恳请专家和读者批评指正！

目录

CONTENTS

Chapter 1 Overview ······ 1

1.1 The nitrogen cycle in nature ······ 1

1.2 Nitrogen contamination risk ······ 2

1.2.1 Global nitrogen pollution situation ······ 3

1.2.2 Nitrogen pollution in China ······ 4

1.3 Nitrogen pollution control ······ 6

1.3.1 Increasingly higher emission standards ······ 6

1.3.2 Improvement of nitrogen pollution in natural water bodies in China ······ 7

1.4 Low carbon source wastewater treatment issue ······ 8

1.4.1 Typical low-carbon source wastewater ······ 8

1.4.2 The challenges of conventional nitrogen removal ······ 9

1.4.3 The challenge of carbon neutrality ······ 9

1.5 New biological nitrogen removal process ······ 10

1.5.1 Shortcut nitrification-denitrification systems ······ 10

1.5.2 Simultaneous nitrification-denitrification systems ······ 11

1.5.3 Anaerobic ammonium oxidation (Anammox) ······ 12

1.5.4 Sulfur-autotrophic denitrification ······ 13

1.5.5 Ferric ammonium oxidation: Feammox ······ 13

1.5.6 Anaerobic oxidation of methane (AOM) ······ 15

1.5.7 Hydrogen autotrophic denitrification ······ 15

Questions ······ 16

References ······ 16

Chapter 2 Partial nitrification and denitrification ······ 17

2.1 Introduction ······ 17

2.2 Advantages of partial nitrification and denitrification ······ 18

2.3 Microorganism involved in nitrification ······ 19
2.3.1 Morphologic and phylogenetic diversity of AOB ······ 19
2.3.2 Morphologic and phylogenetic diversity of NOB ······ 20
2.4 The main influencing parameters of partial nitrification ······ 21
2.4.1 pH, free ammonia (FA) and free nitrous acid (FNA) ······ 21
2.4.2 Temperature ······ 23
2.4.3 DO concentration ······ 24
2.4.4 Sludge retention time ······ 25
2.4.5 Toxic substances ······ 25
2.5 The main challenges for partial nitrification ······ 26
2.6 Application of partial nitrification and denitrification ······ 27
Questions ······ 30
References ······ 30

Chapter 3 Simultaneous nitrification and denitrification ······ 33
3.1 Introduction ······ 33
3.2 The mechanism and advantages of simultaneous nitrification and denitrification ······ 34
3.2.1 The mechanism of SND ······ 34
3.2.2 The advantages of SND ······ 36
3.3 Microorganism involved in SND ······ 37
3.3.1 Nitrifying bacteria within the biofilm ······ 38
3.3.2 Denitrifying bacteria in the biofilm ······ 38
3.3.3 Bacteria capable of heterotrophic nitrification and aerobic denitrification ······ 39
3.4 The main factors affecting SND ······ 40
3.4.1 DO ······ 40
3.4.2 pH ······ 40
3.4.3 C/N ······ 41
3.4.4 Sludge flocs ······ 41
3.5 Applications of SND ······ 41
3.5.1 Moving bed biofilm reactors ······ 42
3.5.2 Hybrid moving bed biofilm reactor–membrane bioreactor systems ······ 46
3.5.3 Aerobic granular sludge systems ······ 47
Questions ······ 51
References ······ 52

Chapter 4　Nitrite-based anaerobic ammonia oxidation (Anammox) ……54
4.1　Discovery of nitrite-based Anammox ……54
4.2　Nitrite-based Anammox stoichiometric ratio ……55
4.3　Nitrite-based Anammox microorganisms and central metabolism ……56
4.3.1　Diversity of functional bacteria for nitrite-based Anammox ……56
4.3.2　Central metabolic mechanism of nitrite-based Anammox ……58
4.4　Factors affecting Anammox of nitrite type ……62
4.4.1　Reactor impact ……62
4.4.2　Influence of environmental factors ……63
4.4.3　Substrate effects ……65
4.5　Nitrite type Anammox main process ……67
4.5.1　SHARON-ANAMMOX ……69
4.5.2　CANON ……70
4.5.3　Oxygen-Limited Autotrophic Nitrification/Denitrification (OLAND) ……72
4.5.4　Simultaneous partial Nitrification, Anammox and Denitrification (SNAD) ……73
4.5.5　Partial Denitrification - Anammox (PD/A) ……74
4.5.6　Denitrifying Anaerobic Methane Oxidation/Anammox (DAMO/A) ……75
Questions ……77
References ……77

Chapter 5　Matching nitrosation for Anammox ……79
5.1　Matched nitrosation reaction ……79
5.2　Process of realization of matched nitrosation ……80
5.2.1　Selection of reactor ……80
5.2.2　Alkalinity ……81
5.2.3　pH ……83
5.2.4　DO content ……85
5.2.5　HRT ……86
5.3　Low temperature for matched nitrosation ……86
5.4　Low substrate concentration for matched nitrosation ……89
5.5　Real-time reactor control ……90
5.6　Microbial populations in matched nitrosation systems ……93
Questions ……96
References ……96

Chapter 6　Start-up of Anammox ……98
6.1　The purpose and significance of quick startup ……98

6.2 Initiation factor control ······ 99
6.2.1 Selection of reactor ······ 99
6.2.2 Selection of inoculated sludge ······ 101
6.2.3 Selection of carrier ······ 103
6.2.4 Start–up load ······ 111
6.2.5 Low temperature start–up control ······ 112
6.3 Start–up process characteristics ······ 113
6.3.1 Start–up stage ······ 113
6.3.2 Stoichiometry ratio ······ 114
6.4 Microbial enrichment status ······ 115
6.4.1 Trends and extent of microbial enrichment ······ 115
6.4.2 Enrichment population categories ······ 116
6.5 Conclusions and prospects ······ 117
Questions ······ 117
References ······ 118

Chapter 7 Promoters and inhibitors of Anammox ······ 120
7.1 Exogenous additives for improving the Anammox process ······ 121
7.1.1 Metals addition ······ 122
7.1.2 Organic matter addition ······ 127
7.1.3 Inorganic matter addition ······ 129
7.1.4 Intermediates addition ······ 131
7.1.5 N–acyl–homoserine lactones addition ······ 132
7.2 Exogenous substances for inhibiting the Anammox process ······ 133
7.2.1 Different inhibitory factors of Anammox process ······ 133
7.2.2 Strategies to control the inhibition ······ 142
Questions ······ 145
References ······ 145

Chapter 8 Coupling of Anammox and denitrification ······ 147
8.1 Reaction mechanism of simultaneous Anammox and denitrification process ··· 147
8.1.1 Stoichiometry in simultaneous Anammox and denitrification ······ 147
8.1.2 Microbiology in simultaneous Anammox and denitrification ······ 149
8.2 Research on the coupling of Anammox and denitrification ······ 151
8.2.1 The coupling reaction of Anammox and denitrification ······ 151
8.2.2 Start–up of coupled Anammox and denitrification reactor ······ 152
8.2.3 Factors influencing the coupling of Anammox and denitrification ······ 153

8.3 Microbial community of the simultaneous Anammox and denitrification process ········ 159
8.4 Emerging extensions of simultaneous Anammox–denitrification process ··· 163
8.4.1 SAND ········ 163
8.4.2 ADSF ········ 167
8.4.3 SDA ········ 169
Questions ········ 171
References ········ 171

Chapter 9 Anammox Granular sludge ········ 173
9.1 Formation mechanism of Anammox granular sludge ········ 173
9.2 Factors influencing the development of Anammox granules ········ 175
9.2.1 Seed sludge ········ 176
9.2.2 Substrate concentration ········ 177
9.2.3 Hydraulic Retention Time (HRT) and Sludge Retention Time (SRT) ········ 178
9.2.4 Temperature and pH ········ 178
9.2.5 Hydraulic shear force and stirring speed ········ 179
9.2.6 Presence of inorganic ions ········ 180
9.2.7 Extracellular polymeric substances (EPS) ········ 180
9.2.8 Influence of nanoparticles present in wastewater ········ 181
9.3 Structure and microecology of Anammox granular sludge ········ 182
9.3.1 Properties of the Anammox granular sludge ········ 182
9.3.2 Microbial ecology of granular sludge ········ 188
9.4 EPS of Anammox granular sludge ········ 190
9.4.1 Compositional characteristics of EPS ········ 190
9.4.2 Influencing factors of EPS ········ 193
9.5 Application of Anammox granular sludge ········ 196
9.6 Main factors affecting EPS secretion and sludge granulation ········ 197
9.6.1 Organic concentration ········ 197
9.6.2 Nitrogen concentration and loading in the reaction system ········ 198
9.6.3 External mediator ········ 199
9.6.4 Other influencing factors ········ 199
9.7 Limitations of Anammox granulation ········ 200
9.7.1 Flotation ········ 200
9.7.2 Storage stability of Anammox granules ········ 201
9.7.3 Susceptibility to heavy metals ········ 202
9.8 Conclusions ········ 203

Questions …… 204
References …… 204

Chapter 10 Application of Anammox …… 205
10.1 Urban domestic sewage …… 205
10.1.1 Realization of a compatible Anammox process …… 205
10.1.2 Influencing factor …… 206
10.1.3 Treatment process …… 210
10.1.4 Engineering practice …… 214
10.2 Industrial wastewater …… 217
10.2.1 Landfill leachate …… 217
10.2.2 Monosodium glutamate wastewater …… 221
10.2.3 Rare earth wastewater …… 223
10.2.4 Pharmaceutical wastewater …… 225
10.3 Agricultural wastewater …… 227
10.3.1 Swine wastewater …… 227
10.3.2 Dairy wastewater …… 230
10.3.3 Aquatic aquaculture wastewater …… 231
Questions …… 234
References …… 234

Chapter 11 Sulfate-reducing ammonium oxidation (sulfammox) …… 236
11.1 Introduction …… 236
11.2 Anthology of sulfammox studies …… 237
11.3 Mechanism of sulfammox …… 238
11.3.1 The presence of organic carbon sources in the influent …… 238
11.3.2 No organic carbon sources in the influent …… 239
11.4 Characteristics of microbes in sulfammox …… 241
11.5 Environmental factors and operational conditions affecting sulfammox …… 243
11.5.1 Process medium and feeding options …… 243
11.5.2 NH_4^+/SO_4^{2-} ratio …… 244
11.5.3 COD addition …… 244
11.5.4 Temperature and pH …… 245
11.5.5 Spontaneity and oxidation-reduction potential …… 246
11.5.6 Other factors …… 246
11.6 Applicable reactors and reported efficiencies …… 246
11.7 Conclusions …… 247

Questions ··· 251
References ··· 251

Chapter 12 Fe(Ⅲ) reduction coupled to anaerobic ammonium oxidation (Feammox) ··· **253**
12.1 Introduction ··· 253
12.2 A collection of existing investigations regarding the Feammox process ··· 254
12.2.1 Anthology of Feammox studies ··· 254
12.2.2 Unveiling of Feammox in the environment and rate measurements ········· 255
12.2.3 Feammox microbial functions ··· 257
12.3 The artificial and natural factors that affect the growth of Feammox microorganisms ··· 260
12.3.1 The ferric iron ··· 260
12.3.2 Soil pH and redox potential ··· 260
12.3.3 Dissolved oxygen ··· 261
12.3.4 Temperature ··· 261
12.3.5 Nitrite and nitrate ··· 262
12.3.6 Carbon sources and electron shuttles ··· 262
12.3.7 In situ soil nutrient characteristics ··· 264
12.4 Prospective: use of Feammox for practical large-scale wastewater treatment ··· 265
12.5 Conclusions ··· 270
Questions ··· 270
References ··· 270

Chapter 13 Anaerobic Methane Oxidation (AOM) ··· **272**
13.1 Discovery and classification of anaerobic methane oxidation ············· 272
13.1.1 Discovery of anaerobic methane oxidation ··· 272
13.1.2 Classification of anaerobic methane oxidation processes ············· 273
13.1.3 Functional microorganisms of anaerobic methane oxidation ············· 274
13.2 Microbial metabolism of anaerobic methane oxidation ············· 277
13.2.1 Sulfate-dependent Anaerobic Methane Oxidation (SAMO) ············· 277
13.2.2 Denitrifying Anaerobic Methane Oxidation (DAMO) ············· 278
13.2.3 Metal-dependent Anaerobic Methane Oxidation (metal-AOM) ············· 280
13.2.4 Anaerobic methane oxidation with other new electron acceptors ············· 281
13.3 Physiological characteristics and ecological distribution of the anaerobic methanotrophic microorganisms ··· 282

13.3.1 Anaerobic methanotrophic archaea 282
13.3.2 Anaerobic methanotrophic bacteria 283
13.4 Enrichment of anaerobic methane-oxidizing microorganisms 285
13.5 Application potential of anaerobic methane oxidation 286
Questions 287
References 287

Chapter 14 Hydrogen-based denitrification 289
14.1 Introduction 289
14.2 Fundamental of H_2-based denitrification 290
14.3 Microorganism involved in H_2-based MBfRs 291
14.4 The key control factors 293
14.4.1 Membrane materials 293
14.4.2 Reactor types 295
14.4.3 Biofilm management 295
14.4.4 H_2 pressure 297
14.4.5 Nitrate loading 297
14.4.6 pH 298
14.5 Applications of H_2-based MBfRs 298
Questions 300
References 300

Chapter 1 Overview

Emphasis: The nitrogen cycle in nature, nitrogen contamination risk, nitrogen pollution control, low carbon source wastewater treatment issue, new biological nitrogen removal process.

Nut: New biological nitrogen removal process.

1.1 The nitrogen cycle in nature

Nitrogen is an integral structural component of proteins and nucleic acids found in all living cells. Although it is an essential building block for life, organisms can only utilize this element in its reactive forms. Molecular nitrogen (N_2) is the most abundant element in the atmosphere and accounts for 78% of all atmospheric gases. Unfortunately, N_2 is unavailable for use by most organisms because it contains a triple bond between the two nitrogen atoms, which very few organisms can break. Therefore, this abundant non-reactive form of nitrogen needs to be converted to useable or "reactive" nitrogen forms (i.e. NH_3 or NO_3^-) in order to support most life. The nitrogen cycle of the world is showed in Figure 1-1.

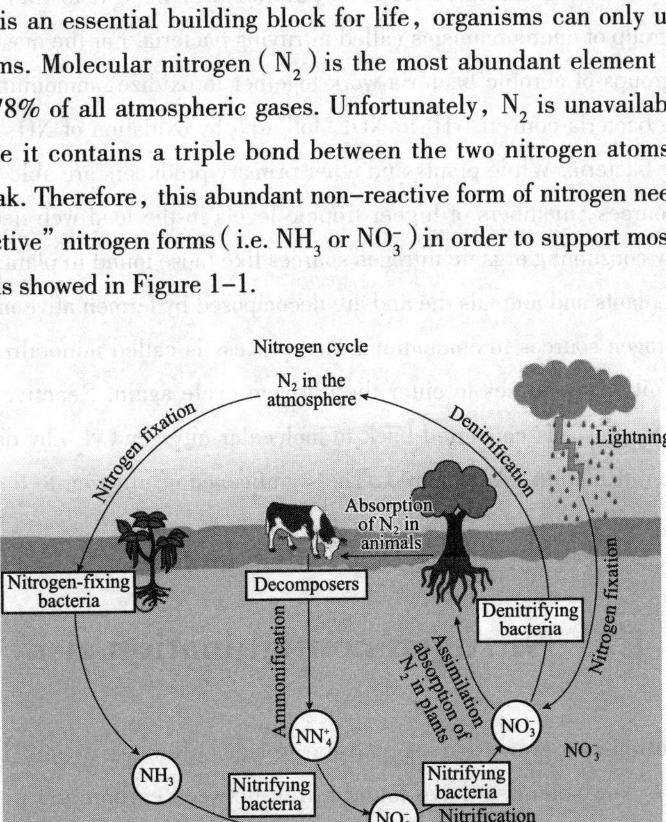

Figure 1-1 Nitrogen cycle diagram

There are a limited number of biological and non-biological ways that N_2 can be converted into reactive nitrogen. The majority of fixed nitrogen is formed biologically by a limited group of prokaryotes (from the domains bacteria and archaea) called diazotrophs that use a unique enzymatic pathway to convert N_2 to NH_3. Some of these prokaryotes form symbiotic partnerships with plants (i.e. legumes and alders) and animals (termites and bivalves), while others live free in terrestrial and aquatic environments. Non-biological sources of nitrogen fixation include lightning, combustion, industrial fixation, and biomass burning.

Each year, approximately 3–10 Tg of fixed nitrogen is formed from lightning. Lightning has enough energy to break the triple bond in N_2 and enables nitrogen atoms to combine with oxygen to form nitrogen dioxide (NO_2) and then nitric acid (HNO_3). Nitric acid is then carried to the ground in the form of rain, snow, or hail and can be utilized by organisms as a reactive inorganic nitrogen source. This source of fixed nitrogen is particularly important in environments where nitrogen-fixing organisms are scarce.

Once non-reactive nitrogen (N_2) has been converted to a reactive form, it can be taken up by plant roots and incorporated into proteins and nucleic acids. Most plants cannot directly utilize ammonia formed by nitrogen-fixing prokaryotes, rather they need it to first be converted into nitrate by another group of microorganisms called nitrifying bacteria. For the most part, nitrification requires that two groups of aerobic bacteria work together to oxidize ammonium to nitrate. First, ammonia-oxidizing bacteria convert NH_3 to NO_2^-, followed by oxidation of NO_2^- to NO_3^- carried out by nitrite-oxidizing bacteria. While plants and other primary producers are able to utilize inorganic reactive nitrogen sources, members of higher trophic levels in the food web need to receive their nitrogen supplies by consuming organic nitrogen sources like those found in plants or animals.

Eventually all plants and animals die and are decomposed by fermentative microorganisms that convert organic nitrogen sources to ammonium. This process is called mineralization and it allows inorganic reactive nitrogen sources to enter the nitrogen cycle again. Reactive nitrogen can also enter the atmosphere after it is converted back to molecular nitrogen (N_2) by denitrification or by anaerobic ammonium oxidation (Anammox). The significance of nitrogen to the biosphere and to cellular life is indisputable.

1.2 Nitrogen contamination risk

Nitrogen pollution is a growing danger to the world's climate, overall human health and the global economy. Now scientists have found a link between carbon and nitrogen cycles that could yield important insights into reducing nitrous oxide emissions and developing strategies for combatting global warming. While greater nitrogen availability has allowed global food production to keep pace with growing human demands, it can also impair water and air quality, reduce plant species diversity and exacerbate global warming. Concerns have been raised about whether the biosphere is able to soak up this extra nitrogen and what that means for the future.

1.2.1 Global nitrogen pollution situation

Although nitrogen is a necessary component of life, an excess of nitrogen can cause serious environmental problems. It is readily transported between air, water and soils and can have deleterious effects on all trophic levels of an ecosystem (Figure 1–2). Dramatic increases in the human population over the past two decades have resulted in production of an excess of nitrogen that is being released into the environment. Currently, humans are producing 210 Tg of nitrogen per year compared to only 7 Tg N $year^{-1}$ produced 2000 years ago through legume cultivation. Nitrogen fertilizer, which is used to increase crop yields, is the largest contributor to human–generated nitrogen (~100 Tg per year). However, fossil fuel combustion, the burning of biomass (when wood is used as fuel and forests and grasslands are cleared for agriculture), and the generation of sewage and industrial wastes have significantly increased nitrogen inputs into the environment.

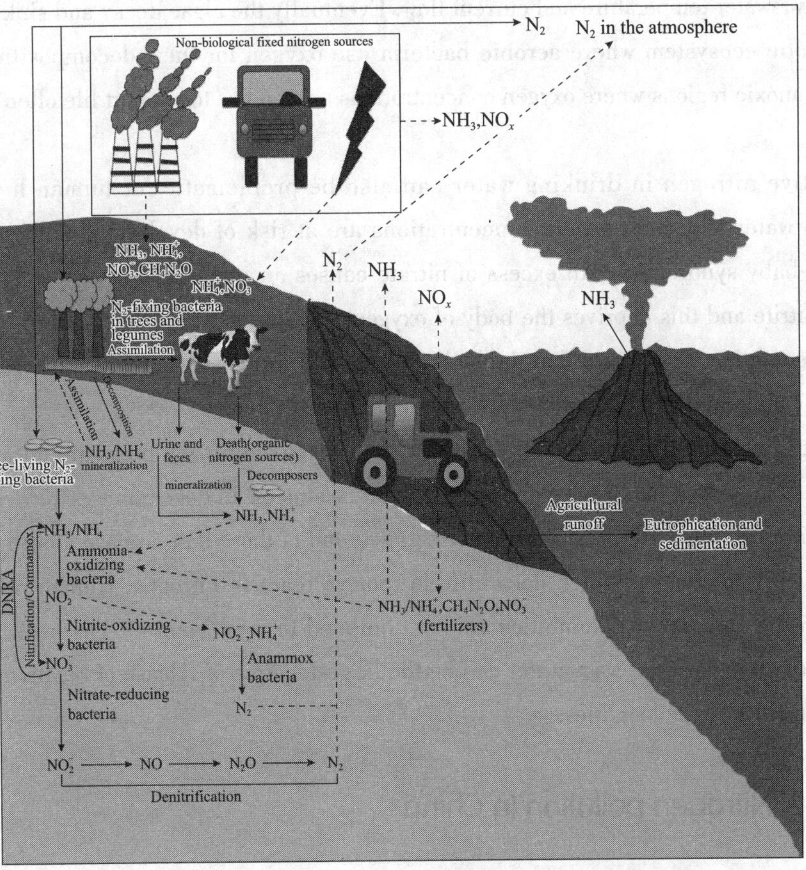

Figure 1–2 Global cycling of nitrogen in the biosphere

An excess of reactive nitrogen in the form of NO_x in the atmosphere promotes the creation of ozone and smog, which can trigger respiratory ailments and damage vegetation. There is also growing evidence showing a connection between reactive nitrogen in the air and several types of cancer and cardiopulmonary diseases. In addition, acid deposition of reactive nitrogen from the atmosphere causes acidification of soils, and inadvertent fertilization of trees and grasslands. This can result in unnatural growth rates, nutrient imbalances and ultimately decrease ecosystem health and biodiversity. In the Northeastern forests of the United States, soil acidification has been linked to a dieback of sugar maples. Acidification of the forest soils depleted the supply of the necessary nutrient Ca^{2+} and increased concentrations of phytotoxic aluminum (Al^{3+}), which can kill sugar maples at various stages of growth. Atmospheric deposition of reactive nitrogen especially in combination with sulfate also contributes to acidification of streams, which leads to a decline in populations of fish, amphibians, and invertebrates.

Reactive forms of nitrogen are highly soluble in water and tend to leach out of the soil and pollute groundwater and aquatic ecosystems. When it reaches aquatic environments like estuaries or other coastal ecosystems, it promotes the growth of algae resulting in eutrophication. Algal blooms formed by this fertilizer effect, block sunlight and interfere with aquatic plant and animal productivity, water temperature and current flow. Eventually the algae decay and sink to the bottom of the aquatic ecosystem where aerobic bacteria use oxygen for their decomposition leading to hypoxic or anoxic regions where oxygen concentrations are too low to support life often referred to as dead zones.

Reactive nitrogen in drinking water can also be problematic for human health. Infants exposed to water with high nitrate concentrations are at risk of developing methemoglobinemia (aka blue-baby syndrome). An excess of nitrate causes an infant's digestive system to convert nitrate to nitrite and this deprives the body of oxygen, leading to digestive and respiratory failures. Recent research has also shown that the body can convert nitrate ingested with drinking water into *N*-nitroso-compounds (NOC), which are potent animal carcinogens.

On a global scale humans excrete about 23 Tg of reactive nitrogen per year and more than half of this waste is released into the environment without treatment. In developing countries, fewer than 35% of their cities have any form of sewage treatment and of those that do have facilities, most only provide primary treatment, which does little to remove reactive nitrogen. Human populations are rapidly growing and many communities are not equipped to implement effective sewage treatment processes which means that wastewater can be the largest source of release of reactive nitrogen into the environment in some localities.

1.2.2 Nitrogen pollution in China

China only has a treatment rate of 45% and a study done by China's Ministry of Construction in 2005 showed that >3.7 billion tons of sewage are discharged daily into rivers and lakes untreated. Based on data from the 2nd Pollution Source Census conducted in 2017, shows that China's total emissions of total nitrogen, ammonia nitrogen and chemical oxygen demand in 2017 were 304.14, 96.34 and 2143.98 million tons. The emissions of total nitrogen, ammonia nitrogen and chemical oxygen demand from the seven river basins were 272.27, 85.64 and 1957.48 million tons (Figure 1–3). The seven river basins are the main sources of nitrogen pollution in China, with their emissions of total nitrogen, ammonia nitrogen and chemical oxygen demand accounting for 89.5%, 88.9% and 91.3% of China's total water pollutant emissions respectively. Therefore, the control of nitrogen pollutants in rivers needs to be further strengthened. In addition, China's aquaculture, plantation, livestock and poultry farming, industrial and industrial wastewater, rural domestic wastewater and urban domestic wastewater are the main sources of nitrogen emissions from human production activities. Total of 45.41 million tons of ammonia nitrogen and a total of 101.87 million tons of total nitrogen are emitted from urban domestic sewage pollutants, with the highest emissions for both indicators. The aquaculture industry discharges relatively less total nitrogen and ammonia nitrogen than other industries at 9.91 million tons and 2.23 million tons respectively, while rural domestic wastewater discharges a total of 24.5 million tons of ammonia nitrogen, second only to urban domestic wastewater in terms of ammonia nitrogen emissions (Figure 1–3). Industrial enterprises discharge less total nitrogen and ammonia nitrogen than rural domestic sewage and urban domestic sewage, while in the production of industrial enterprises, chemical products manufacturing, agricultural and food processing industries, textile industry are the top three industries in terms of ammonia nitrogen and total nitrogen emissions. In addition, there is a huge difference in total and ammonia nitrogen emissions from both the plantation and livestock farming industries, indicating that the organic or nitrate nitrogen content of pollutants discharged from livestock farming and plantation industries is high. Only industrial enterprise

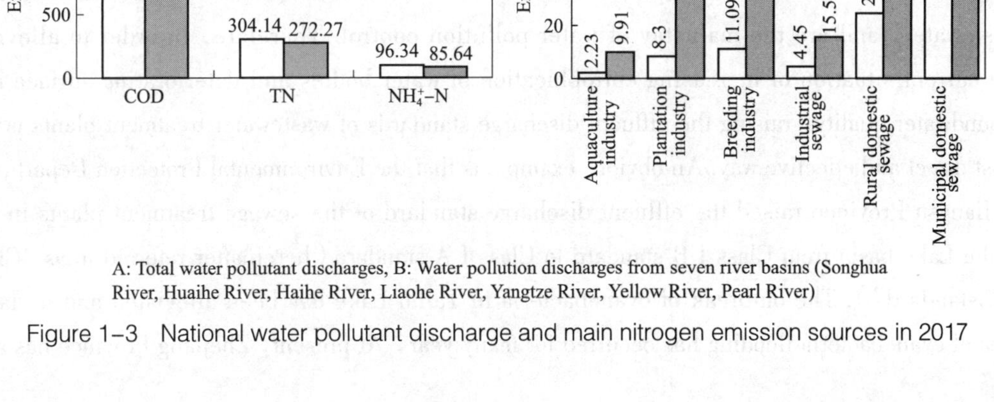

A: Total water pollutant discharges, B: Water pollution discharges from seven river basins (Songhua River, Huaihe River, Haihe River, Liaohe River, Yangtze River, Yellow River, Pearl River)

Figure 1–3 National water pollutant discharge and main nitrogen emission sources in 2017

sewage, rural domestic sewage and urban domestic sewage three total emissions of total nitrogen 162.09 million tons, sewage denitrification treatment can be a large reduction in nitrogen emissions is also an urgent problem, sewage denitrification treatment and new efficient denitrification process technology research and development, there is still a huge potential space.

1.3 Nitrogen pollution control

The total amount of naturally occurring reactive nitrogen has remained largely stable at the global scale since the industrial era, while anthropogenic reactive nitrogen production has continued to rise due to the impact of human activities, particularly the dramatic increase in the consumption of chemical nitrogen fertilizers and fossil fuels. In the global system, between the 1860s and the 1990s, the total amount of reactive nitrogen released into the environment from mineral combustion and fertilizer use increased from 15.3 to 156.0 Tg/y, exceeding the total reactive nitrogen naturally produced by terrestrial and marine ecosystems. The increase in anthropogenic reactive nitrogen has increased the transfer of reactive nitrogen to the atmosphere and water bodies, and the excess nitrogen flux has led to decreasing rates of denitrification removal and plant uptake, and a continued increase in nitrogen retention in environmental water bodies. Based on global trends in population growth, food demand, agricultural activity and energy use the total anthropogenic reactive nitrogen is predicted to reach 270 Tg/y by 2050, and human activities will likely lead to a doubling of the rate of the terrestrial nitrogen cycle and the overall global nitrogen cycle, which could lead to eutrophication of oceans, rivers, lakes and other water bodies and nitrate pollution of the groundwater environment caused by reactive nitrogen, becoming a major environmental problem that is difficult to solve globally. This will become a major environmental problem that is difficult to solve globally.

1.3.1 Increasingly higher emission standards

Urban wastewater treatment plants collect and treat most of the urban domestic and industrial wastewater, and are the mainstay of water pollution control. Therefore, in order to alleviate the current situation of increasing eutrophication of water bodies and deteriorating surface and groundwater quality, raising the effluent discharge standards of wastewater treatment plants is the most direct and effective way. An obvious example is that the Environmental Protection Department of Jiangsu Province raised the effluent discharge standard of the sewage treatment plants in the Taihu Lake basin from Class I B standard to Class I A standard (hereinafter referred to as "Class I A standard"). The outbreak of cyanobacteria in Taihu Lake has been alleviated and no large area of cyanobacteria flooding has occurred for many years. At present, Zhejiang Province has also proposed to implement Class A standard for the effluent quality of sewage treatment plants in the Taihu Lake and Qiantang River basins by the end of 2015, and the mandatory requirement for all urban sewage treatment plants in the province to implement Class A standard by 2017. Although, at present, only the Jiangsu and Zhejiang areas of the sewage treatment plant to upgrade the standard of the mandatory requirements, in the national scope of the sewage treatment plant to upgrade the standard is the general trend, which is an inevitable requirement. Compared with the Class B standard used by most wastewater treatment plants, the Class A standard sets higher discharge standards for total nitrogen, ammonia nitrogen and total phosphorus, e.g. TN, NH_4^+–N and TP are reduced from 20 mg/L, 8 mg/L and 1 mg/L for Class B standard to 15 mg/L, 5 mg/L and 0.5 mg/L for Class A standard. This places a higher demand on wastewater treatment plants for nitrogen removal and phosphorus removal. requirements.

1.3.2 Improvement of nitrogen pollution in natural water bodies in China

In China, the critical surface-water quality standard (1.0 milligrams of nitrogen per litre) was being exceeded in most provinces by the mid-1980s, and that current rates of anthropogenic nitrogen discharge (14.5 ± 3.1 megatonnes of nitrogen per year) to fresh water are about 2.8 times the estimated "safe" nitrogen discharge threshold (5.2 ± 0.7 megatonnes of nitrogen per year). Excessive nitrogen content poses a high risk of eutrophication in Chinese water bodies. Eutrophication is the process in which a water body becomes overly enriched with nutrients, leading to the plentiful growth of simple plant life. The excessive growth (or bloom) of algae and plankton in a water body are indicators of this process. Eutrophication is considered to be a serious environmental concern since it often results in the deterioration of water quality and the depletion of dissolved oxygen in water bodies. Eutrophic waters can eventually become "dead zones" that are incapable of supporting life, see Figure1–4. Eutrophication is common in Chinese lakes, especially in economically developed areas, see Figure 1–5.

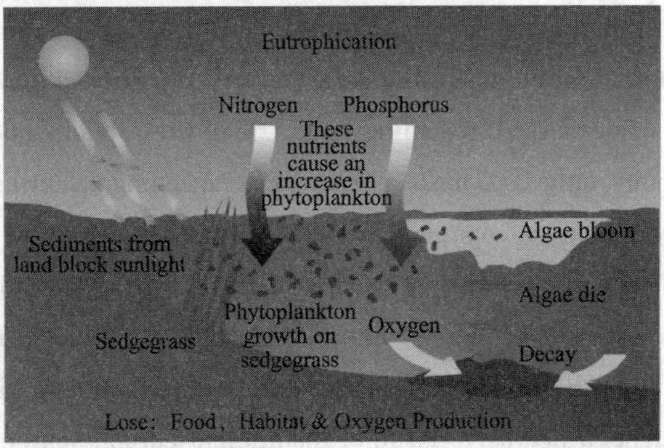

Figure 1–4 Process and mechanism of water eutrophication

Figure 1-5 Eutrophication is common in Chinese lakes and sea

1.4 Low carbon source wastewater treatment issue

1.4.1 Typical low-carbon source wastewater

Inadequate raw water carbon sources and low carbon to nitrogen ratios are common in all regions of China. For example, common southern urban wastewater is typical of low carbon source wastewater, with low organic matter concentrations (COD<200 mg/L) and low carbon to nitrogen ratios. According to statistical analysis, the influent water quality at 127 wastewater treatment facilities across China was assessed, revealing that 80% of these plants operated with a BOD_5/TN ratio below 3.6. Conversely, only 10% maintained a ratio exceeding 4. Further examination of urban wastewater treatment systems data, compiled by the Ministry of Housing and Construction, indicated that approximately 70% of the incoming wastewater at urban treatment plants nationwide exhibited a BOD_5/TN ratio under 4. Notably, there was a discernible geographic pattern, with northern city treatment plants recording a BOD_5/TN ratio of 4, while those in southern cities averaged a ratio of 3. This disparity highlights a pervasive issue of insufficient carbon sources in the influent water, impacting treatment efficacy across the country.

1.4.2 The challenges of conventional nitrogen removal

China's sewage plants are mostly adopted the traditional biological/chemical process to treat low-carbon source sewage, although it can meet the urban sewage discharge standards, the sewage plant operating costs are high, while the feed water is complex, and difficult to manipulate, but also to the sewage plant to deal with low-carbon source sewage increased the difficulty, there are several major problems.

1. High operating costs of sewage plants

As the traditional nitrification-denitrification process requires the consumption of a large amount of organic matter, to meet the denitrification, a large number of carbon sources need to be added to increase C/N. Sewage plants usually choose acetic acid, ethanol, glucose and other organic matter as an additional carbon source, its average market price is more than 2500 yuan. In recent years, the natural solid organic matter-based, non-toxic and inexpensive additional carbon source has become a hot spot. The researcher used rice husk, corn cob and aged rice as additional carbon sources to study the effects of different natural organic carbon sources on biological denitrification. In addition, the corn cob solid carbon source bioreactor was used to study the effect of SND on the treatment of low carbon to nitrogen ratio wastewater. If the valence conversion process of nitrogen can be reduced, the consumption of carbon sources can be theoretically reduced, thus reducing the waste of organic matter.

2. Complexity of feed water

To improve the efficiency of biological denitrification, most wastewater treatment plants use segmented influent and periodic changes in influent. On the one hand, the improved segmented feed has the advantages of making full use of carbon sources, high denitrification efficiency and easy operation and management; on the other hand, there are also the shortcomings of complex operation and difficult operation and control of the segmented feed process. In addition, the process requires multiple reactors to operate in series, with a large footprint and correspondingly higher operating costs.

1.4.3 The challenge of carbon neutrality

Currently, more than 86% of global energy consumption is still mainly dependent on traditional fossil fuels, which causes resource scarcity and even emission of high amounts of carbon dioxide (CO_2), resulting in a severe "greenhouse effect". Considering this situation, the concept of "carbon neutrality" has been put forward by 125 countries one after another. To achieve the goals of "carbon neutrality", two main strategies to reduce CO_2 emissions and develop sustainable clean energy can be adopted. China has also put forward a clear carbon neutral target, with CO_2 emissions peaking before 2030 and striving to achieve carbon neutrality by 2060 (Figure 1-6).

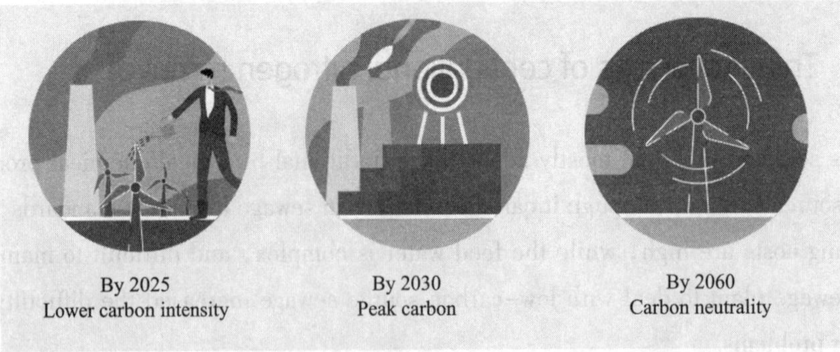

Figure 1-6 China's big climate goals

Carbon neutralization is a very complex system engineering, which requires a variety of technical channels and various efforts to reduce carbon. In the sewage treatment process, due to a large number of chemicals, as well as aeration, sludge dewatering equipment, water pumps, etc., the power consumption is very large. Therefore, while protecting the water environment, the sewage treatment industry is also a high energy consuming industry. At the same time, some sewage treatment processes are also accompanied by CH_4, CO_2 and other direct greenhouse gas emissions, and the problem of carbon emissions in the sewage treatment process cannot be underestimated. At present, nitrogen removal of wastewater with low carbon nitrogen ratio is generally an additional carbon source, which is contrary to the need for low-carbon treatment in the wastewater treatment industry. Therefore, it is urgent to develop the research and development of low-carbon biological nitrogen removal technology in wastewater.

1.5 New biological nitrogen removal process

A traditional wastewater treatment plant treats waste in three distinct steps prior to release into the environment. During primary treatment, solids are separated from the wastewater resulting in the formation of "sludge" which contains nearly 50% suspended solids. In the second step of treatment, biological and abiotic oxidation processes further purify sludge generated during the first step. There are three different secondary treatment approaches (filtration, aeration, and oxidation). Aeration is a commonly used method with the added benefit that it activates microorganisms that can help carry out the third step of treatment, the removal of phosphates and nitrogen from the wastewater. The processes referred to in the following sections are used for tertiary treatment to remove reactive nitrogen from wastewater prior to release.

1.5.1 Shortcut nitrification-denitrification systems

Shortcut nitrification-denitrification (also called partial nitrification-denitrification) systems, do not require as much energy or supplementation. In these systems, ammonia is only oxidized to nitrite, which significantly reduces energy needed for the aeration step. In addition, concentrations of electron donor required for nitrite reduction to N_2 are ~40% lower than those needed for complete denitrification from nitrate. In order for the shortcut nitrification-denitrification process to operate maximally, AOB need to be the dominant nitrifiers while NOB should be eliminated from the system. The most common method to control AOB and NOB concentrations is to keep free ammonia (FA) concentrations at levels that are too high for NOB but not AOB. NOB are significantly more sensitive to FA than AOB. For example, an ammonia-oxidizing *Nitrosomonas* species could grow at FA concentrations of 10–150 mg/L, while a nitrite-oxidizing *Nitrobacter* species was inhibited at FA concentrations as low as 0.1–1.0 mg/L. As a result, shortcut nitrification-denitrification processes are usually successful at treating waste water with high ammonia concentrations.

One of the most famous shortcut nitrification-denitrification processes is called the SHARON (single reactor high activity ammonia removal over nitrite) process. SHARON was originally developed for the removal of ammonia via the so called "nitrite route". In this process (patented), both autotrophic nitrification and heterotrophic denitrification take place in a single SHARON reactor system using intermittent aeration. The stoichiometry of the SHARON reaction is represented by the following equations:

$$2NH_4^+ + 3O_2 \longrightarrow 2NO_2^- + 4H^+ + 2H_2O$$

$$2NO_2^- + 4.8g\ COD + 2H^+ \longrightarrow N_2 + 1.8g\ sludge$$

$$2NH_4^+ + 3O_2 + 4.8g\ COD \longrightarrow N_2 + 2H^+ + 1.8g\ sludge$$

Methanol is added to stimulate denitrification, which helps keep the pH of the system neutral by compensating for acidification caused during the nitrification phase. Although the process is not suitable for all wastewater due to a high temperature dependency, the SHARON process is ideally suited to remove nitrogen from waste streams with high ammonia concentrations (>500 mg/L) and elevated temperatures.

1.5.2 Simultaneous nitrification-denitrification systems

Although shortcut nitrification-denitrification processes are quite effective in certain instances, the costs associated with external carbon addition for wastewaters lacking sufficient organics for complete denitrification, as well as the problems associated with the unstable operational conditions and the limitation of wastewater characteristics (e.g. high ammonia concentrations and high temperatures), have driven the development of alternative processes.

Many studies have shown that nitrification and denitrification can occur concurrently in one reactor when dissolved oxygen (DO) concentrations are low in simultaneous nitrification and denitrification (SND) systems. In these reactors, DO concentrations (usually <0.5 mg/L) are kept at concentrations that permit the simultaneous activity of denitrifying and nitrifying bacteria. There are two groups of denitrifiers in these systems: aerobic denitrifiers and microzone denitrifiers. Aerobic denitrifers can use nitrate and oxygen simultaneously as terminal electron acceptors, and microzone denitrifiers carry out anaerobic denitrification in anoxic zones within biological flocs or biofilms.

Many previous studies have shown that certain heterotrophic microorganisms can carry out denitrification in aerobic environments with relatively high DO concentrations. Complete denitrification at high DO concentrations was first observed in *Paracoccus* denitrificans when it produced dinitrogen gas from ammonium and/or nitrite and nitrate in well-mixed aerobic cultures. Other heterotrophic bacteria that have been shown to denitrify under aerobic or microaerophilic conditions include *Magnetospirillum magnetotacticum* and *Pseudomonas stutzeri* SU2. The aerobic denitrification process may offer several potential advantages because it occurs directly in aerated bioreactors in the presence of readily biodegradable organics. However, the respiratory pathway has yet to be studied in depth.

Some autotrophic nitrifiers can also carry out simultaneous nitrification and denitrification. For example, *Nitrosomonas eutropha* can use the nitrite it forms during nitrification as an electron acceptor for growth with hydrogen. Studies have shown that denitrification by *N. eutropha* can also be stimulated by adding gaseous N_2O under anaerobic conditions. Hydroxylamine is formed as an intermediate during ammonia oxidation to nitrite and *N. europaea* can grow mixotrophically on ammonia and hydroxylamine. In the absence of oxygen, hydroxylamine is oxidized with nitrite as electron acceptor, while nitrous oxide is produced.

1.5.3 Anaerobic ammonium oxidation (Anammox)

The Anammox process, like the traditional nitrogen removal processes (nitrification and denitrification), is mediated by a complex group of microorganisms. Since Anammox bacteria are obligate anaerobes, they need an anaerobic environment with a low oxidant-reduction potential (ORP). An increase of approximately 50% in the basic ammonia-oxidation bacteria (AOB), ammonia-oxidizing archaea (AOA) and nitrite-oxidizing bacteria are the key bacteria involved in the Anammox process.

Under anaerobic conditions, the Anammox bacteria oxidize ammonia to nitrogen gas by using nitrite as an electron acceptor, as shown in the equation below.

$$NH_4^+ + 1.32NO_2^- + 0.066HCO_3^- + 0.13H^+ \longrightarrow 1.02N_2 + 0.26NO_3^- + 0.066CH_2O_{0.5}N_{0.15} + 2.03H_2O$$

The Anammox process, which is a relatively novel biological nitrogen removal (BNR) process has attracted a lot of attention over the years for good reason. This is largely due to its high rate cost efficiency and energy conservation, and its ability to treat wastewaters with low carbon to nitrogen ratio (>2.5) without any external carbon source supply. Similarly, it ensures low energy costs for aeration due to the low oxygen consumption, results in lower N-oxides production, saves 90% of operational costs in sludge disposal, creates a 90% reduction in greenhouse gas emissions and has relatively small space requirement. In addition, the Anammox bacteria has the ability of reactivation after 2–5 days of being re-exposed to suitable operating conditions after it had been dormant due to adverse conditions. As such, the Anammox process presents a more sustainable way of treating wastewater.

1.5.4 Sulfur-autotrophic denitrification

In sulfur-autotrophic denitrification, reduced sulfur compounds (i.e., sulfide, elemental sulfur, thiosulfate, tetrathionate or sulfite) are used as electron donors for nitrate reduction. Stoichiometric equations of denitrification with sulfide and thiosulfate as electron donors are:

$$14NO_3^-+5FeS_2+4H^+ \longrightarrow 7N_2+10SO_4^{2-}+5Fe^{2+}+2H_2O$$

$$8NO_3^-+5S_2O_3^{2-}+H_2O \longrightarrow 4N_2+10SO_4^{2-}+2H^+$$

$$5S^0+6NO_3^-+2H_2O \longrightarrow 3N_2+5SO_4^{2-}+4H^+$$

Autotrophic denitrification with elemental sulfur has been studied extensively and its high denitrification efficiency compares well with that of heterotrophic denitrification. As a result, several pilot-scale experiments in China have started using this approach to remove nitrogen from municipal wastewater. However, the low solubility of sulfur compounds, the production of sulfates and the need to add limestone for pH adjustment limit its applicability (Figure 1-7).

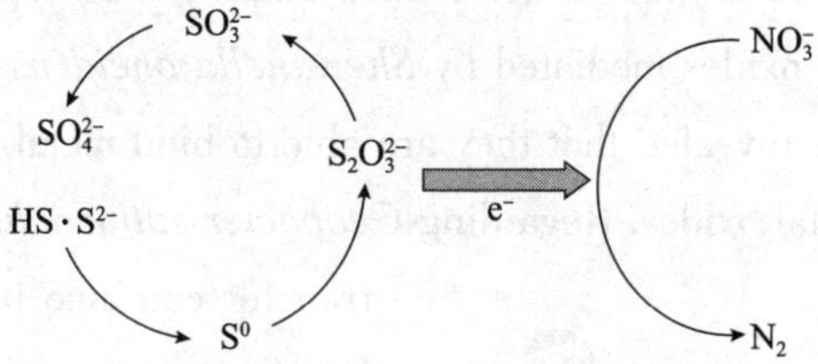

Figure 1-7 The sulfur-autotrophic denitrification process

1.5.5 Ferric ammonium oxidation: Feammox

In 2005, the oxidation of NH_4 to NO_2^- coupling with dissimilatory iron reduction anaerobically in wetland soils was firstly reported, and the authors pointed out that this biological process potentially plays a significant part in the nitrogen cycle in soils. Subsequently, Sawayama found that NH_4^+ was converted to NO_2^- in a fixed-bed anaerobic reactor added with Fe(Ⅲ)-EDTA and termed this process "Feammox". During the Feammox process, NH_4^+ is oxidized to N_2, NO_2^-, or NO_3^- and Fe(Ⅲ) is reduced to Fe(Ⅱ), as shown in the equation below. Among them,

Feammox to N_2 can proceed in a broad pH range while the production of NO_2^- and NO_3^- only takes place at pH <6.5. Additionally, compared with the Feammox reactions to form NO_2^- and NO_3^-, Feammox to N_2 yields more energy while requiring significantly less Fe(Ⅲ) to oxidize per mole NH_4^+. Thermodynamically, the oxidation of NH_4^+ to N_2 is more preferential than to NO_2^-/NO_3^-. Some researchers have verified this by applying the isotope labeling method to track the conversion pathways of NH_4^+ during the incubation of tropical highland soil slurries in the presence of Fe^{3+}, and demonstrated that the direct generation of N_2 through Feammox could be responsible for 47%–72% of the total N loss. Similarly, further research revealed that N_2 accounted for 67.0%–93.2% of the products of Feammox reaction. Additionally, NO_x^- produced by Feammox has been shown to have the ability to oxidize Fe(Ⅱ) to Fe(Ⅲ) under anaerobic conditions, termed NO_x^--dependent Fe(Ⅱ) oxidation (NDFO), which is crucial for the recycling of Fe in the ecosystems.

$$3Fe(OH)_3+5H^++NH_4^+ = 3Fe^{2+}+9H_2O+0.5N_2 \quad \Delta G=-245kJ/mol$$

$$6Fe(OH)_3+10H^++NH_4^+ = 6Fe^{2+}+16H_2O+NO_2^- \quad \Delta G=-164kJ/mol$$

$$8Fe(OH)_3+14H^++NH_4^+ = 8Fe^{2+}+21H_2O+NO_3^- \quad \Delta G=-207kJ/mol$$

To date, the biochemical mechanisms of ammonium oxidation in Feammox remain not fully understood, but several pathways of Fe(Ⅲ) reduction by Feammox microorganisms have been proposed. One pathway is electron transfer via direct cell-to-mineral contact. Many iron-reducing bacteria (IRB) rely on multiple cytochromes or some redox-sensitive proteins to transfer extracellular electrons through direct contact with Fe(Ⅲ) minerals, forming an extracellular electron transfer pathway that connects the intracellular respiratory chain with external Fe(Ⅲ) minerals. Several cytochromes have been found to transfer electrons directly to the Fe(Ⅲ) mineral surface. The OMC MtrC (OmcB) and OmcA, for example, were reported to directly participate in the reduction of Fe(Ⅲ) oxides mediated by *Shewanella oneidensis* MR-1. The investigations of purified MtrC and OmcA revealed that they are able to bind metal ox ides, like α-Fe_2O_3, and transfer electrons to the metal oxides. Regarding *Geobacter sulfurreducens*, extracellular electron transfer can also be mediated by OmcE, and OmcS. It is considered that OmcE and OmcS can loosely attach to the cell surface and act as intermediates for the transfer of electrons to electrically conductive type Ⅳ pili (T4P). And T4P is believed to transfer electrons directly to the iron oxide. In spite of these distinctions, both *Shewanella oneidensis* and *Geobacter sulfurreducens* could reduce metal oxides directly through cell surface-localized cytochromes (Figure 1-8).

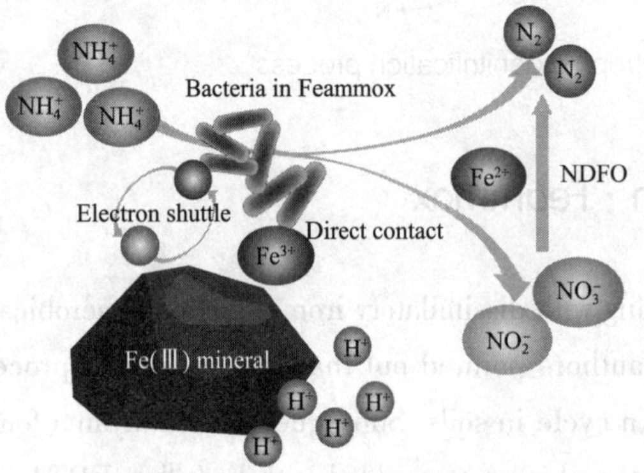

Figure 1-8 The Feammox process

1.5.6 Anaerobic oxidation of methane (AOM)

Anaerobic oxidation of methane (AOM) coupled with denitrification, also called nitrite/nitrate-denitrifying anaerobic methane oxidation (N-DAMO) process, is a novel bioprocess which uses methane as electron acceptor donor and nitrate or nitrite as electron acceptor. The N-DAMO process is considered as a widespread process. Two microorganisms, NC10 bacteria (N-DAMO bacteria) and ANME archaea (N-DAMO archaea), were detected in a NDAMO culture which was obtained by Raghoebarsing et al. after 16 months of enrichment in 2006. A year later, Ettwing et al. found that N-DAMO bacteria could achieve nitrite-dependent DAMO (Nitrite-DAMO) process without N-DAMO archaea, which made the role of N-DAMO archaea in the N-DAMO enrichment unclear. The selective effect of nitrogen on the N-DAMO microbial community structure was subsequently confirmed. It is widely believed that N-DAMO archaea was significant for nitrate reduction, while N-DAMO bacteria played an important role in nitrite reduction. N-DAMO bacteria could achieve Nitrite-DAMO reaction with nitrite as electron acceptor in a new "intra-aerobic" pathway. DAMO bacteria reduced nitrite to nitric oxide and then achieved methane oxidation via using the in situ produced oxygen through the dismutation of nitric oxide to oxygen and nitrogen. The finding provided a reasonable explanation for the pathway of N-DAMO process.

The microbial process of denitrifying anaerobic methane oxidation (DAMO) where methane is oxidized anaerobically to provide electrons for denitrification links the nitrogen and carbon cycles. DAMO is a new sustainable wastewater treatment process that does not require any external dosing of electron donors for denitrification. Rather, the electron donor, methane, is produced by methanogens during the anaerobic digestion of disposal sewage sludge. One obvious advantage to this approach is that less sludge is generated during wastewater treatment. This process is still in its infancy but its future applications look very promising.

1.5.7 Hydrogen autotrophic denitrification

Autotrophic denitrification using hydrogen as an electron donor is another promising method being studied by numerous researchers. Hydrogen is a clean energy gas and hydrogenotrophic denitrification does not generate any toxic intermediates. The process proceeds by the following reaction:

$$2NO_3^- + 5H_2 \longrightarrow N_2 + 4H_2O + 2OH^-$$

Hydrogen is one of the most thermodynamically favorable electron donors for nitrate-based respiration, and its high diffusivity through biofilms allows enhanced nitrate removal. Nevertheless, the use of hydrogen in engineered denitrification systems is limited by its relatively high cost, low solubility, and explosive properties during handling and storage.

Questions

[1] What is the main nitrogen cycle in nature?

[2] Nitrogen pollution has become an important environmental problem facing humanity. What is the current situation of nitrogen pollution facing the world and China?

[3] What are the problems faced by deep denitrification of low carbon to nitrogen ratio wastewater?

[4] What are the common new biological nitrogen removal processes currently used for low carbon to nitrogen ratio wastewater? What are the advantages and disadvantages of each?

[5] Briefly introduce the principle, advantages and disadvantages of the Anammox process.

References

[1] Chen C J, Wang J F, Zhang H Q, et al. Research progress on anaerobic ammonia oxidation wastewater treatment process and its practical application [J] . Journal of Ecology and Environment, 2014, 23 (3) : 521-527.

[2] Adams M, Xie J X, Kabore A J W, et al. Research advances in Anammox Granular Sludge : a review. Critical Reviews in Environmental Science and Technology, 2022, 52 (5) : 631-674.

[3] Adams M, Xie J X, Xie J W, et al. The effect of carrier addition on Anammox start-up and microbial community : a review [J] . Reviews in Environmental Science and Bio/Technology, 2020, 19 : 355-368.

[4] Galloway, James N. The global nitrogen cycle : changes and consequences [J] . Environmental Pollution, 1998, 102 (1) : 15-24.

[5] Hu K, Li W, Wang Y, et al. Novel biological nitrogen removal process for the treatment of wastewater with low carbon to nitrogen ratio : a review [J] . Journal of Water Process Engineering, 2023, 53 : 103673.

[6] Sancho I, Lopez-Palau S, Arespacochaga N, et al. New concepts on carbon redirection in wastewater treatment plants : a review [J] . Science of the Total Environment, 2019, 647 : 1373-1384.

[7] Fu X, Hou R, Yang P, et al. Application of external carbon source in heterotrophic denitrification of domestic sewage : a review [J] . Science of the Total Environment, 2022, 817 : 153061.

Chapter 2　Partial nitrification and denitrification

Emphasis: Partial nitrification, denitrification, free ammonia, free nitrous acid, DO, sludge retention time, application.

Nut: Control technology.

2.1　Introduction

The biological nitrogen removal (BNR) process, which includes nitrification and denitrification, has proven to be a more effective and cost-efficient method for reducing nitrogenous compounds in wastewater compared to physicochemical processes. In recent years, there have been significant advancements in BNR processes aimed at reducing energy input. These advancements include partial nitrification and denitrification, simultaneous nitrification and denitrification (SND), anaerobic ammonium oxidation (Anammox), and completely autotrophic nitrogen removal over nitrite (CANON).

Nitrification, a vital component of BNR, is a two-step biological oxidation process [Eq. (2-1)]. The conversion of NH_4^+-N to NO_2^--N is the rate-limiting step carried out by ammonia-oxidizing bacteria (AOB), while the subsequent oxidation of NO_2^- to NO_3^- is rapidly performed by nitrite-oxidizing bacteria (NOB) in the presence of molecular oxygen [Eq. (2-2)] (Figure 2-1).

$$NH_4^+\text{-}N + 1.5O_2 \xrightarrow{AOB} NO_2^-\text{-}N + 2H^+ + H_2O \qquad (2\text{-}1)$$

$$NO_2^-\text{-}N + 0.5O_2 \xrightarrow{NOB} NO_3^-\text{-}N \qquad (2\text{-}2)$$

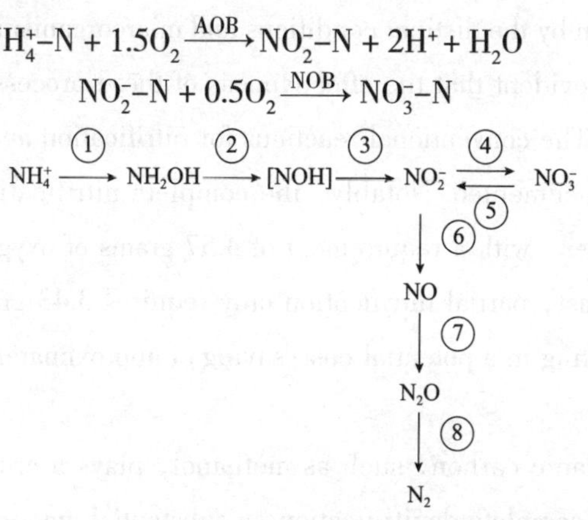

Figure 2-1　Possible metabolic pathway for nitrification and denitrification

During nitrification, ammonium is oxidized to hydroxylamine (step 1) . Hydroxylamine is oxidized to nitrite (steps 2 and 3) . Nitrite is converted to nitrate (step 4) . During denitrification, nitrate is reduced to nitrite (step 5) which is converted to gaseous NO, N_2O and N_2 (steps 6, 7 and 8) .

The conversion of NH_4^+–N to NO_2^-–N involves two steps mediated by hydroxylamine (NH_2OH) as an intermediate species [Eqs. (2–3), (2–4)]. In the initial step, NH_3 rather than NH_4^+–N, serves as the primary substrate, undergoing oxidation to NH_2OH facilitated by ammonia monooxygenase (AMO) catalysis. Subsequently, NH_2OH is further oxidized to NO_2^-–N through the catalytic action of hydroxylamine oxidoreductase (HAO) [Eq. (2–4)]. Notably, one of the oxygen atoms in NO_2^-–N originates from water, while the other is sourced from an oxygen molecule [Eq. (2–5)]. The second step released four electrons, of which two compensate for the electron input of the first step, while the remaining two are passed to the terminal oxidase via an electron transport chain, thereby generating a proton motive force.

$$NH_3 + O_2 + 2H^+ + 2e^- \xrightarrow{AMP} NH_2OH + H_2O \qquad -120\ kJ/mol \qquad (2\text{–}3)$$

$$NH_2OH + \tfrac{1}{2}O_2 \xrightarrow{HAO} HNO_2 + 2H^+ + 2e^- \qquad -114\ kJ/mol \qquad (2\text{–}4)$$

$$NH_2OH + H_2O \longrightarrow HNO_2 + 4H^+ + 4e^- \qquad +114\ kJ/mol \qquad (2\text{–}5)$$

Regarding denitrification, the process of biological nitrogen removal encompasses the sequential conversion of NH_4^+–N to NO_3^-–N during nitrification, followed by the subsequent denitrification of nitrate nitrogen to gaseous nitrogen (N_2), which is ultimately released into the atmosphere. It has been established that nitrification is an aerobic process executed by autotrophic bacteria, whereas denitrification occurs under anaerobic conditions and is conducted by heterotrophic bacteria.

2.2 Advantages of partial nitrification and denitrification

The separation of nitrification and denitrification processes in either time or space is a well–established practice, driven by the distinct conditions and microorganisms involved in each process. Over time, it has become evident that the effectiveness of these processes is enhanced when they are carried out separately. The conventional reactions for nitrification and denitrification have been extensively studied and documented. Notably, the complete nitrification reaction necessitates a substantial amount of oxygen, with a requirement of 4.57 grams of oxygen per gram of ammonium nitrogen nitrified. In contrast, partial nitrification only requires 3.43 grams of oxygen per gram of ammonium nitrogen, resulting in a potential cost–saving of approximately 25% in terms of aeration expenses.

The utilization of organic carbon, such as methanol, plays a crucial role in the process of denitrification. To achieve complete denitrification, a substantial amount of methanol, specifically 2.47 g per gram of nitrate, is required. However, the addition of electron donors like methanol significantly contributes to the overall cost of implementing denitrification on a large scale. Notably, nitrite serves as an intermediate product in both nitrification and denitrification processes. In the context of nitrification, inhibiting the oxidation of nitrite (partial nitrification) can effectively shorten the denitrification step from nitrate to nitrite. In comparison to denitrification with nitrate, partial nitrification offers a potential reduction of up to 40% in the demand for electron donors. Additionally, denitrification rates for nitrite are 1.5 to 2 times faster than those for nitrate denitrification, potentially resulting in a theoretical reduction of sludge production by approximately 33% for partial nitrification and approximately 55% for denitrification.

2.3 Microorganism involved in nitrification

Nitrification involves two phylogenetically distinct groups of autotrophic bacteria, AOB and NOB. It is important to note that the role of Ammonia-Oxidizing Archaea (AOA) was not discussed in the context of partial nitrification. AOA are another group of ammonia-oxidizing microorganisms that are phylogenetically distinct from AOB and have been recognized for their contribution to ammonia oxidation in various environments. While AOB and NOB are typically the focus when discussing nitrification, AOA can also play a role in the nitrogen cycle and contribute to ammonia oxidation in certain conditions.

2.3.1 Morphologic and phylogenetic diversity of AOB

AOB can be distinguished by their cell morphologies, Gram-negative multilayered cell walls, and motility facilitated by flagella. Since the first isolation of AOB in 1890, five recognized genera of AOB have been reported, belonging to two phylogenetically distinct groups: the β- and γ-subclass of *Proteobacteria*. The β-subclass includes four genera of AOB, such as *Nitrosomonas* (e.g., *Nitrosococcus mobilis*), *Nitrosospira*, *Nitrosovibrio*, and *Nitrosolobus*, while the γ-subclass consists of only one cluster of *Nitrosococcus*. To date, a total of 25 AOB species have been cultured from various environments, with *Nitrosomonas* and *Nitrosospira* being the most extensively studied.

The characterized autotrophic AOB found in freshwater exclusively belong to the β-subdivision of *Proteobacteria* and are exemplified by *Nitrosomonas europaea*. These bacteria form a distinct group within the β-subdivision and are closely related to an iron-oxidizing bacterium (*Gallionella ferruginea*), the photosynthetic bacterium *Rhodocyclus purpureus*, and methylotrophic bacteria. On the other hand, *Nitrosococcus oceanus*, an ammonia oxidizer, is a marine species belonging to the γ-*Proteobacteria*, lineage.

Among the autotrophic nitrite-oxidizing bacteria, those most commonly studied belong to the α-subdivision of *Proteobacteria*, with *Nitrobacter winogradskyi* being a representative species. Other characterized chemolithoautotrophic NOB are phylogenetically widespread within the class *Proteobacteria*, occurring in the α, δ, and γ subdivisions. Phylogenetic analysis of the α–subdivision of *Proteobacteria* has revealed that *Nitrobacter winogradskyi* is most closely related to *Bradyrhizobium japonicum* and *Rhodopseudomonas palustris*.

2.3.2 Morphologic and phylogenetic diversity of NOB

While the discussion above focused on AOB for partial nitrification, NOB is also important for the process. NOB are more phylogenetically distinct and widespread than AOB within the *Proteobacteria*. Eight species of NOB have been cultured, and they are grouped into four phylogenetically distinct groups. The genera *Nitrococcus* and *Nitrobacter* are assigned to the α– and γ–subclass of *Proteobacteria*, respectively. The genus *Nitrospira* is in its own subdivision phylum *Nitrospira* and is closely related to the δ–subclass, with the first–deciphered complete genome of *Candidatus Nitrospira* defluvii.

NOB are Gram–negative and have four primary shapes. *Nitrobacter* was previously considered the most important NOB, but recent studies have shown that *Nitrospira* is more dominant and specialized in most WWTPs, drinking water, and soil systems. *Nitrospira* is considered a K–strategist with high substrate affinities and low maximum activity for nitrite and oxygen, while *Nitrobacter* represents R–strategists under limited substrate conditions. *Nitrobacter* is also able to utilize organic energy sources as part of its facultative autotrophs and anaerobes. Generally, *Nitrospira* was dominant under high inorganic carbon conditions, while *Nitrobacter* was dominant under low inorganic carbon conditions (Figure 2–2).

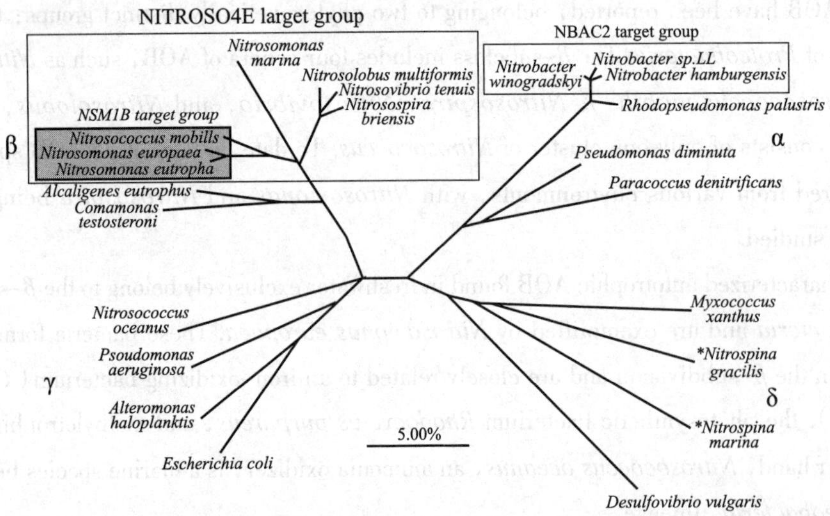

Figure 2–2 Phylogenetic relationships of the chemolithoautotrophic ammonia and nitrite–oxidizing bacteria

2.4 The main influencing parameters of partial nitrification

The key to successful partial nitrification and denitrification, or "short-cut" BNR, is indeed to enrich AOB while inhibiting NOB. Various strategies have been developed to achieve this, including regulating environmental parameters such as free ammonia (FA, NH_3) and free nitrous acid (FNA, HNO_2) concentrations, temperature, pH, and dissolved oxygen (DO) concentration. Process control strategies such as alternating anoxic and oxic conditions, aggressive SRT (sludge retention time) control, and intermittent aeration are also important.

The sensitivity of AOB and NOB to various environmental factors plays a crucial role in determining nitrite accumulation in a nitrifying system. In general, NOB are more sensitive to adverse environmental conditions compared to AOB. Certain compounds, such as hydroxylamine and chlorate, have the ability to inhibit NOB irreversibly without affecting AOB. These compounds can inactivate NOB, while substances like ammonia and nitrous acid can lead to the adaptation of NOB. This differential sensitivity of AOB and NOB to specific influences contributes to the potential for nitrite accumulation in nitrification systems.

Economic feasibility should also be considered when using temperature, pH, and inhibitors as regulation parameters. By applying lower oxygen concentrations, aeration can be saved, and nitrite oxidizers cannot proliferate in the system. This comprehensive approach to process control and environmental parameter manipulation is crucial for achieving successful partial nitrification and denitrification in BNR systems.

2.4.1 pH, free ammonia (FA) and free nitrous acid (FNA)

The pH directly influences the activities of AOB and NOB by altering enzymatic reaction mechanisms, or indirectly through FA and FNA. The concentrations of FA and FNA play a significant role, as these uncharged nitrogen forms serve as the actual substrate/inhibitor for ammonium and nitrite oxidation, rather than the ionized forms (NH_4^+ and NO_2^-). The balance between charged and uncharged forms is determined by the pH and temperature (℃) within the reactor and can be calculated based on the acid-base equilibrium. The concentrations of FA and FNA can be determined using a specific equation. Increasing pH leads to higher levels of ammonia and lower levels of nitrous acid, which in turn promotes the activity of AOB while inhibiting NOB.

$$\text{FA}(\text{mg/L}) = \frac{17}{14} \times \frac{[\text{NH}_4{}^+ - \text{N}] \times 10^{\text{pH}}}{K_\text{a} / K_\text{w} + 10^{\text{pH}}} \quad (2\text{-}6)$$

$$\text{FNA}(\text{mg/L}) = \frac{47}{14} \times \frac{[\text{NO}_2{}^-]}{\exp[-2300/(273+T) \times 10^{\text{pH}}]} \quad (2\text{-}7)$$

Here, [NH_4^+–N] and [NO_2^-–N] are ammonium and nitrite concentrations.

The optimum pH for both AOB and NOB lies between 7 and 8. This is due to the influence of pH on the NH_4^+/NH_3 and HNO_2/NO_2^- equilibria, as well as direct pH effects on activity. Below pH 7, nitrification rate will decrease due to carbon limitation. However, high nitrification rates at low pH were detected.

FA and FNA have been used as core inhibitory agents to achieve partial nitrification. On one hand, FA can serve as a substrate for AOB to promote the ammonia oxidation process; on the other hand, AOB has a higher FA inhibition threshold and a faster activity recovery rate compared to NOB, making FA an effective inhibitor for selectively eliminating NOB. Different threshold values for nitrification inhibition have been proposed in the literature, but these are sensitive to bacterial adaptation (Figure 2–3). It has been reported that the inhibitory effect of FA on NOB becomes effective when it is below 1 mg NH_3–N/L, and it is likely to completely halt NOB growth when it exceeds 6 mg NH_3–N/L. In comparison, the FA inhibition threshold for AOB can reach 16 mg NH_3–N/L. When exposed nitrifying sludge to FA, followed by alternating exposure to high FA and FNA, would result in the development of resistance in the AOB community to FA and FNA, while the presence of NOB was not detected in the system, effectively achieving complete NOB inhibition. There is evidence to suggest that once NOB are controlled to a minimum level, it is difficult to re-establish their population in a short period of time.

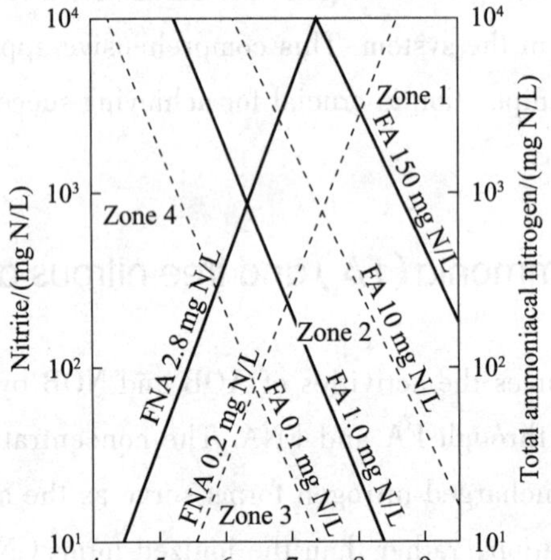

Figure 2–3 Relationship between concentrations of FA and FNA and inhibition to nitrifiers

The dashed lines mark the lower limit and the solid lines mark the upper limit of the range of boundary conditions of zones of nitrification inhibition. Zone 1 = Inhibition of nitritation and nitratation by FA ; Zone 2 = Inhibition of nitratation by FA ; Zone 3 = Complete nitrification ; Zone 4 = Inhibition of nitratation by FNA.

Additionally, different NOB species exhibit varying levels of tolerance to FA, with *Nitrotoga* showing higher FA tolerance than *Nitrospira*, particularly when the FA concentration is below 30 μg N/L, where the dominant species in wastewater, *Nitrospira*, is more sensitive than *Nitrobacter*. The inhibitory effect of FA on NOB will be more pronounced when employing partial nitrification for treating high-ammonia nitrogen wastewater. However, NOB have also been shown to have a certain level of adaptability to FA. For example, when exposed aerobic granules in a reactor to FA [(0.77 ± 0.55) mg N/L] continuously for nearly 18 months, strong resistance of the NOB fixed in the larger granules to FA gradually transitioning from partial nitrification to complete nitrification. Research on wastewater treatment systems indicates that NOB species are more sensitive to FNA than AOB. The threshold range for the FNA inhibitory activity of NOB to decrease by 50% is 0.42–1.72 mg N/L; lower concentrations of FNA (0.011–0.07 mg N/L) have an inhibitory effect on NOB activity, while a FNA concentration of 0.026–0.22 mg N/L will completely inhibit NOB. The inhibitory effect of FNA is particularly significant in high-ammonia nitrogen wastewater (such as anaerobic digestion effluent, landfill leachate, and fecal wastewater) and depends on the oxidation of high-strength ammonia.

Furthermore, some scholars have innovatively operated bioreactors in acidic environments, utilizing the acid tolerance characteristics of AOB to maintain FNA in the system to achieve stable accumulation of nitrite in the effluent, thereby expanding the application scope of partial nitrification.

2.4.2 Temperature

Temperature is an important parameter in the nitrification process, but the influence of temperature on the nitrification process is complex and involves interactions with mass transfer, chemical equilibria, and growth rate. A rise in temperature leads to increased NH_3 inhibition and increased activity of the organisms, according to the Arrhenius principle. However, this increased activity is only effective up to a certain critical temperature, above which biological activity decreases.

It has been observed that temperatures above 25 ℃ led to an increase in the specific growth rate of AOB, surpassing that of NOB. This principle forms the basis of the SHARON (Single reactor High activity Ammonia Removal Over Nitrite) process, where nitrification of ammonium to nitrite is established by working at high temperatures (above 25 ℃) and maintaining an appropriate SRT of 1–1.5 days to retain AOB while washing out NOB. The SRT was operated at value that retains the AOB and falls short of the minimum SRT required for NOB. Thus, NOB are washed out and further nitrification of nitrite to nitrate is prevented (Figure 2–4).

It is essential to highlight that partial nitrification can still be achieved at low temperatures if the operational conditions favor AOB outcompeting NOB, such as maintaining adequate levels of DO, pH, FNA, and FA. The partial nitritation process has demonstrated successful startup and maintenance at lower temperatures ranging between 15 and 30 ℃. These findings suggest that partial nitritation is not limited to effluents with temperatures around 30 ℃, such as those from methanogenic reactors, but can be applicable to various types of industrial wastewater treatments.

However, it is crucial to note that the performance of partial nitrification significantly declines below 15 ℃.

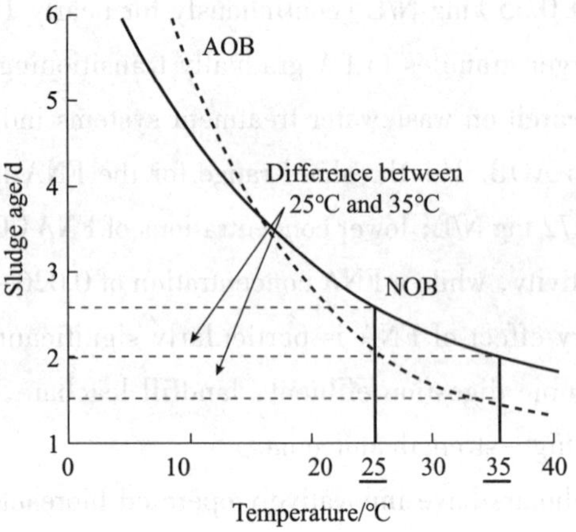

Figure 2-4 Effect of temperature on the minimal required sludge age for ammonia and nitrite oxidation

Above 25 ℃ it is possible to wash out the nitrite oxidizers (–) while maintaining the ammonium oxidizers.

2.4.3 DO concentration

DO serves as a substrate and the terminal electron acceptor in the oxidation-reduction chains for both AOB and NOB. DO plays a crucial role in the ecological separation of AOB and NOB, playing a vital role in ecologically separating these bacterial groups.

Low DO concentrations result in reduced growth rates for both AOB and NOB. It is widely acknowledged that the oxygen half-saturation constant for AOB is lower than that of NOB. Studies have reported that the half-saturation coefficient (K_0) of AOB to DO ranges from 0.2 to 0.4 mg/L, significantly lower than that of NOB ranging from 1.5 to 12 mg/L. This suggests that AOB have a higher affinity for oxygen than NOB, providing them with a competitive advantage in oxygen utilization. As a result, AOB are often considered to have a competitive advantage over NOB in terms of oxygen utilization. However, AOB do not always possess an absolute advantage in oxygen competition, as *Nitrospira* in NOB can adapt to sustained low DO environments, leading to a shift in the NOB community towards internal species with higher oxygen affinity. For example, when DO is below 0.5 mg/L, the growth rate of *Nitrospira* is two orders of magnitude higher than that of *Nitrobacter*.

Additionally, researchers have found that the oxygen affinity of AOB is significantly higher than that of NOB species with cytoplasmic-oriented NXR, but similar to that of NOB species with periplasmic-oriented NXR. In short, in partial nitrification systems, when *Nitrobacter* is the dominant species in the NOB community, AOB will dominate the competition for oxygen; whereas when *Nitrospira* dominates the NOB community, the competition for oxygen between NOB and AOB becomes more intense, providing important insights for the microbial structure regulation of partial nitrification systems. Therefore, controlling *Nitrobacter* as the dominant species of NOB will help achieve partial nitrification. In fact, studies have shown that reactors operated under anaerobic/aerobic alternating conditions are prone to achieve *Nitrobacter* dominance, resulting in nitrification rates twice as high as under strict aerobic conditions. Bao, et al. (2017) found that a sudden increase in DO levels (from 0.3 to 1.8 mg/L) also rapidly leads to *Nitrobacter* becoming the dominant species, thereby achieving a complete transformation from complete nitrification to partial nitrification, which differs significantly from conventional control strategies. Furthermore, NOB have been observed to exhibit a longer lag time compared to AOB when transitioning from anoxic to oxic conditions. This delay is attributed to the deactivation of a crucial enzyme that inhibits NOB following anoxic disturbance, or the inhibition of NOB by intermediate products like hydroxylamine. To suppress NOB activity, an intermittent aeration-based operational strategy has been suggested. This approach involves creating transient anoxic conditions in SBRs or continuous step-feed reactors to suppress NOB activity effectively.

2.4.4 Sludge retention time

In suspended growth systems, the selective cultivation of AOB and the washout of NOB can be achieved by adjusting the SRT due to their distinct minimum required sludge ages. AOB have a minimum doubling time of 7–8 hours, while NOB have a minimum doubling time of 10–13 hours. Full-scale experience has indicated that maintaining an SRT between 1 and 2.5 days leads to optimal performance. In the SHARON process, AOB are selected based on their different growth rates. This process operates at a HRT equal to SRT of 1 day at high temperatures and elevated oxygen levels to promote AOB growth and washout of NOB. Successful partial nitrification has also been achieved with longer sludge ages. Studies by Pollice et al. (2002) demonstrated stable partial nitrification under oxygen limitation regardless of the SRT (10, 14, and 40 days), while Peng and Zhu (2006) achieved stable nitrite accumulation at normal or low temperature (<13 ℃) with extended SRT of 30 days.

2.4.5 Toxic substances

Nitrifying bacteria are very sensitive to some organic and inorganic compounds. When the concentration of these compounds increases, the nitrification rate will decrease and even the nitrifying bacteria die. The organic substances that have been reported to be toxic to nitrifying bacteria include solvent organic chemicals, amines, phenolic compounds, cyanates, carbamate and benzene.

If the industrial wastewater contains the above organic compounds, it is necessary to pay attention to their impact on the activity of nitrifying bacteria. Because the quality of industrial wastewater is relatively complex, even if it is confirmed that it has inhibitory effect on nitrifying bacteria, it is difficult to find out which organic substances produce toxicity. If it is necessary to find the source of toxic substances, it is necessary to conduct a large-scale survey and sampling of the service area of the wastewater collection system.

2.5 The main challenges for partial nitrification

Partial nitrification, while offering numerous advantages, presents significant challenges within the realm of wastewater treatment. An essential obstacle in this process is the inhibition of NOB, particularly evident in mainstream conditions characterized by low influent ammonia levels and reduced temperatures, which promote the preferential growth of NOB over AOB. The precise regulation of aeration duration stands out as a key strategy for achieving partial nitrification in mainstream wastewater treatment, demanding precise process control and automation, especially in systems with minimal DO concentrations in suspended growth configurations. Notably, aeration duration control has predominantly found application in SBRs, less common in municipal wastewater treatment plants. Consequently, the imperative need arises for the development of alternative control methodologies to realize partial nitrification, particularly in non-SBR systems. Additionally, there is a pressing requirement for further investigation into the influence of the COD/N ratio on the performance and microbial dynamics (e.g., heterotrophs, AOB, NOB, and Anammox bacteria) in partial nitrification across diverse configurations (e.g., suspended growth versus attached growth) at varying temperatures.

In partial nitrification systems, the accumulation of elevated nitrite levels has been observed to have inhibitory effects on Polyphosphate Accumulating Organisms (PAOs). Both aerobic and anaerobic phosphate uptake processes have shown susceptibility to inhibition by nitrite, with reported threshold inhibition concentrations of 0.4 mg N/g VSS and approximately 2 mg N/g VSS, respectively. Research has indicated complete inhibition of both anaerobic and aerobic phosphate uptake at nitrite concentrations surpassing 8 mg NO_2^--N/L. Furthermore, fundamental aerobic metabolic processes of PAOs, including Polyhydroxyalkanoate (PHA) oxidation, phosphorus uptake, and growth, have been reported to be impeded by FNA. In addition, FNA exerts lower inhibitory effects on Glycogen Accumulating Organisms (GAOs) compared to PAOs, leading to decreased phosphorus removal due to a reduced aerobic glycogen synthesis rate. Therefore, further investigations are essential to comprehend the inhibitory impact of nitrite on Enhanced Biological Phosphorus Removal (EBPR) in conjunction with partial nitrification systems. Recently, exploration into low DO operation has emerged as a potential strategy for integrating mainstream partial nitrification with biological phosphorus removal. Furthermore, the effects of low DO concentration and the ratio of Denitrifying Phosphorus Accumulating Organisms (DPAOs) to nitrifiers, revealed that DPAOs held competitive advantages over nitrifiers for oxygen under DO-limited conditions. Additionally, the utilization of the granular sludge process has demonstrated efficacy in enabling simultaneous partial nitrification and phosphorus removal.

An additional obstacle to attain consistent mainstream partial nitrification is the presence of nitrifying bacteria (AOB and NOB) in untreated wastewater. A recent investigation by Duan et al. delved into the impact of influent NOB on the effectiveness of NOB inhibition in a mainstream nitrite shunt system, facilitated through sidestream sludge treatment employing FA or FNA. The findings revealed an augmented resistance of NOB to sidestream sludge treatment and potentially to alternative NOB suppression techniques, consequently leading to the inadequacy of NOB inhibition measures.

2.6 Application of partial nitrification and denitrification

Recent review papers highlight the global implementation of over 140 full-scale partial nitrification processes, with outcomes detailed in Table 2-1. These processes have been predominantly employed in sidestream applications, encompassing anaerobic digestate, recycled dewatering streams, leachate, and industrial wastewater treatment, with minimal mainstream utilization. Among the installations, suspended growth processes and biofilm growth processes accounted for 65% and 29% of the total respectively. Within suspended growth processes, the DEMON process emerged as the predominant technology (63%), followed by SHARON and other commercial processes. Among the various partial nitrification systems, Sequential Batch Reactors (SBRs) stood out as the most widely adopted technology. While the median volumetric Nitrogen Loading Rate (NLR) for SBRs [0.35 kg N/(m^3 · d)] was lower compared to Completely Stirred Tank Reactors (CSTRs) with settlers [0.58 kg N/(m^3 · d)], Moving Bed Biofilm Reactors (MBBRs) [0.62 kg N/(m^3 · d)], and biofilm processes [0.8 kg N/(m^3 · d)], SBRs demonstrated easier NOB suppression due to elevated ammonia and nitrite concentrations post-fill and early in aeration, leading to increased FA and FNA inhibition of NOB. Remarkably, Fluidized Bed Biofilm Reactors (FBBRs) show promise from a loading perspective, with two studies documenting NLRs as high as 1.2-4.8 kg N/(m^3 · d).

While the treatment systems mentioned operate at extended SRTs and primarily rely on DO control for NOB suppression, the SHARON process functions at notably short SRTs (1-3 days) and elevated DO levels. No comparative studies directly comparing SHARON with other processes were identified in the accessible literature. The DEMON® configuration, a development by the University of Innsbruck and the Achental-Inntal-Zillertal Wastewater Treatment Association (Austria), is integrated into over 80% of SBR processes. It incorporates a hydro-cyclone to regulate the SRT for AOB and Anammox bacteria separately and to segregate the slow-growing Anammox bacteria from incoming solids. The system operates at a pressure drop of 2 bar, achieving a 95% capture efficiency of Anammox granules with high specific gravity at the bottom.

Table 2-1 Full-scale partial nitrification/ammonification systems

Process	Number of Installations	Design load /(kg N/d) Min-Max (average)	Design load /[kg N/(m^3 · d)] Min-Max (average)
DEMON	61	45-2400(603)	0.4-1.0(0.6)
SHARON	10	408-3500(1500)	0.02-0.65(0.39)
PN/A process	5	56-1300(400)	0.23-0.72(0.44)
Cleargreen™	2	600-1900(1250)	0.55-0.7(0.63)
SNAD	3	180-500(340)	0.5
NAS	5	320-890(520)	0.02-0.59(0.2)
Other SBRs	6	270-990(480)	0.15-0.99(0.5)

SHARON processes developed by the Delft University of Technology (Netherlands) are also a typical partial nitrification technology. The typical characteristics of the SHARON process are: partial nitrification and partial denitrification are implemented in the same reactor, resulting in a shorter process. The reactor does not retain activated sludge, and has a simple structure. The operating temperature is relatively high (30-38 ℃), leading to good treatment efficiency. By utilizing denitrification, it regulates the acidity and alkalinity (pH 7-8) without the need for alkali neutralization. From a microbiological perspective, the SHARON process does not seem to be anything new. However, in the field of wastewater biological deammonification, the concept and practice of full nitrification and full denitrification have always been popular. Breaking free from traditional concepts has become an innovation. In fact, the SHARON process does have microbiological characteristics that are different from traditional biological deammonification processes. The main reasons why full nitrification and full denitrification have been used for long-term wastewater biological deammonification are: initially, the purpose of applying nitrification processes was to eliminate the adverse effects of ammonia on water bodies. If nitrification is incomplete, the formation of nitrite is a "triple threat" substance, posing a threat to aquatic life and human safety. Nitrite still has oxygen-consuming capabilities, depleting the DO in water bodies. Nitrite bacteria and nitrate bacteria commonly coexist and live together in open wastewater treatment systems, making it difficult to separate them in engineering terms. In nature or wastewater treatment systems, ammonia oxidation is often the limiting step in the entire nitrification process, with little accumulation of nitrite. Due to the fact that the substrate provided by the nitrification process is nitrate, the denitrification process can only proceed as full denitrification.

Before the successful development of the SHARON process, many researchers made beneficial attempts at partial nitrification (such as pH control, DO control, temperature control, ammonia concentration control, nitrite control, etc.) in an attempt to terminate nitrification at the nitrite stage, but all were unsuccessful. The reasons for this may be multifaceted: ① From a guiding ideology perspective, they were dominated by the idea of meeting effluent standards, and when enriching nitrifying bacteria, they consciously or unconsciously screened some nitrite bacteria with high substrate affinity but slow growth, making it difficult to improve reaction activity. ② From a microbiological perspective, they generally only considered the differences in a certain physiological characteristic (such as sensitivity to ammonia concentration, sensitivity to oxygen concentration, acid resistance, etc.) between two types of nitrifying bacteria, and it is difficult to separate the two bacterial populations in engineering by using only one physiological characteristic difference, even if they can be separated, they often lack practical value. ③ From a reactor analysis perspective, they often separated the nitrification process from the entire denitrification process (separated spatially or temporally), and did not fully utilize the alkalinity produced by denitrification. Since external alkalinity is needed to maintain normal ammonia oxidation reaction, the economic advantage is not obvious, and promotion is difficult. ④ From an operational perspective, they used reactors that could retain biomass and tried to increase the biomass concentration in the device, causing nitrate bacteria to accumulate in the device, making it difficult to achieve short-term nitrification.

Full-scale SHARON processes are in operation in various countries such as Netherlands, France, US, England, Sweden, and Switzerland for sidestream nitrogen removal, typically processing between 400 and 3500 kg of nitrogen per day. In a typical SHARON system, an anoxic tank with a Hydraulic Retention Time (HRT) of 0.75 days is employed to utilize biodegradable carbon for denitritation, aiding in the removal of nitrogen compounds efficiently. This system is designed to convert high levels of ammonium into nitrites to achieve nitrogen removal in wastewater treatment processes (Table 2-2).

Table 2-2 Full-scale application of SHARON process for nitrogen removal

Full-scale plants	Capacity (pe)	Influent	N removal/%	N load /(kg N/d)	Operational year
Utrecht	400000	Rejection water	>95	900	1997
Rotterdam	470000	Rejection water	>95	850	1999
Zwolle	200000	Rejection water	>95	410	2003
Beverwijk	320000	Rejection water and condensate	>95	1200	2003
Garmerwolde	300000	Filtrate and condensate	>95	2500	2005
Hague	930000	Centrate	>95	1200	2005

The Dokhaven wastewater treatment plant in Rotterdam, the Netherlands, is located in the city center, and the main structures (aeration tank and sedimentation tank) are built underground, with no expansion space around it. The plant uses the A/B process to treat wastewater equivalent to 470000 population. Although this process is suitable for BOD removal and nitrification, the denitrification effect of the B-level reactor is poor, mainly due to the lack of electron donors (organic matter). Nitrogen balance analysis of the entire wastewater treatment plant found that the amount of ammonia nitrogen in the anaerobic digestion sludge separation liquid accounts for about 15% of the nitrogen content. If this part of the ammonia nitrogen is removed and not allowed to return to the A/B system, the total ammonia emission from wastewater treatment can be significantly reduced. After the SHARON process operated normally, the effluent quality of the Dokhaven wastewater treatment plant improved significantly, with TKN concentration and ammonia concentration decreasing from 7.5 mg/L and 6.2 mg/L to 3.8 mg/L and 2.2 mg/L, respectively, and the total nitrogen discharge reduced by nearly half.

Questions

[1] What are the advantages of partial nitrification and denitrification?

[2] Please describe how to achieve partial nitrification and denitrification in the mainstream wastewater treatment.

[3] Please describe the SHARON process.

[4] Please analyze the application potential of partial nitrification and denitrification process.

References

[1] Hovanec T A, Dnlong E F. Comparative analysis of nitrifying bacteria associated with freshwater and marine aquaria [J]. Applied and Environmental Microbiology, 1996, 62(8): 2888–2896.

[2] Almeida J S, Reis M A M, Carrondo M J T. Competition between nitrate and nitrite reduction in denitrification by *Pseudomonas fluorescens* [J]. Biotechnology and Bioengineering, 1995, 46(5): 476–484.

[3] An Z, Kent T R, Sun Y, et al. Free ammonia resistance of nitrite-oxidizing bacteria developed in aerobic granular sludge cultivated in continuous upflow airlift reactors performing partial nitritation [J]. Water Environment Research, 2021, 93(3): 421–432.

[4] Anthonisen A, Loehr R, Prakasam T, et al. Inhibition of nitrification by ammonia and nitrous acid [J]. Journal Water Pollution Control Federation, 1976, 48(5): 835–852.

[5] Bao P, Wang S, Ma B, et al. Achieving partial nitrification by inhibiting the activity of *Nitrospira*-like bacteria under high-DO conditions in an intermittent aeration reactor [J]. Journal of Environmental Sciences, 2017, 56: 71–78.

[6] Barak Y, Van Rijn J. Relationship between nitrite reduction and active phosphate uptake in the phosphate-accumulating denitrifier *Pseudomonas sp.* strain JR 12 [J]. Applied and Environmental Microbiology, 2000, 66(12): 5236–5240.

[7] Baumann B, Snozzi M, Van Der Meer J R, et al. Development of stable denitrifying cultures during repeated aerobic-anaerobic transient periods [J]. Water Research, 1997, 31(8): 1947–1954.

[8] Belser L W, Mays E L. Specific inhibition of nitrite oxidation by chlorate and its use in assessing nitrification in soils and sediments [J]. Applied and Environmental Microbiology, 1980, 39(3): 505–510.

[9] Garrido J M, Van Benthum W A J, Van Loosdrecht M C M, et al. Influence of dissolved oxygen concentration on nitrite accumulation in a biofilm airlift suspension reactor [J]. Biotechnology and Bioengineering, 1997,

53(2): 168–178.

[10] Hellinga C, Schellen A, Mulder J W, et al. The SHARON process: an innovative method for nitrogen removal from ammonium-rich wastewater [J]. Water Science and Technology, 1998, 37(9): 135–142.

[11] Hovanec T A, Delong E F. Comparative analysis of nitrifying bacteria associated with freshwater and marine aquaria [J]. Applied and Environmental Microbiology, 1996, 62(8): 2888–2896.

[12] Khin T, Annachhatre A P. Novel microbial nitrogen removal processes [J]. Biotechnology Advances, 2004, 22(7): 519–532.

[13] Liu X, Kim M, Nakhla G, et al. Partial nitrification-reactor configurations, and operational conditions: performance analysis [J]. Journal of Environmental Chemical Engineering, 2020, 8(4): 103984.

[14] Meinhold J, Arnold E, Isaacs S. Effect of nitrite on anoxic phosphate uptake in biological phosphorus removal activated sludge [J]. Water Research, 1999, 33(8): 1871–1883.

[15] Mulder J W, Van Loosdrecht M C M, Hellinga C, et al. Full-scale application of the SHARON process for treatment of rejection water of digested sludge dewatering [J]. Water Science and Technology, 2001, 43(11): 127–134.

[16] Neufeld R, Greenfield J, Rieder B. Temperature, cyanide and phenolic nitrification inhibition [J]. Water Research, 1986, 20(5): 633–642.

[17] Peng Y, Zhu G. Biological nitrogen removal with nitrification and denitrification via nitrite pathway [J]. Applied Microbiology and Biotechnology, 2006, 73(1): 15–26.

[18] Peng Y Z, Chen Y, Peng C Y, et al. Nitrite accumulation by aeration controlled in sequencing batch reactors treating domestic wastewater [J]. Water Science and Technology, 2004, 50(10): 35–43.

[19] Pollice A, Tandoi V, Lestingi C. Influence of aeration and sludge retention time on ammonium oxidation to nitrite and nitrate [J]. Water Research, 2002, 36(10): 2541–2546.

[20] Regan J M, Harrington G W, Noguera D R. Ammonia- and nitrite-oxidizing bacterial communities in a pilot-scale chloraminated drinking water distribution system [J]. Applied and Environmental Microbiology, 2002, 68(1): 73–81.

[21] Siegrist H, Reithaar S, Lais P. Nitrogen loss in a nitrifying rotating contactor treating ammonium rich leachate without organic carbon [J]. Water Science and Technology, 1998, 37(4-5): 589–591.

[22] Sijbesma W F H, Almeida J S, Reis M A M, et al. Uncoupling effect of nitrite during denitrification by *Pseudomonas fluorescens*: an in vivo 31P-NMR study [J]. Biotechnology and Bioengineering, 1996, 52(1): 176–182.

[23] Stein L Y, Arp D J. Loss of ammonia monooxygenase activity in *Nitrosomonas europaea* upon exposure to nitrite [J]. Applied and Environmental Microbiology, 1998, 64(10): 4098–4102.

[24] Van Dongen U, Jetten M S M, Van Loosdrecht M C M. The SHARON®-Anammox® process for treatment of ammonium rich wastewater [J]. Water Science and Technology, 2001, 44(1): 153–160.

[25] Van Hulle S W H, Vandeweyer H J P, Meesschaert B D, et al. Engineering aspects and practical application of autotrophic nitrogen removal from nitrogen rich streams [J]. Chemical Engineering Journal, 2010, 162(1): 1–20.

[26] Van Rijn J, Tal Y, Barak Y. Influence of volatile fatty acids on nitrite accumulation by a *Pseudomonas stutzeri* strain isolated from a denitrifying fluidized bed reactor [J]. Applied and Environmental Microbiology, 1996, 62(7): 2615–2620.

[27] Wong-Chong G M, Loehr R C. Kinetics of microbial nitrification: nitrite-nitrogen oxidation [J]. Water

Research, 1978, 12(8): 605–609.

[28] Ye L, Zhang T. Estimation of nitrifier abundances in a partial nitrification reactor treating ammonium–rich saline wastewater using DGGE, T–RFLP and mathematical modeling [J]. Applied Microbiology and Biotechnology, 2010, 88(6): 1403–1412.

[29] Yoo H, Ahn K H, Lee H J, et al. Nitrogen removal from synthetic wastewater by simultaneous nitrification and denitrification (SND) via nitrite in an intermittently–aerated reactor [J]. Water Research, 1999, 33(1): 145–154.

[30] Zaman M M, Liu X, Nakhla G. Impact of dissolved oxygen concentration and DPAOs: nitrifiers population ratio on nutrient removal in EBPR process [J]. International Journal of Environmental Science and Development, 2019, 10(9): 257–260.

[31] Zhang M, Wang S, Ji B, et al. Towards mainstream deammonification of municipal wastewater: partial nitrification–Anammox versus partial denitrification–Anammox [J]. Science of the Total Environment, 2019, 692: 393–401.

Chapter 3 Simultaneous nitrification and denitrification

Emphasis: Simultaneous nitrification and denitrification, heterotrophic nitrification and aerobic denitrification, influencing factors, microbial community, aerobic granular sludge, moving bed biofilm reactors.

Nut: Control technology.

3.1 Introduction

Nitrogen is an important pollutant causing eutrophication and environmental contamination in water bodies. The control of nitrogen and phosphorus emissions in wastewater has gained increasing attention, and the research on processes with efficient nitrogen and phosphorus removal capabilities is becoming increasingly important.

According to traditional denitrification theory, the removal of ammonia nitrogen is achieved through two independent processes: nitrification and denitrification. Due to different environmental requirements, these two processes cannot co-occur. Current biological denitrification processes arrange nitrification and denitrification as two separate stages in different reactors (spatially) or utilize intermittent aerobic and anaerobic conditions (temporally) to achieve nitrogen removal, often resulting in complex systems, high energy consumption, and inconvenient operation and management.

Simultaneous nitrification and denitrification (SND) refers to the simultaneous occurrence of nitrification and denitrification processes in the same reactor under the same operating conditions. The implementation of SND in the same treatment system allows the products of nitrification to directly serve as substrates for denitrification, avoiding the accumulation of NO_3^- during nitrification that inhibits the nitrification reaction and accelerates the nitrification rate. Additionally, the alkalinity released during denitrification can partially compensate for the alkalinity consumed during nitrification, stabilizing the pH value in the system. Furthermore, nitrification and denitrification are performed in the same system under the same operating conditions, tremendously simplifying the operational complexity. The implementation of SND and the achievement of dynamic equilibrium in both processes will greatly simplify the biological denitrification process and improve denitrification efficiency, thereby saving investment and enhancing treatment efficiency. Therefore, a great advance has been increasingly investigated on the new denitrification technology for maximizing the efficiency of nitrification and denitrification in the same reactor during domestic and industrial wastewater treatment.

The SND phenomenon was first discovered in 1974, when Starr, et al. (1974) observed and proposed this phenomenon in soil water leaching processes. The nitrogen loss observed in the aeration system of the wastewater treatment process without obvious anaerobic or anoxic conditions also demonstrated the possibility of the SND phenomenon in wastewater denitrification.

3.2 The mechanism and advantages of simultaneous nitrification and denitrification

3.2.1 The mechanism of SND

Research on the mechanism of SND reactions shows that there is currently no direct evidence to provide a complete and unified explanation for the reaction mechanism. This section summarizes the three theories of SND reaction mechanism that have been more concentrated and recognized in recent years, including the macro-environment theory and micro-environment theory from a physical perspective, and the microbial theory from the perspective of nitrifying and denitrifying bacteria.

1. Macroscopic environment theory

In activated sludge processes without obvious anaerobic or anoxic stages, non-assimilatory nitrogen loss in the aeration system was frequently observed, with a loss rate of approximately 10% to 20% depending on the control conditions.

Generally, even in activated sludge systems with predominantly aerobic conditions, especially when using point-source aeration devices or uneven aeration, a large range of local hypoxic environments often occur, which is the macro-environment of the bioreactor. For example, in a biofilm reactor, anoxic zones are formed inside the membrane due to the influence of substrate concentration and membrane thickness. Similar phenomena also exist in RBC, SBR reactors, oxidation ditches, and so on. In fact, completely uniform mixing states do not exist in bioreactors at production scale; therefore, it is entirely possible to have some degree of denitrification during the aeration phase, that is, SND. This theory suggests that the formation of SND is due to the uneven distribution of DO in different areas of the reactor caused by its internal structure or aeration devices. This leads to the coexistence of aerobic, anaerobic, or anoxic zones within the reactor, allowing aerobic nitrifying bacteria and anaerobic denitrifying bacteria to occupy adaptive areas simultaneously and carry out nitrification and denitrification. The nitrifiers and denitrifiers mainly localize in higher (more than 1–2 mg/L) and lower concentration of DO (less than 0.5 mg/L), respectively.

2. Micro–environment theory

The micro–environment theory explains the phenomenon of SND from a physical perspective. This theory considers the transfer and changes of various ecological factors (such as DO, organic matter, and other nutrients) in the microenvironment of activated sludge and biofilms, the metabolic activities and interrelationships of various microorganisms, and the changes in the physical, chemical, and biological conditions or states of the micro–environment.

The micro–environment theory suggests that due to limitations in oxygen diffusion, there is a gradient of dissolved oxygen within microbial aggregates or biofilms. This means that the outer surface of microbial flocs or biofilms has a high DO concentration, favoring aerobic nitrifying bacteria and AOB. However, as one moves deeper into the big aggregates, oxygen transfer is hindered and there is significant external oxygen consumption, leading to the formation of hypoxic zones where denitrifying bacteria dominate, creating a micro–environment conducive to SND (Figure 3–1). Therefore, SND can occur in aerobic granular sludge or biofilm reactors, where denitrifying bacteria mainly confined in the deeper anoxic layer.

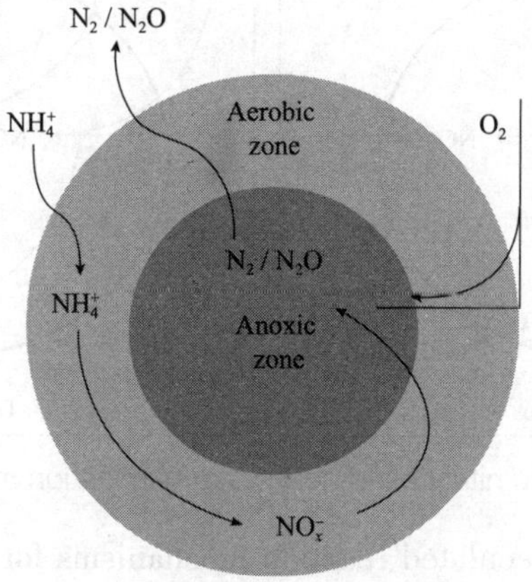

Figure 3–1 Schematic of activated sludge floc having internal aerobic and anoxic zones

Granular sludge of a certain size (larger than 200 μm), especially aerobic granular sludge, can also form aerobic and anaerobic zones internally due to limitations in oxygen diffusion, providing a micro–environment for SND (Figure 3–2). The presence of different redox conditions in a single granule allow occurrence of SND process even in the case of aeration. The advantages of using aerobic granular sludge for biological denitrification include its dense structure, excellent microbial activity and settling ability, as well as high biomass retention. In recent years, successful cultivation of aerobic granular sludge using hydraulic sieving methods has been reported. The results have shown high removal rates for NH_4^+–N and TN in SBR reactors, reaching up to 95% and 60% respectively, with the NH_4^+–N load of approximately 0.18 kg NH_4^+–N/(m^3 d). However, reports on the achievement of SND using aerobic granular sludge are still limited and currently in the exploratory stage.

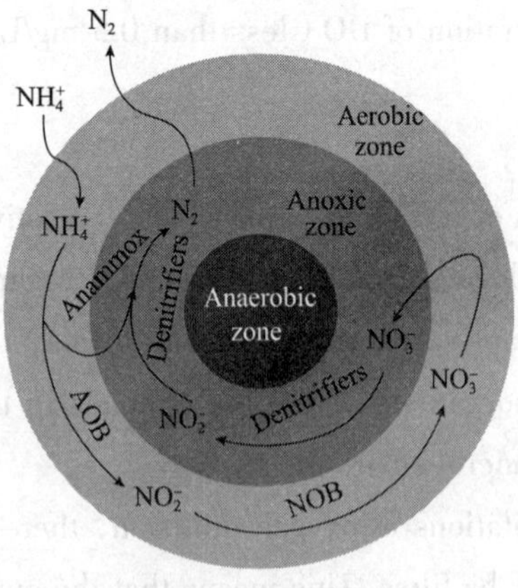

Figure 3–2 Nitrogen removal pathway in aerobic granular sludge

3. Microbial theory

The discovery of aerobic denitrifying bacteria and heterotrophic nitrifying bacteria has challenged the traditional theory that nitrification reactions can only be carried out by autotrophic bacteria and denitrification can only occur under anaerobic conditions. The microbial mechanism of SND is based on the activity of heterotrophic nitrifiers and aerobic denitrifiers with the capacity of denitrification even at oxygen–saturated condition. It has been found that many heterotrophic nitrifying bacteria can carry out aerobic denitrification reactions, directly converting NH_4^+–N into gaseous end products while reducing NO_3^-–N and NO_2^-–N (Figure 3–3). Therefore, from a microbiological perspective, SND for biological denitrification is possible.

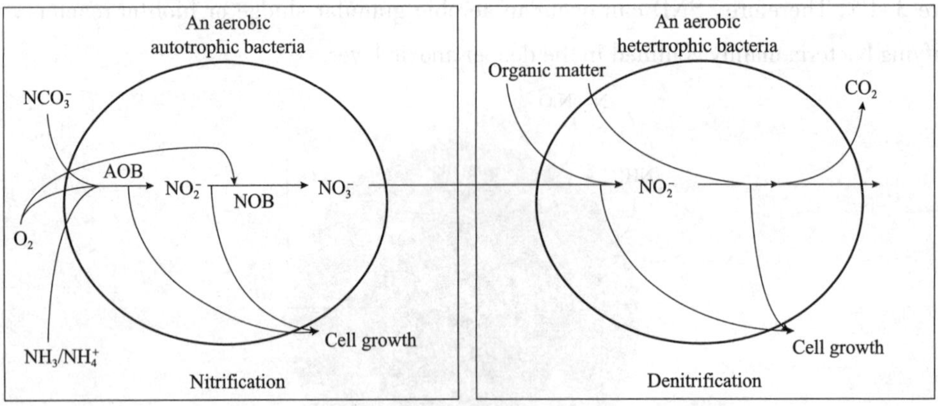

Figure 3–3 Heterotrophic nitrification aerobic denitrification mediated nitrogen removal

Among the three main speculated reaction mechanisms for SND, the micro–environment theory is currently more widely accepted by researchers. However, this theory cannot fully explain all observed SND phenomena, such as the issue of organic carbon sources for endogenous anoxic denitrification, indicating that its mechanism still requires further study.

3.2.2 The advantages of SND

The advantages of SND reactions include:

(1) SND in a single reactor without the need for sludge return, saving construction space and infrastructure costs.

(2) Relatively stable pH in the system without the demand for additional acid/alkali sources.

(3) More efficient utilization of carbon sources in raw water during the SND process, reducing

the need for additional carbon sources.

(4) Lower general DO concentration control during the SND reaction process compared to traditional nitrification and denitrification, leading to energy savings and operating costs reduction.

(5) Direct consumption of nitrate nitrogen produced during nitrification by the denitrification process, avoiding the accumulation of nitrate nitrogen that could inhibit the system's nitrification process and speeding up reaction times.

(6) Lower sludge production, especially for aerobic granular sludge reactor, reducing the treatment cost of surplus sludge.

3.3 Microorganism involved in SND

In biofilm reactors, the efficiency of removing a specific substrate is determined by the composition and growth of the biofilm, which aids in degrading the substrate. The inner surfaces of bio-carriers exhibit a diverse microbial community with distinct morphologies such as cocci, bacilli, and rods. This microbial community forms a biofilm held together by the secretion of extracellular polymeric substances (EPS) by microorganisms, creating a gel-like matrix that surrounds the cells. The EPS acts as a binding agent, contributing to the structural integrity of the biofilm. Within the biofilm, concentration gradients for dissolved compounds like DO, nitrates, and ammonium are established due to diffusion into the biofilms. Studies on biofilm dynamics have shown that the spatial distribution of different bacterial species in the biofilm depends on the C/N ratio. Both groups of microorganisms, responsible for nitrogen removal, can coexist in the outer layers of the biofilm even in the absence of sufficient carbon sources. The microorganisms involved in nitrogen removal processes are illustrated in Figure 3-4, highlighting their roles within the biofilm structure.

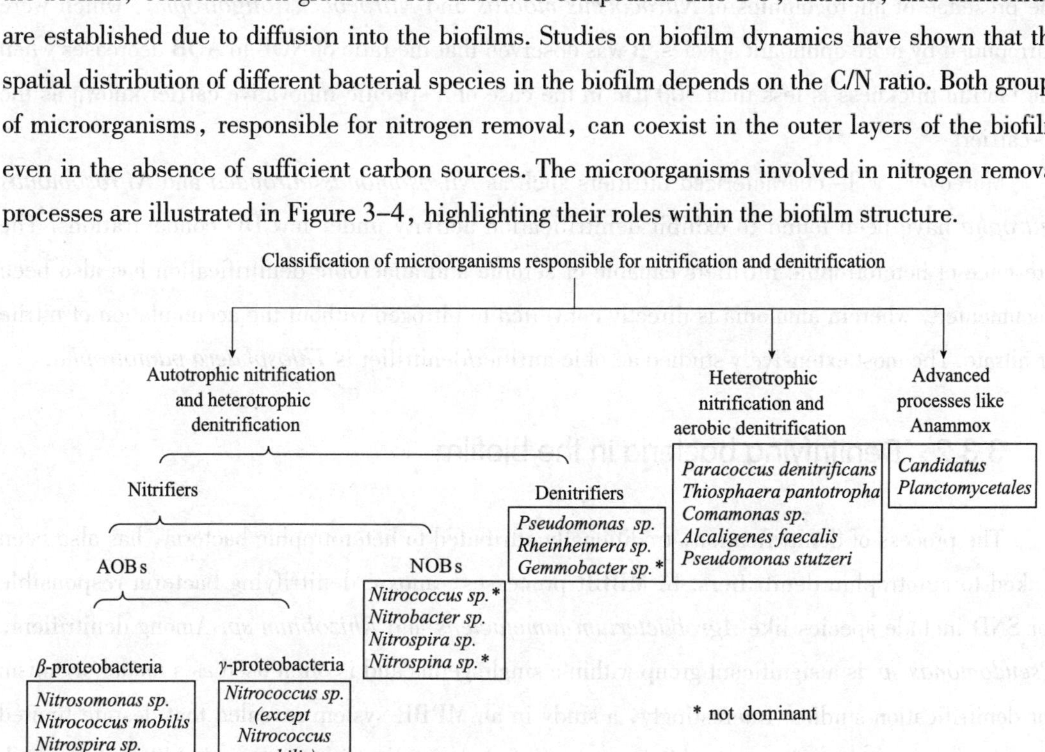

Figure 3-4 Classification of microorganisms responsible for nitrification and denitrification

3.3.1 Nitrifying bacteria within the biofilm

The nitrification process involves two distinct groups of microorganisms, namely AOB and NOB. Genetic analysis has revealed that AOB can be further classified into two monophyletic groups, one within the *beta-Proteobacteria*, which includes *Nitrosomonas* and *Nitrosospira*, and another within the *gamma-Proteobacteria*, which encompasses *Nitrococcus*, with *Nitrococcus mobilis* being an exception classified under *beta-Proteobacteria*. Additionally, NOB are categorized into four groups, including *Nitrobacter*, *Nitrococcus*, *Nitrospina*, and *Nitrospira*. Notably, species within the *Nitrococcus* and *Nitrospina* genera are not predominant in municipal wastewater treatment plants. In an experiment involving varying ammonia concentrations up to 175 mg/L, despite that *Nitrospira*-like microorganisms dominated during most of the investigation, they were occasionally outnumbered by *Nitrobacter*.

The biomass yield of slow-growing nitrifiers is relatively low in comparison to the amount of NH_4^+-N oxidation. In lab-scale investigations of nitrifying communities treating landfill leachate, the biofilm predominantly consisted of AOB, with *Nitrosomonas europaea* and *Nitrosomonas eutropha* were observed to be the most abundant species, suggesting that the growth of these strains was facilitated by high ammonia concentrations. A study on nitrifying bacteria in SBBR revealed the presence of microcolonies of *Nitrococcus mobilis* and *Nitrococcus oligotropha*, which were surrounded by more abundant species. It was observed that the ratio of NOB to AOB decreases when the biofilm thickness is less than 300 μm in the case of a specific innovative carrier known as the Z-carrier.

Moreover, well-characterized nitrifiers such as *Nitrosomonas europaea* and *Nitrosomonas eutropha* have been found to exhibit denitrification activity under low DO concentrations. The presence of heterotrophic nitrifiers capable of aerobic and anaerobic denitrification has also been documented, wherein ammonia is directly converted to nitrogen without the accumulation of nitrite or nitrate. The most extensively studied aerobic nitrifier/denitrifier is *Thiosphaera pantotropha*.

3.3.2 Denitrifying bacteria in the biofilm

The process of denitrification, traditionally attributed to heterotrophic bacteria, has also been linked to autotrophic denitrifiers. In MBBR processes, anoxic denitrifying bacteria responsible for SND include species like *Agrobacterium tumefaciens* and *Rhizobium sp.* Among denitrifiers, *Pseudomonas sp.* is a significant group within a single genus and is often used as a model organism for denitrification studies. Interestingly, a study in an MBBR system revealed that despite limited organic carbon sources, heterotrophic bacteria prevailed, challenging the long-held belief that high C/N ratios favor the growth of denitrifying species.

Furthermore, in environments where DO levels are adequate to reach denitrifying bacteria, certain species have been observed to undergo aerobic denitrification, utilizing nitrate and oxygen as electron acceptors. This process results in the formation of intermediates like nitric oxide and nitrous oxide before eventual conversion to nitrogen gas.

Apart from conventional autotrophic nitrifiers and heterotrophic denitrifiers, other species have been identified for their denitrification capabilities under different conditions. For instance, *Magnetospirillum magnetotacticum* and *Pseudomonas stutzeri* U2 have shown denitrification abilities in aerobic settings. Additionally, species like *Thiosphaera pantotropha* and *Nitrosomonas*-like microorganisms, including *Bacillus cereus*, *Bacillus subtilis*, and *Bacillus licheniformis*, have demonstrated the capacity for SND, even in fully aerobic or anoxic conditions. These findings highlight the diverse metabolic potential of microbial communities involved in nitrogen cycling processes.

Denitrifying strains isolated from laboratory-scale bioreactors are often closely related to *Hyphomicrobium*, *Pseudomonas* and *Comamonas* species within the *Proteobacteria* phylum. However, there is a discrepancy between these frequently isolated strains and the dominant species found in real wastewater denitrification systems. In actual wastewater treatment systems, species like *Azoarcus*, *Zoogloea* and *Comamonadaceae* family are commonly found to dominate the denitrifying microbial communities. These species are predominantly affiliated with the *Proteobacteria* and *Bacteroidetes* phyla.

The abundance of dominant populations in denitrifying communities can vary significantly depending on factors such as influent characteristics, treatment configurations, and operating conditions. The complex interactions between microbial species and environmental conditions influence the composition and dominance of specific taxa in denitrification processes within wastewater treatment systems.

3.3.3 Bacteria capable of heterotrophic nitrification and aerobic denitrification

Heterotrophic nitrification and aerobic denitrification bacteria widely distribute and have been found in many environment conditions. Autotrophic nitrifying microorganisms mainly exist in bacteria genera, while heterotrophic nitrification and aerobic denitrification bacteria exist in the genera of fungi and bacteria. Bacterial genus like *Arthrobacter*, *Paracoccus*, *Pseudomonas*, *Klebsiella pneumoniae*, *Bacillus*, *Acinetobacter*, *Rhodococcus*, *Agrobacterium*, *Aeromonas*, *Microvirgula*, *Alcaligenes faecalis*, *Diaphorobacter*, *Chryseobacterium*, *Marinobacter*, *Bacillus cereus*, etc, are observed to exhibit heterotrophic nitrification and aerobic denitrification. Heterotrophic nitrification and aerobic denitrification fungi were often found in soil, such as *Basidiomycetes*, *Aspergillus*, *Penicillium*, *Mortierella*, *Trichoderma*, *Verticillium* and

Penicillium tropicum, while few were reported in water environment.

3.4 The main factors affecting SND

3.4.1 DO

DO is a crucial factor affecting nitrification and denitrification processes. High DO levels can enhance nitrification efficiency but inhibit denitrification and lead to inadequate use of carbon sources. Low DO levels can result in incomplete nitrification and influence the total nitrogen removal. The threshold oxygen inhibition concentration can be as low as 0.1 mg O_2/L in wastewater denitrification systems. Expression and activity of nearly all N reductases are suppressed in the presence of oxygen. It has been reported that oxygen has an immediate and reversible inhibitory impact on nitrate respiration, with the maximal inhibition reached at as low as 0.2% oxygen saturation. Nitrite reductase was found to be less sensitive to O_2 than nitrate reductase, with a threshold inhibitory concentration of 2.5 mg O_2/L. Therefore, it is essential to control the DO concentration at an appropriate level in the SND reactor.

The most suitable DO value for SND reactions varies for different wastewater treatment reactors or structures. For example, it has been reported that the optimal DO is 0.3–0.8 mg/L for four-channel oxidation ditches, approximately 0.5 mg/L for activated sludge, 1–2 mg/L for biofilm processes, and <1.0 mg/L for intermittent aeration oxidation ditches. Pochana, et al. (1999) successfully achieved SND in an SBBR reactor operating at a DO of 2 mg/L.

3.4.2 pH

pH is a critical factor in the SND system, which has a significant impact on the growth and metabolism of microorganisms. The appropriate pH level was responsible for the excellent denitrification efficiency of the system, and both nitrification and denitrification reactions have their optimal pH values. The best removal efficiency of NH_4^+–N and TN by SND was found to be in the pH range of 7.0–7.5, while nitrification efficiency is severely inhibited when the pH is higher than 9 and lower than 5. For denitrification, the pH should be maintained at a slightly alkaline level (7.5–9.2). Otherwise, the denitrification efficiency would be greatly reduced. For example, the final product of denitrification will be the potent greenhouse gas N_2O instead of N_2 when the pH drops below 7.3. Therefore, pH 7.5 is beneficial for SND process compared to acidic (pH 4.8) and alkaline (pH 9.7) conditions.

3.4.3 C/N

The C/N ratio mainly affects the SND system by influencing the supply of organic carbon sources for the growth and metabolism of heterotrophic microorganisms and as electron donors for denitrification. A low C/N ratio can significantly inhibit denitrification and affect TN removal. However, a high C/N ratio caused by an excessive amount of organic carbon sources can promote the excessive proliferation of heterotrophic microorganisms, which inhibits the activity of the autotrophic microorganisms (e.g. nitrifying bacteria), leading to reduction in nitrification efficiency. Therefore, a suitable range of C/N ratio for SND is important for maintaining the SND performance.

The optimal C/N ratio for SND process exhibited a high degree of variability in different bioreactors, which depends on the differences in carrier materials, biofilm composition, types of influent, and treatment strategies. For instance, the effluent ammonia, nitrite, and nitrate were observed to be low when the C/N ratio was stabilized at 6.3–9.3 in a continuous flow activated sludge system. However, a feed C/N ratio of 5.6 in a microaerobic moving bed biofilm reactor (mMBBR) can have a negative impact on the efficiency of both nitrification and denitrification. In comparison, a feed C/N ratio of 2.7 in the mMBBR can lead to a stronger nitrification process than those reported at higher C/N ratio.

3.4.4 Sludge flocs

The structure of sludge flocs, including their density and size, plays a crucial role in the SND process. This is because the structure of sludge flocs was responsible for the transfer of substances such as DO and organic matter within the flocs, the formation of anaerobic micro-environments, the size of anaerobic and aerobic zones, and the ease with which microorganisms can obtain nutrients and carbon sources. It can be difficult or even impossible to form anaerobic micro-environments for denitrification within the flocs with loose structure and tiny particle size. It was found that sludge flocs with a diameter of 155 μm exhibited higher denitrification activity in the SND process compared to those with a diameter of 77 μm. However, as aggregate increase in size, substrate removal typically becomes mass transfer-limited and organisms compete for substrates and space within the aggregate, and thus having a negative influence on SND process.

3.5 Applications of SND

SND has been the subject of investigation across diverse operational conditions and reactor configurations. This section critically describes the configurations and essential features of the most utilized biofilm reactors for SND, namely Moving Bed Biofilm Reactors (MBBR), Hybrid Moving Bed Biofilm Reactor–Membrane Bioreactor (hybrid MBBR–MBR), and Aerobic Granular Sludge (AGS), with the aim of offering guidance for successful and cost–effective operation.

3.5.1 Moving bed biofilm reactors

MBBR has been widely adopted as an effective manner for the wastewater treatment. The presence of mobile carriers within the reactor facilitates the development of a stratified biofilm (Figure 3–5). The microbial community in biofilm is influenced by the wastewater composition and operational conditions. Substrates are transported from the bulk into a co–diffusional biofilm through diffusions, allowing for the formation of anaerobic, anoxic, and aerobic layers within the biofilm. In comparison to fixed–bed biofilm systems, MBBR is more advantageous due to low head loss, high specific surface area, non–clogging, and no backwashing.

Biofilm systems and hybrid processes are increasingly becoming successful alternatives to suspended growth systems for nutrient removal on a broader scale. This shift helps avoid limitations associated with increased hydraulic and organic loadings. Furthermore, these systems allow for the incorporation of a higher amount of active biomass into the treatment plant, negating the need for sludge recycle. Recent developments include the establishment of new MBBR plants, not only for new setups but also for economically enhancing existing conventional treatment facilities for both municipal and industrial wastewater containing high nitrogen concentrations, such as cheese and dairy wastewater, refinery effluent, slaughterhouse wastewater, and effluents from the pulp and paper industry. Various factors like carrier type, C/N ratio, DO concentration, and filling ratio have been studied concerning the SND process in MBBR systems. Table 3–1 and Table 3–2 present the performance and operating conditions for the SND process in continuous–flow MBBRs and MBBR–SBRs respectively.

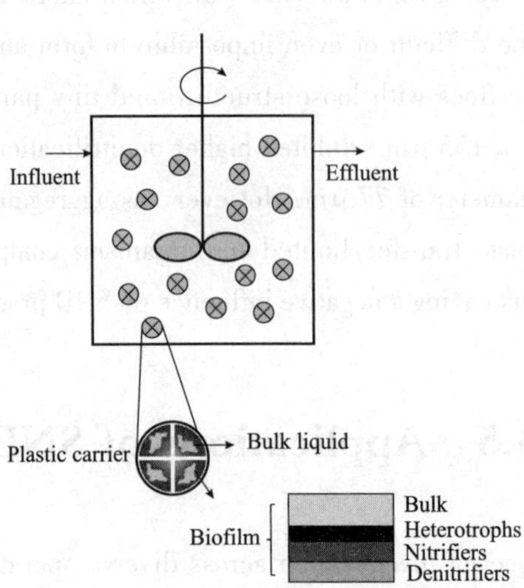

Figure 3–5 Biofilm development and stratification in MBBR applied for SND

Table 3–1 SND performance and operating conditions of continuous–flow MBBRs treating municipal wastewater

Carrier type	Specific surface/ (m^2/m^3)	Materials	Filling ratio/ %	HRT/h	DO/ (mg/L)	Carbon source	Feed TN/ (mg/L)	Feed C/N	COD RE /%	TN RE /%
Novel carrier	560–600	PE–PQAS 10–Fe_2O_3	30	8	0.6–0.8	BOD_5 Glucose	30–70	≤ 5[b]	79–86	62–76
Conventional carrier	560–600	PE	30	8	0.6–0.8	BOD_5 Glucose	30–70	≤ 5[b]	—	45–72
Novel carrier	560–600	PE–PQAS 10–Fe_2O_3	35	8	0.75	Glucose	40	5–9[b]	86–91	45–78
Conventional carrier	560–600	PE	35	8	0.75	Glucose	40	5–9[b]	85	43–62
PE plastic	500	PE	20–40	5	3.0	Glucose	40 ± 10	5	81	34–42
PU foam	280	PU	20–40	5	3.0	Glucose	40 ± 10	5	83	47–59
Zeolite powder–based PU sponges	0.846[a]	PU–zeolite (SiO_2)	10	12	5.0–6.5	—	28–33	3.5	>94[c]	84
Sponges biocarriers	0.846[a]	PU	10	12	5.0–6.5	—	28–33	3.5	>94[c]	75
Bio–carriers	1200	—	30	10	—	Glucose	51	6.3[b]	<90	≤30
Cubic–shaped sponges	—	PU	10–30	12	5.0–6.5	Glucose	16	7.1	94[c]	85–95
Kaldnes–3	500	PE	50	4–12	2.0–4.0	BOD_5	—	5[b]	50–88	26–46
Ring R2	500	PE	50	4–12	2.0–4.0	BOD_5	—	5[b]	54–85	26–50
PUF carrier	1120	PU	20–40	5–7	4.0–6.0	Glucose	25 (± 3)	~11[b]	81	25–54
PU carrier	900	PU	20	14	—	Glucose	44 (± 5)	3.7[b], 8.1[b]	90[c]	20–55
Biodegradable polymer	0.346[a]	PCL	16.7	14	—	Glucose	44 (± 5)	3.7[b], 8.1[b]	72[c]	59
Polyethylene carriers	500	PE	30	10	2.0–4.0	—	100	4.5–13.4	94–96	26–51

[a] reported as m^2/g.
[b] available as COD/N.
[c] measured as total organic carbon (TOC) RE.

Table 3–2 SND in MBBR–SBR systems

Carrier type	Specific surface/ (m^2/m^3)	Materials	Filling ratio/ %	Cycle Length/ min	AN phase/ min	O phase/ min	A phase/ min	Settling/ min	Idle/ min	Discharge/ min	DO/ (mg/L)	Carbon source	Feed TN/ (mg/L)	Feed C/N[b]	COD RE/ %	TN RE/ %
PU foam	—	PU	30	—	—	600	—	—	120	—	1.5–5.5	$(C_6H_{10}O_5)_n$	25	12	–92	42–85
AC CBMC	2.70[a]	AC, PE	40, 55	360	—	350	—	—	—	5	6.0	Acetate	55	5	82–94	68–88
Conventional carrier	0.2478[a]	PE	40, 55	360	—	350	—	—	—	5	6.0	Acetate	55	5	82–94	59–76
Combined fibers	—	PP	—	—	175	90	175	30	—	5	2.0	Glucose	100	3.8	94	73
Novel carrier	650	PQAS–10+ Fe_2O_3+PE	30	240	—	200	40	—	—	—	1.0–1.5	—	50	5	81–84	78–80
PUF	—	PU	30	600, 480	—	—	—	—	—	—	0.5–1.5	Acetate	89	~4.5	85	60
HX9KL, BIOMASTER BCN 012KLS	500	PE	40, 60	240, 360	—	90, 50	150, 90	0, 10	—	0, 3	0.8–1.1	—	35 (±7) 47 (±6)	5.5–6	78–95	50–76
Polyurethane filler + rice husk	382	PU + rice husk	15.8	720	—	240, 60	120–480	—	—	30	0.1–2.0	Acetate, yeast, sucrose	35 (±4)	1.8–7.6	64–83	41–89
Foam carrier	—	PU	30	720	—	600	—	—	120	—	1.5–5.5	$(C_6H_{10}O_5)_n$	22–31	~10.2	> 90	42–84
Plastic fiber	—	—	12	720	50	190–270	380–460	—	—	—	1.5–2.2	Glucose	50	1–4	50–90	83–98
Cylindrical carrier	—	PP	30	—	90	210	90	30	40	—	0.2–2.5	—	35	11.4	95	94
PU foam	1621	PU	—	360	—	297	—	50	—	—	0.1–0.9	Glucose	65–227	1.8–10.5	—	55–79
PU foam	—	PU	8	—	60	60	—	90	570	—	0–7	Peptone, sucrose, acetate	48	5	65	57–100
Bio–carrier K3	500	PE	30	480	60	240	120	30	—	—	—	Acetate	35	15.7	—	71.5

AN= Anaerobic phase.

O= Aerobic phase.

A= Anoxic phase.

[a]reported as m^2/g.

[b]available as COD/N.

1. Carrier type

The choice of carrier material and its physical properties are critical factors for the adhesion and reproduction of microorganisms. The characteristics of carriers, such as shape, surface area, porosity, and roughness, can affect biofilm development and modulate gradients of parameters within the biofilm, including DO. Various carrier types have been investigated for supporting SND in MBBRs, including polyethylene (PE), polyurethane (PU) sponge, high-density PE (HDPE) carriers, ceramic carriers, and zeolite powders combined with PU sponge (Z-PU).

Sponges, in particular, exhibit a large specific surface area and high porosity, which are conducive to microbial growth and biofilm development. Comparative studies have shown that MBBRs filled with PU foam carriers can achieve higher TN removal efficiencies due to the higher nitrifying activity and increased biofilm-liquid contact area of PU foam compared to those with conventional PE carriers. Besides, the proper pore size facilitates sufficient transfer of mass in the interior of the pores and thus resulting in the attachment of nitrifiers/denitrifiers and the removal of nitrogen. It has been reported that the sponge of the pore size 46 ppi exhibited excellent performances for mass transfer, with a high biomass concentration of 2135.6 mg/L, a COD uptake rate of 150.2 mg/h, and a nitrogen uptake rate of 17.5 mg/h. Moreover, the size of the sponge carriers has also been found to impact SND performance, with smaller cubes facilitating a favorable DO gradient for the denitrification process. The biofilm carriers harbor the potential for NO_x^- reduction in MBBR even in the absence of an external carbon source, possibly due to carbon storage within the deep biofilm layers. Overall, considering the effect of carrier type on biofilm thickness, depth, and diffusion, it is important to design and select the optimal carrier for maximizing SND in MBBRs.

2. Filling ratio

The filling ratio in the operation of MBBRs typically ranges between 20% and 70%. A value <70% is generally recommended for adequate mixing, while a ratio >20% for sufficient concentration of attached-growth biomass within the MBBR. Mixing and aeration maintain a relative balance between the growth and detachment of biofilms in order to sustain the growth of biofilms under steady-state conditions and a constant biofilm thickness. The filling degree has a significant impact on the overall performance and operational costs, such as energy consumption and carrier procurement. Feng, et al. (2012) evaluated the impact of different polyurethane foam (PUF) carrier filling ratios (20%, 30%, and 40%) on SND performance in three parallel MBBRs. As a result, the TN removal efficiency in MBBR filled with 40% PUF carriers was much higher than that of MBBR filled with 20% PUF carriers (Table 3-1). Furthermore, the filling ratio is responsible for biofilm structure and nitrification activity. A dense biofilm was observed in MBBR with a 20% filling ratio, whereas a thinner biofilm and hollow structure were found on the carrier in MBBR with high filling ratios. According to microprofile analysis, the dense biofilm can restrict the diffusion of DO into the inner layer for the inhibition of nitrifier growth, as evidenced by the higher relative abundance of nitrifying bacteria *Nitrosomonas* and *Nitrospira* in MBBRs with filling ratios of 30% and 40% compared to those with a 20% filling ratio. Similar results was observed by Ferrentino, et al. (2018), where the filling ratio from 40% to 60% in a Sequential Batch Reactor (SBR) –MBBR performing SND led to an enhanced nitrification activity, corresponding in an obvious increase in TN removal.

Zhang, et al. (2017) investigated the SND process in three MBBRs filled with PUF carriers at filling ratios of 10%, 20%, and 30%, where the MBBR filled at 30% achieved the highest average TN removal efficiency of 87% with the highest rates of NH_4^+ oxidation and denitrification. There was no significant difference in maximum TN removal efficiency (94%–95%) and SND efficiency (98%–100%) for the MBBRs filled with 20% and 30% carriers. By comparison, Lim, et al. (2012) observed that the increase of the filling ratio from 20% to 30% and 40% in three Integrated Anoxic MBBR–Sequential Batch Reactors (IA MBBR–SBRs) could reduce suspended biomass concentration and thus resulting in a decline of NH_4^+ removal. Nevertheless, clearly visible was an enhanced denitrification efficiency due to the increase of anoxic zone and carbon storage in the biofilm. Accordingly, the optimal filling ratio in the IA MBBR–SBRs system was determined to be 30%, which was responsible for the maximized TN removal by SND. Overall, despite variations in biomass dynamics in different MBBR systems, filling ratios ⩾30% might be an optimal ratio to enhance SND performance and nitrogen removal.

3.5.2 Hybrid moving bed biofilm reactor–membrane bioreactor systems

The integration of hybrid moving bed biofilm reactor and MBR, called hybrid MBBR–MBR, has been shown to improve pollutants removal and alleviate membrane biofouling (Figure 3–6). Specifically, on the one hand, the addition of carriers in MBR systems can cause a strong scouring effect on membrane surface, and hence the loose surface of the biofilm could be detached under continuously hydraulic shear force. On the other hand, the carriers in system are beneficial for the growth of microorganisms in the mixed liquor on their surface, forming a biofilm with hypoxic zone, which can enhance nitrification and denitrification. It has been reported that the formation of anoxic and aerobic micro–environments within the carrier–attached biofilm was responsible for an around 8% enhancement of the TN removal efficiency in the hybrid MBBR–MBR compared to the conventional MBR. Besides, a slower increase of the transmembrane pressure (TMP) and 1.5 times longer operational time in the hybrid MBBR–MBR were observed compared to those in conventional MBR. However, the excessive addition of biofilm carriers could cause the breaking up of sludge flocs, resulting in the increase of small particles and SMP content. The above would has a negative impact on membrane fouling control. Therefore, maintaining an appropriate carrier–to–suspended biomass ratio is important for SND performance and membrane fouling mitigation. Despite excellent performances in hybrid MBBR–MBR, the operation of HMBRs is rather expensive due to the high cost of membranes and carriers, and hence a cost–benefit evaluation for full–scale application should be addressed.

3.5.3 Aerobic granular sludge systems

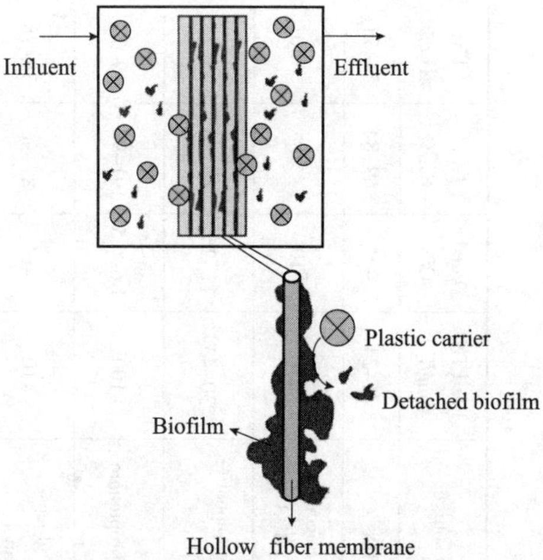

Figure 3–6 Biofilm development and stratification in Hybrid biofilm–membrane bioreactor applied for SND

Aerobic granular sludge (AGS) is much favored by synergistic virtue of activated sludge and biofilm systems. Aerobic granules are near–spherical bacterial aggregates with larger size and denser structure than sludge flocs. Therefore, the AGS is more advantageous in terms of structure, biomass retention, settleability and biodegradation ability compared with sludge flocs smaller than 0.2 mm. In general, hydrodynamic shear force and feast–famine feeding regimes involving alternating anaerobic and aerobic conditions can promote granules formation (Figure 3–7). The micro–environments consisting of aerobic, anoxic, and anaerobic zones within the granule enable the coexistence of different bacterial populations performing specific roles in the biodegradation of water pollutants, such as nitrifying and denitrifying bacteria, phosphate accumulating organisms (PAOs), and glycogen accumulating organisms (GAOs), which is conducive to the simultaneous removal of C, N, and P.

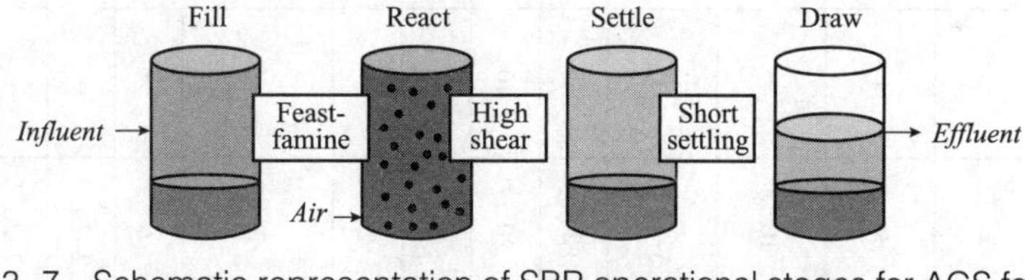

Figure 3–7 Schematic representation of SBR operational stages for AGS formation

Table 3–3 presents the operational parameters and SND performance of AGS reactors. The most common setup to enrich for aerobic granules performing SND and to maximize C, N, and P removal is to cultivate AGS in a SBR. SND in AGS–SBRs has been successfully established in lab–scale SBRs with a height/diameter ratio (H/D) between 5 and 10. Despite the excellent TN removal in AGS system, the poor long–term stability of aerobic granules remains a challenging issue. The stability of AGS is affected by the various operational factors, including feed C/N ratio, substrate type, aeration rate, organic loading rate, biomass loading rate, toxic substances, temperature, as well as intermittent feast–famine and aerobic and non–aerated conditions. These operational factors could cause the variations in granule size, microbial composition, filamentous population and EPS composition, and thus impacting on AGS stability and SND process. The following are several factors closely related to the SND process.

Table 3-3 SND applications in AGS systems

Cycle length/h	Cycle	AN phase/ min	O phase/ min	A phase/ min	Settling/ min	Discharge/ min	Granule diameter/mm	H/D	DO/(mg/L) (O)	Carbon source	Feed TN/ (mg/L)	Feed C/N	COD RE/%	TN RE/%
3-6	AN-O	20-60	30-250	—	13.3-25	3.3-6.7	1.5	18.75	1.0-2.5	BOD_5	31-34	-5[a]	69-84	31-71
6	AN-O-A	120	90	144	2	—	—	—	—	Acetate, Succinate (1 : 1, 1 : 3, 3 : 1)	20	-10[a]	90	-80
5.6	AN-O	90	240	—	5	1	>0.3	8.4	2	Acetate-propionate-BOD_5	30-43	11-26[a]	83-93	45-77
4	AN-O-A, AN-O, O	0-45-55	104-220	0-30	10	—	0.2	10	8-10	Acetate, propionate	100	1.6-6.5	91-98	19-75
6	AN-O	54	270-285	—	5-20	1	0.7-1.5	10	—	Acetate, glucose, ethanol	110	-6[a]	87-96	34-78
6	AN-O-A	120	90	144	2	—	—	5.2	7-8	Acetate	2-55	4-20[a]	94-95	75-95
3.6	O	—	150	—	20	40	0.09-0.4	—	2	BOD_5	91	-14.7[a]	92	87
6	AN-O-A	120	90	144	2	—	—	5	5	Acetate	20	-10[a]	89.2	88.5
6	AN-O-A	120	90	144	2	—	—	5	0.8-1.2	Acetate, glucose	20	-10[a]	81-91	81-94
6	AN-O-A	120	90	144	2	—	1.4 ± 0.6 1.8 ± 0.9	5	3.5-4.5 1.7- 2.3 0.8-1.2	Acetate	20	-10[a]	95	64-94
6	AN-O-A	120	90	144	2	—	—	5	5	Acetate	19	10.7[a]	94	94
8	AN-O-A	180	140-180	90-148	2-20	—	0.6-1.0	6.3	5	Acetate	40-50	-7[a]	80-90	88-97
8	AN-O-A	180	150-180	90-138	2-20	—	1.2	5.2	5	Acetate	15	-10[a]	87	97
6	AN-O	130	160	—	60	—	0.3-1.6	4.2	0.8-1.6	Acetate	50	-16[a]	—	65
4	O	—	180	—	5	2	0.5-1.7	8	>4.0	Acetate	250-330	-1[a]	100	33-77
4	AN-O	10-20	184-219	—	10-35	—	0.5	9.7	—	BOD_5	83	-7[a]	92	61

Continued

Cycle length/h	Cycle	AN phase/ min	O phase/ min	A phase/ min	Settling/ min	Discharge/ min	Granule diameter/mm	H/D	DO/(mg/L) (O)	Carbon source	Feed TN/ (mg/L)	Feed C/N	COD RE/%	TN RE/%
6.5–3	AN–O–A	—	60–300	—	30	—	>0.2	—	1.8–2.5	BOD_5	50	−10[a]	87	86
3	O	—	162–224	—	2–7	6	>1.5	11.1	7–8	Acetate	77–154	8–16	>90	>90
3–4	O	—	112	—	15	0.5	0.3–0.7	22.2	—	BOD_5	76–91	4.7–11.2[a]	78–92	66–97
3	AN–O	60	35–160	3–40	3	5	0.9–1.1	—	20%–80% of saturation	Propionate	50	8–1.2[a]	100	40–95
6	AN–O–A	45–100	114	—	2–40	—	0.2	—	2	BOD_5+ Acetate	67	−5[a]	100	75–84
5.4	O	—	213	10	3	3	2.4	7.2	6–8	Sucrose, acetate, propionate	40	−10[a]	>80	>75
4	A–O	—	130–167	—	2	5	0.7–1.5	20	2	—	60	−10[a]	80–85	68–96
4	O	—	240	5	8–45	5	0.8	5	8	BOD_5	60	−16.7[a]	80	50
2.5–4	A–O	90	90	30	—	—	—	3.2	2–3	Acetate–BOD_5	43–147	0.4–2.3	64–89	47
—	AN–O–A	30	120	—	—	—	>1.0	—	—	Acetate	60	2.5	−100	−100
6	AN–O	90	169	—	0.5	—	1.0	—	2–5	Acetate	60	−10[a]	−100	−100
3	O	—	230	0–120	3	5	1.0	—	10%–100% or saturation	Acetate	—	8.3[a]	—	8–45
4–6	O, AN–O, AN–A	0–120			2	4	0.4–1.9	13.3	0.5–2.0	Ethanol	25–150	3–5[a]	92	20–40

AN= Anaerobic phase.

O= Aerobic phase.

A= Anoxic phase.

[a]available as COD/N.

1. Granule size

Granule size is a major factor affecting the removal of nitrogen. Nutrient accessibility and DO penetration are reported to associated to granule size. Limited anoxic zones in smaller granules have no significant impact on SND efficiency. For example, the granules with small sizes (0.25 mm < d < 0.63 mm) exhibited a low efficiency SND due to limited extension and carbon availability in the anoxic zone, while a diameter of 0.9 mm could significantly enhance TN removal efficiency. Chen, et al. (2011) reported an increase in TN removal efficiency from 68% to 72% with an increase in granule size from 0.7 to 1.5 mm. However, excessive granule size also hinders SND process. Both filamentous outgrowth and pore clogging in large AGS have a negative impact on the effective diffusion of nutrients and DO down to the deeper layers within granule. The consequent starvation conditions and the anaerobic micro-environment within the core of granule result in the decline of microbial activity, endogenous respiration, cell lysis and even granule disintegration. Accordingly, this is not conducive to SND. In general, aerobic granules exhibit diameters ranging from 0.2 to 3.0 mm. The size of granule cultivated with real municipal wastewater varies between 0.2 and 1.3 mm, while granules developed with synthetic wastewater in lab-scale SBRs have a larger size (> 2.0 mm). It is well-known that 1-3 mm is regarded as the appropriate granule size range for large-scale SBR operation. To maximize SND efficiency, the optimal average diameter of AGS can be maintained approximately 2.0 mm.

2. Substrate type

Different type of substrate can cultivate granules with different structures. Typically, carbon source highly affects the microbial community composition and granule density in AGS systems, and thus influencing nitrogen removal. Common substrates include glucose, acetate, phenol, starch, ethanol, molasses and sucrose. For instance, acetate feeding can form large and compact granules (average diameter of 1.5 mm) with high microbial diversity, excellent pollutant removal efficiency (TN: 72%, TP: 42%). In comparison, glucose-fed AGSs harbor a loose morphology and fluffy surface due to a dominance of filamentous bacteria, corresponding in the low TN and TP removal efficiencies (44% and 21%, respectively), and low microbial diversity. Besides, ethanol-fed AGSs exhibit moderate TN removal efficiency (53%) and low TP removal efficiency (31%). Taken together, energy-rich substrates (e.g., glucose, sucrose) might induce the proliferation of filamentous bacteria, while simple carbon substrates but with a large molecule (e.g., acetate) select for dense and uniform microstructures. Regarding complex organic substrates such as carbohydrates and peptone, they can be degraded via multistep processes, which induces the formation of AGS with multi-layered microstructures and greater microbial diversity.

Real wastewaters collected from domestic or industrial origin contain diverse carbon sources along with a variety of other organic and inorganic compounds. This complex composition and the different substrate degradation rates can influence the structure and microbial community in AGS. For example, the granules fed with domestic wastewater were observed less dense structure and lower stability compared to the acetate-fed AGS, which might possess a poor SND performance.

3. Aeration strategy

Aeration strategy plays a significant role in AGS formation and stability via oxygen supply and hydraulic shear force, which can affect SND performance in AGS system. Rapid penetration of oxygen can diminish the duration of the anoxic phase and thus inhibiting denitrification, whereas extreme low aeration rate can induce the overgrowth of filamentous bacteria at the granule surface and subsequently disintegrate leading to biomass washout. As such, the oxygen supply should be optimized for minimizing energy consumption and ensuring efficient N-removal in AGS system. Intermittent aeration (IA) during the aerobic phase was reported to enhance the persistence of anoxic zones and then facilitate denitrification. As discussed by Layer et al. (2020), the 2-step aeration maintains DO levels at 2.0 mg/L for 15 min and at 0.5 mg/L for 270 min in the aerobic phase, while IA was conducted by activating aeration every 15 min that DO range is 0–2.0 mg/L. As a result, a substantial increase was observed in denitrification efficiency from 14%–37% to 65%–79% and in TN removal efficiency from 13% to over 65%.

Considering the differences in granule cell physiology and metabolic kinetics between the feast and famine periods in the SBR operational cycle, reducing appropriately the aeration rate during the famine period is a strategy to effectively decrease the total energy input of the AGS system without compromising the granule stability. It has been reported that the aeration rate with a superficial velocity of 0.55 cm/s during famine conditions can maintain the stable structure of granule for more than 3 months.

Questions

[1] What is the mechanism of SND?

[2] How many environmental factors can affect SND performance? Please list four factors and describe them.

[3] In the MBBRs, what are the strategies to achieve high SND performance?

[4] In the aerobic granular sludge systems, what are the strategies to achieve high SND performance?

[5] Please analyze the application potential of SND process.

References

[1] Bhattacharya R, Mazumder D. Simultaneous nitrification and denitrification in moving bed bioreactor and other biological systems [J]. Bioprocess and Biosystems Engineering, 2021, 44(4): 635–652.

[2] Chen F Y, Liu Y Q, Tay J H, et al. Operational strategies for nitrogen removal in granular sequencing batch reactor [J]. Journal of Hazardous Materials, 2011, 189(1): 342–348.

[3] Di Capua F, Iannacone F, Sabba F, et al. Simultaneous nitrification–denitrification in biofilm systems for wastewater treatment: key factors, potential routes, and engineered applications [J]. Bioresource Technology, 2022, 361: 127702.

[4] Feng Q, Wang Y, Wang T, et al. Effects of packing rates of cubic–shaped polyurethane foam carriers on the microbial community and the removal of organics and nitrogen in moving bed biofilm reactors [J]. Bioresource Technology, 2012, 117: 201–207.

[5] Ferrentino R, Ferraro A, Mattei M R, et al. Process performance optimization and mathematical modelling of a SBR–MBBR treatment at low oxygen concentration [J]. Process Biochemistry, 2018, 75: 230–239.

[6] Franca R D G, Pinheiro H M, Van Loosdrecht M C M, et al. Stability of aerobic granules during long–term bioreactor operation [J]. Biotechnology Advances, 2018, 36(1): 228–246.

[7] Körner H, Zumft W G. Expression of denitrification enzymes in response to the dissolved oxygen level and respiratory substrate in continuous culture of *Pseudomonas stutzeri* [J]. Applied and Environmental Microbiology, 1989, 55(7): 1670–1676.

[8] Layer M, Villodres M G, Hernandez A, et al. Limited simultaneous nitrification–denitrification (SND) in aerobic granular sludge systems treating municipal wastewater: mechanisms and practical implications [J]. Water Research X, 2020, 7: 100048.

[9] Lim J W, Lim P E, Seng C E. Enhancement of nitrogen removal in moving bed sequencing batch reactor with intermittent aeration during REACT period [J]. Chemical Engineering Journal, 2012, 197: 199–203.

[10] Liu Y Q, Tay J H. Cultivation of aerobic granules in a bubble column and an airlift reactor with divided draft tubes at low aeration rate [J]. Biochemical Engineering Journal, 2007, 34(1): 1–7.

[11] Liu Y Q, Tay J H. Variable aeration in sequencing batch reactor with aerobic granular sludge [J]. Journal of Biotechnology, 2006, 124(2): 338–346.

[12] Nancharaiah Y V, Kiran Kumar Reddy G. Aerobic granular sludge technology: mechanisms of granulation and biotechnological applications [J]. Bioresource Technology, 2018, 247: 1128–1143.

[13] Pochana K, Keller J, Lant P. Model development for simultaneous nitrification and denitrification [J]. Water Science and Technology, 1999, 39(1): 235–243.

[14] Rout P R, Shahid M K, Dash R R, et al. Nutrient removal from domestic wastewater: a comprehensive review on conventional and advanced technologies [J]. Journal of Environmental Management, 2021, 296: 113246.

[15] Sarma S J, Tay J H. Carbon, nitrogen and phosphorus removal mechanisms of aerobic granules [J]. Critical Reviews in Biotechnology, 2018, 38(7): 1077–1088.

[16] Starr J L, Broadbent F E, Nielsen D R. Nitrogen transformations during continuous leaching [J]. Soil Science Society of America Journal, 1974, 38(2): 283–289.

[17] Wang K M, Jiang S F, Zhang Z H, et al. Impact of static biocarriers on the microbial community, nitrogen removal and membrane fouling in submerged membrane bioreactor at different COD : N ratios [J]. Bioresource Technology, 2020, 301 : 122798.

[18] Zeng R J, Lemaire R, Yuan Z, et al. Simultaneous nitrification, denitrification, and phosphorus removal in a lab-scale sequencing batch reactor [J]. Biotechnology and Bioengineering, 2003, 84(2): 170–178.

[19] Zhang X, Song Z, Guo W, et al. Behavior of nitrogen removal in an aerobic sponge based moving bed biofilm reactor [J]. Bioresource Technology, 2017, 245 : 1282–1285.

Chapter 4 Nitrite-based anaerobic ammonia oxidation (Anammox)

Emphasis: Nitrite-based Anammox; Nitrite-based Anammox microorganisms and central metabolism; Factors affecting Anammox of nitrite type; Nitrite type Anammox main process.

Nut: Factors affecting Anammox of nitrite type.

4.1 Discovery of nitrite-based Anammox

Scientists can combine different electron donors and acceptors based on thermodynamic information to hypothesize possible reactions on which microorganisms depend. An Austrian chemist proposed two inorganic trophic reactions based on thermochemistry, one of which was a microbial reaction utilizing oxidation, as shown in Eq. However, this microbial reaction was not found in nature for ten years after it was proposed. Scientists at the time believed that ammonia was chemically stable and that oxygen and the versatile ammonia monooxygenase enzyme must be involved in ammonia oxidation, so it was considered impossible to oxidize ammonia and nitrogen using nitrite under anaerobic conditions.

$$NO_2^- + NH_4^+ \longrightarrow N_2 + 2H_2O$$

In the 1980s, ammonia loss was found in denitrification pilot reactors, and further studies revealed that ammonia nitrogen loss was accompanied by nitrate loss. And this possible reaction process was named anaerobic ammonia oxidation, abbreviated as Anammox. Subsequently, we took the sludge from the above reactor and found that ammonia nitrogen and nitrate were consumed synchronously, and the oxidation of ammonia nitrogen stopped when nitrate was consumed, and the reaction occurred again after nitrate was added again. This process is accompanied by the production of gas, the gas is mainly nitrogen, so that the above reactor process is considered to be a microbial reaction rather than a chemical reaction. The substance flow and functional genes in the nitrogen cycle are shown in Figure 4-1.

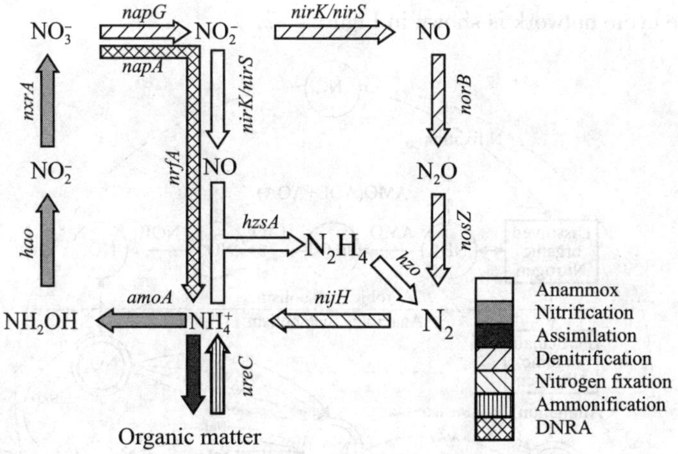

Figure 4-1 The substance flow and functional genes in the nitrogen cycle

4.2 Nitrite-based Anammox stoichiometric ratio

As a key factor that determine the operational stability of Anammox process, the stoichiometry of the Anammox metabolism is important for process design and control. The most widely acknowledged Anammox stoichiometric equation was proposed on the basis of a SBR reactor. In that study, mass balance was enabled for the first time under stable operational conditions. In recent year, a new stoichiometric equation was proposed with a smaller error in the electron balance and a higher accuracy in the carbon balance.

Thus, maintaining a balance between the different available microbial groups involved is critical. The Anammox process involves "the oxidation of NH_4^+ to N_2 gas using inorganic carbon (such as CO_2) as carbon source, nitric oxide (NO) and hydrazine (N_2H_4) as intermediates and NO_2^- as the electron acceptor under anoxic conditions", represented by the Eq. below. This stoichiometric equation has been revised, further stressing the importance of NO_2^-/NH_4^+ ratio for Anammox process design and optimization. Implementing and achieving a stable and effective treatment process requires a thorough understanding of the stoichiometry of biological reactions.

$$NH_4^+ + 1.32NO_2^- + 0.066HCO_3^- + 0.13H^+ \longrightarrow 1.02N_2 + 0.26NO_3^- + 0.066CH_2O_{0.5}N_{0.15} + 2.03H_2O$$

$$NH_4^+ + 1.146NO_2^- + 0.071HCO_3^- + 0.057H^+ \longrightarrow 1.02N_2 + 0.986NO_3^- + 0.071CH_2O_{0.5}N_{0.15} + 2.002H_2O$$

The latest stoichiometry, indicates a positive correlation with NLR. Elevated NLRs results in high cell yield, causing excess nitrate production. It was surmised that Lotti's stoichiometry was valid only at low NLR while Strous' was also valid under high NLR conditions. To this day, no foolproof stoichiometry has been offered. As a result, comparative examination of the various postulates is required, as it is critical in maintaining a favorable NO_2^-/NH_4^+ ratio, even though careful adjustment contingent on the prevalent nitrogen species in the effluent seems plausible.

Nitrogen biosphere cycle network is shown in Figure 4-2.

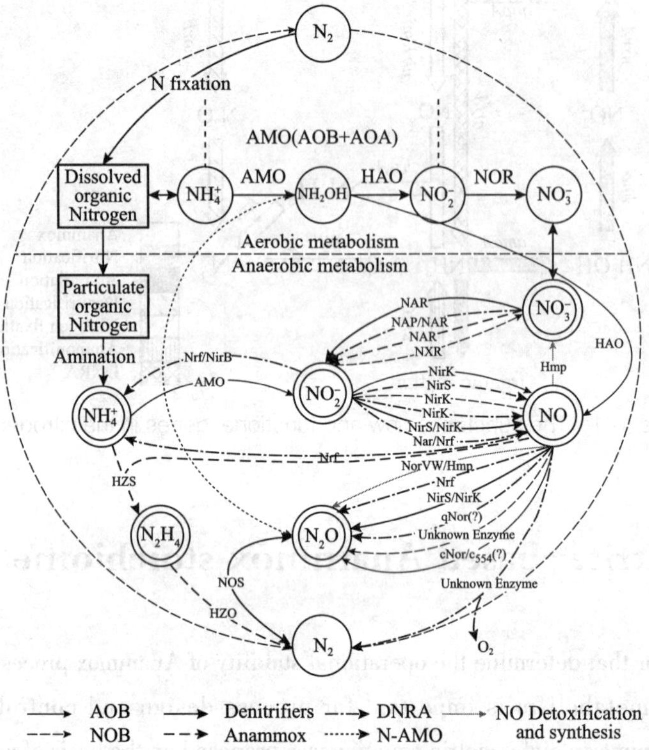

Figure 4-2 Nitrogen biosphere cycle network

4.3 Nitrite-based Anammox microorganisms and central metabolism

4.3.1 Diversity of functional bacteria for nitrite-based Anammox

Based on 16S rRNA sequencing technology, anaerobic ammonia-oxidizing bacteria are classified as deep-branching bacterioplankton, and 26 species of anaerobic ammonia-oxidizing bacteria belonging to six genera have been identified so far with the continuous research (see Table 4-1). Since the first discovery of anaerobic ammonia-oxidizing bacteria in artificial bioreactors, researchers have found anaerobic ammonia-oxidizing bacteria in natural environments such as marine sediments, oil fields, estuarine sediments, mangrove forests, urban inland waterways, oceans, freshwater lakes, paddy soils, lake sediments, and submarine hot springs. *Candidatus Brocadia*, *Candidatus Kuenenia*, *Candidatus Anammoxoglobus*, *Candidatus Jettenia*, *Candidatus Anammoximicrobium* were mostly found in freshwater environments or artificial bioreactors. *Candidatus Scalindua*, on the other hand, is found mainly in the sea, especially in some marine sediments and in the oxygen minimum zone (OMZ). *Brocadia* and *Jettenia* dominate the soil. It indicates that anaerobic ammonia-oxidizing bacteria have a wide ecological range and diverse living environments, and are widely distributed in nature. These findings will contribute to the wide application of anaerobic ammonia-oxidation process, and the treatment of high temperature, high salt and toxic and hazardous high concentration nitrogenous wastewater will be expected.

Table 4–1 The anaerobic ammonia-oxidizing bacteria reported currently

Genus	Species	Source	Year of discovery	Doubling time
Anammoxoglobus	*Candidatus Anammoxoglobus propionicus*	Delft Laboratories SBR Strainer, Netherlands	2007	—
	Candidatus Anammoxoglobus sulfate	China Laboratory Biological Carousel Reactor	2008	—
Anammoximicrobium	*Candidatus Anammoximicrobium moscowii*	Moscow River Bottom Mud	2013	32
Brocadia	*Candidatus Brocadia Anammoxidans*	Gist-Brocades Wastewater Treatment Plant, Netherlands	1999	10.7
	Candidatus Brocadia fulgida	Rotterdam Wastewater Treatment Plant, Netherlands	2001	18
	Candidatus Brocadia sinica	China Denitrogenation Reactor	2010	7
	Candidatus Brocadia brasiliensis	Brazilian Wastewater Treatment Plant SBR Reactor	2011	—
	Candidatus Brocadia caroliniensis	U. S. Hog Farm Wastewater Denitrification System	2011	—
	Candidatus Brocadia sapporoensis	Japan MBR Nitrogen Removal Reactor	2017	3.5
Jettenia	*Candidatus Jettenia asiatica*	China Laboratory Biofilm Reactor	2008	—
	Candidatus Jettenia ecosi	Moscow Laboratory Vertical Anaerobic Upflow Bioreactor	2018	13
	Candidatus Jettenia caeni	Japan Sewage Treatment Plant	2012	14.2
	Candidatus Jettenia moscovienalis	Moscow Digestive Sludge Bioreactor	2015	28
Kuenenia	*Candidatus Kuenenia stuttgartiensis*	Biofilter, Stuttgart, Germany	2000	8.3–11
Scalindua	*Candidatus Scalindua brodiae*	Pitsea Landfill Leachate, UK	2003	—
	Candidatus Scalindua sorokinii	Sub-oxygenated Region of the Western Black Sea Basin	2003	—
	Candidatus Scalindua wagneri	Pitsea Landfill Leachate, UK	2003	—
	Candidatus Scalindua sinooifed	Shengli High Temperature Oilfield, China	2010	—
	Candidatus Scalindua marina	Marine sediments in the Gulf of Gullmar, Sweden	2011	—
	Candidatus Scalindua arabica	Black Sea Suboxic Region	2008	—
	Candidatus Scalindua profunda	Marine sediments in Gullmar Bay, Sweden	2008	—
	Candidatus Scalindua zhenghei	Seafloor sediments in the South China Sea	2011	—
	Candidatus Scalindua richardsii	Black Sea Suboxic Region	2012	—
	Candidatus Scalindua rubra	Red Sea	2017	—
	Candidatus Scalindua flavia	East China Sea and Yellow Sea, China	2017	—
	Candidatus Scalindua pacifica	Bohai Sea Floor Sediments	2013	—

Anaerobic ammonia–oxidizing bacteria are chemoenergetic autotrophic specialized anaerobic bacteria, most of them oxidize ammonia nitrogen/nitrite nitrogen under anaerobic conditions to obtain energy and use CO_2 as the source. Slow growth, long doubling time (3.5–32 d), suitable growth temperature of 20–43 ℃, the optimal growth temperature of 40 ℃ ; pH value range of 6.7–8.3, the best 8.0; sensitive to oxygen, can only be in the partial pressure of oxygen is less than 5% of the oxygen saturation concentration to survive. It is generally believed that anaerobic ammonia–oxidizing bacteria need to be enriched to a certain number (10^{11}) in order to play an active role, and mature anaerobic ammonia–oxidizing bacteria are generally red or brick red. Anaerobic ammonia–oxidizing bacteria have been shown to be widespread in marine, wetland, terrestrial and specialized ecosystems (geothermal reservoirs, deep–sea hydrothermal fluids, polar regions). It has also been found that bacteria of the genus *Pseudomonas* in the order *Pseudomonas spp.* also have some anaerobic ammonia–oxidizing activity, which means that bacteria with anaerobic ammonia–oxidizing function are not necessarily genera of the order *Pseudomonas aeruginosa*, and should be explored and studied further.

4.3.2 Central metabolic mechanism of nitrite–based Anammox

Currently, metabolic pathway models proposed were generally recognized. Briefly, NO_2^-–N was converted to hydroxylamine (NH_2OH) which was considered as the intermediate product. Then, NH_2OH reacted with NH_3 to form hydrazine (N_2H_4), and then, N_2H_4 was finally oxidized to N_2 by hydrazine–oxidizing enzyme (HZO) on the external side of anammoxosome. It proposed that NO_2^-–N was reduced to NO which was the intermediate product. Anammox bacteria generally coexisted with other microorganisms in the whole community, such as ammonia–oxidizing bacteria (AOB), heterotrophic microorganisms, nitrite oxidizing bacteria (NOB) and so on. The interaction of microbial communities in Anammox system was also a pivotal issue affecting the metabolism activities. Metabolic exchanges as well as cross–feeding are ubiquitous in microbial communities. So far, six genera of Anammox bacteria have been discovered, i.e., *Candidatus Kuenenia*, *Candidatus Scalindua*, *Candidatus Jettenia*, *Candidatus Anammoxoglobus*, *Candidatus Brocadia* and *Candidatus Anammoximicrobium*.

Quorum sensing (QS) is a chemical communication process that bacteria use to orchestrate the behavior of groups. Meanwhile, it could mediate the metabolism of various life activities of microorganisms. QS signaling molecules in the Anammox system, including acyl homoserine lactone (AHLs) type, autoinducer 2 (AI–2) type, autoinducer peptide (AIP) type, diffusion signaling factor (DSF) and c–di–GMP type were observed. Various studies have found that QS mechanism could regulate the metabolism of Anammox system, such as the synthesis and hydrolysis of extracellular polymeric substance (EPS) and the information exchange among microorganisms.

The effects of conventional environmental factors on metabolic mechanisms of the Anammox system is shown in Figure 4-3.

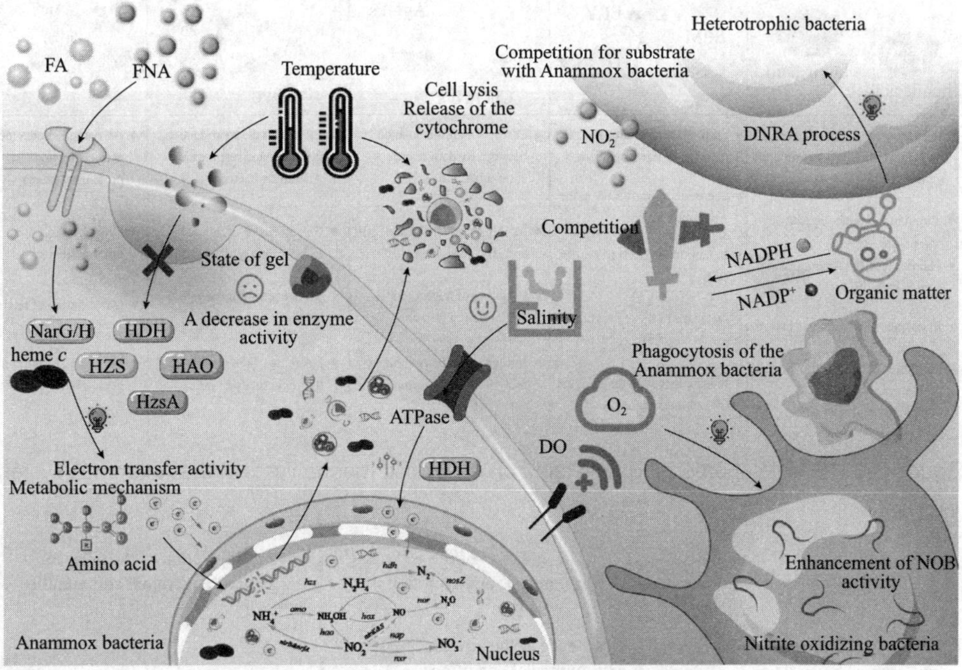

Figure 4-3 Effects of conventional environmental factors on metabolic mechanisms of the Anammox system

Several natural organic matters such as humic substances, including fulvic acid and humic acid, contained various functional groups, which could serve as electron mediators to mediate the activity and metabolic processes of microorganisms, shown in Figure 4-4.

As a density-dependent microbial communication mechanism, QS could regulate a variety of bacterial behaviors. QS bacteria would regulate the expression of relevant genes and trigger corresponding behaviors when the concentration of signaling molecules reached a threshold in the microenvironment. The up-regulation of QS gene expression level could contribute to improve coordinated interactions in microorganisms with the enhancement of the adaptability of Anammox bacteria to the complex dynamic environment, so as to ensure the stability and function of nitrogen removal system.

Studies have reported that QS promoted the growth and coordinated interactions of microorganisms under adverse conditions, such as the rapid start-up of Anammox system at low temperatures, the improving tolerance under the salinity stress. It was found that AHLs could affect the expression of Anammox metabolite-related genes, which further affect the activity of Anammox bacteria, and play an important role in regulating the mutual aid balance between microbial communities. The above contents mainly summarized the metabolic process regulated by QS mechanisms in the Anammox system. The role of QS on metabolic mechanisms of the Anammox system is shown in Figure 4-5.

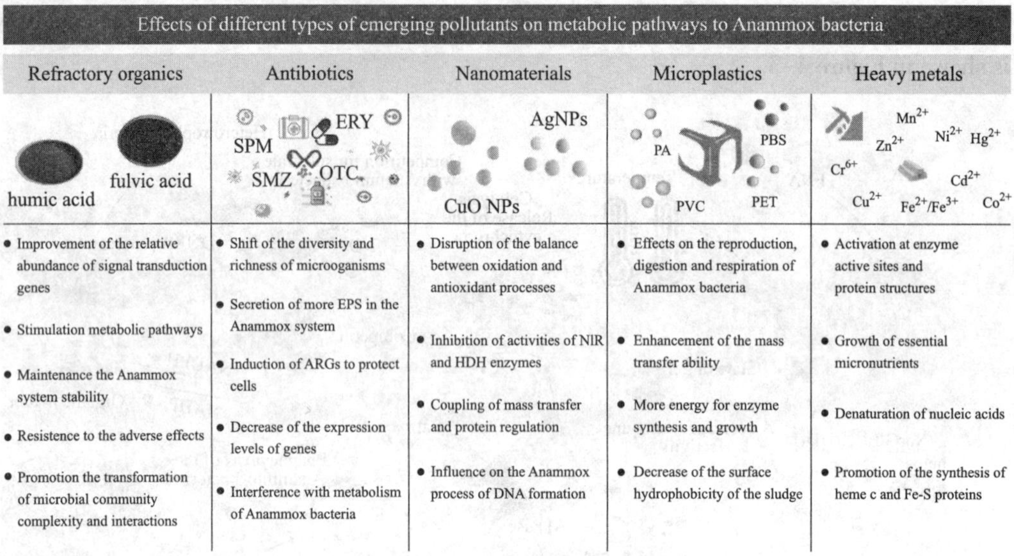

Figure 4-4 Effects of emerging pollutants on metabolic mechanisms of the Anammox system

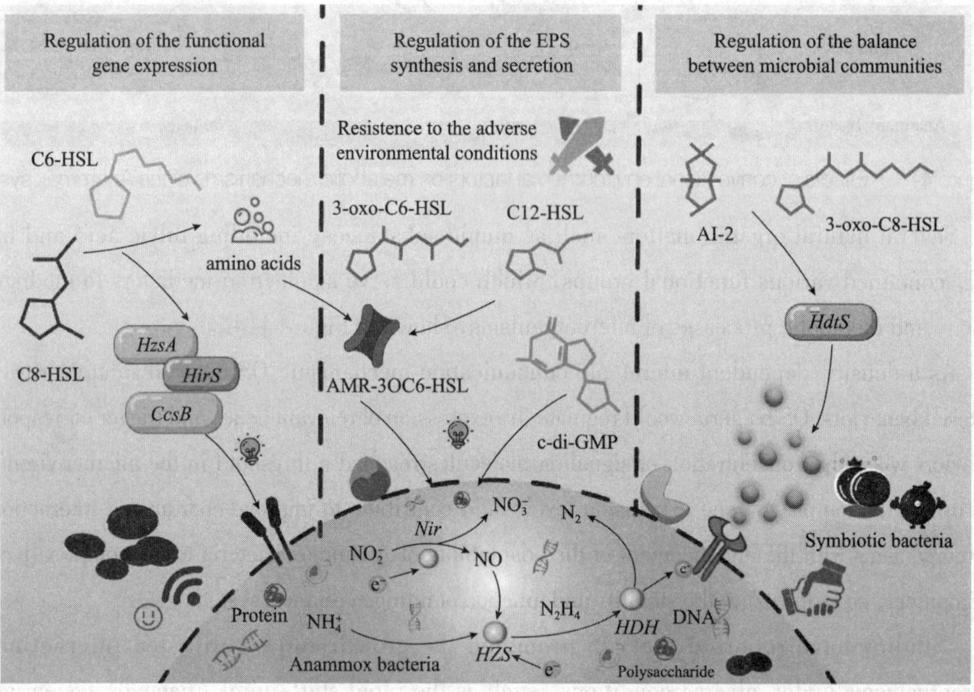

Figure 4-5 Role of QS on metabolic mechanisms of the Anammox system

Metabolic exchanges are ubiquitous in microbial communities. In particular, diverse microbial communities coexisted in the Anammox system, and metabolic exchanges would influence the metabolic pathway of Anammox bacteria. It was found that most heterotrophic organisms in anaerobic ammonia oxidized particles encode nitrate breathing capacity through partial denitrification to remove excess nitrate during the Anammox process. The heterotrophic bacteria might help remove detritus and peptides produced during the metabolism of Anammox bacteria and provide fuel for energy metabolism of Anammox bacteria by alternative electron donors such as acetate, and formate. Moreover, heterotrophic bacteria in the Anammox system could utilize organic matter as electron donors to reduce nitrate to nitrite, and establish a cross-intake relationship with them, further improve the nitrogen removal efficiency.

Cross-feeding is one of the significant nutritional interactions between Anammox bacteria and other symbiotic bacteria. For example, amino acid, the decomposition product of bacteria EPS, was considered to be the main organic carbon source that supported the heterotrophic bacteria growth in the Anammox system. Also, the cross-feeding of Anammox bacteria increased the levels of molybdopterin cofactors and folate, as well as reduced the metabolic cost of carbon fixation to support anabolism, which maintained the contents of heme c and extracellular polymers. It was also found that glycogen and vitamins might be synthesized in the Anammox metabolism to feed other symbiotic bacteria which lacked pathways of glycogen and vitamins synthesis. The metabolic exchanges of microbial communities in the Anammox system are shown in Figure 4-6.

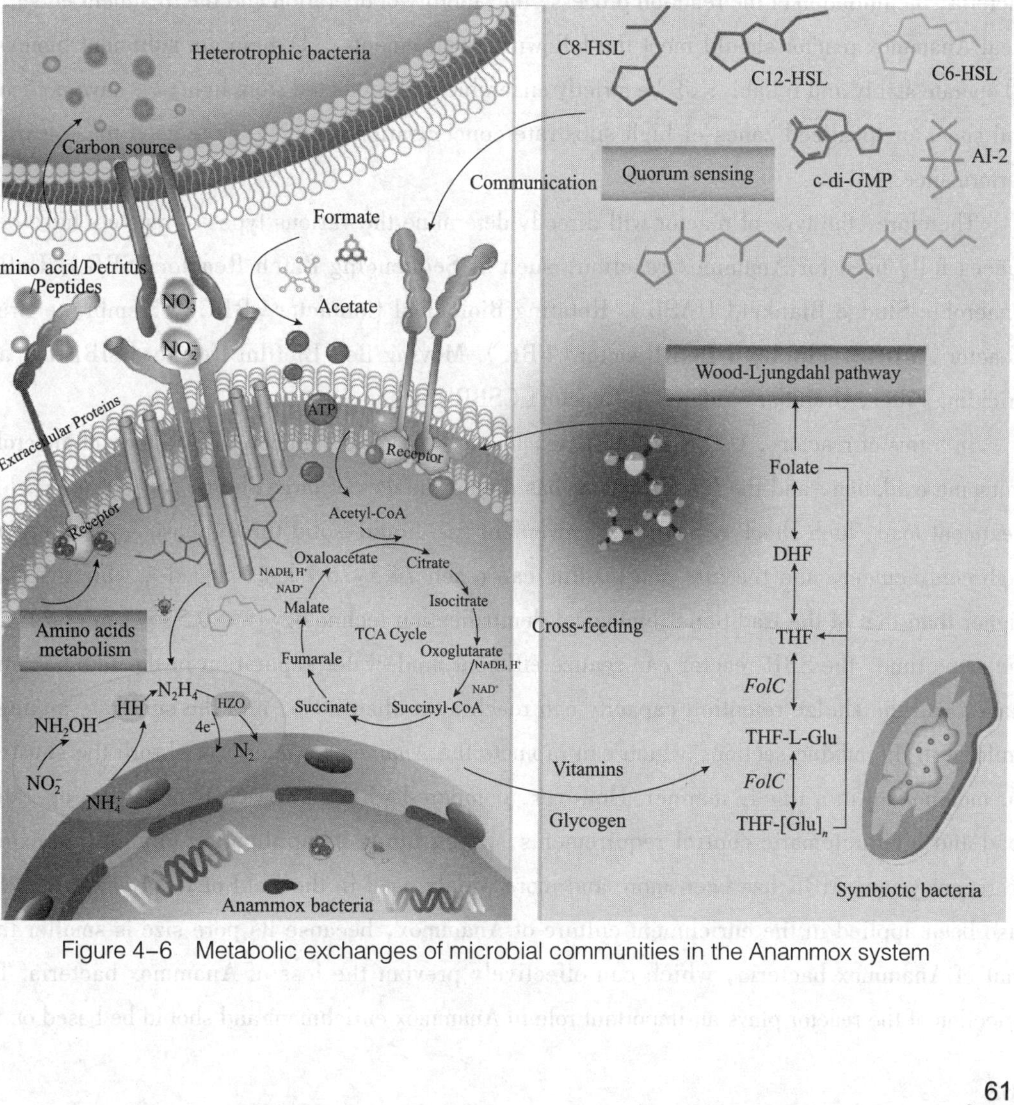

Figure 4-6 Metabolic exchanges of microbial communities in the Anammox system

4.4 Factors affecting Anammox of nitrite type

The influencing factors of Anammox are mainly focused on reactor structure and conditions such as Anammox reactor type, sludge properties, packing, environmental factors such as temperature, pH, oxygen, electromagnetic field, and substrate conditions such as organic matter, nitrogen, phosphate, salinity, and metal ions.

4.4.1 Reactor impact

1. Reactor type

The type and configuration of the Anammox reactor will affect the enrichment of Anammox bacteria, the initiation of the reaction process, the stability of operation and the treatment effect. An ideal Anammox reactor should meet the following requirements: ① maintain sufficient biomass; ② operate stably and reliably; ③ be strictly anaerobic and protected from light; ④ have no dead-end zones or localized zones of high substrate concentration; and ⑤ have good mass transfer performance.

Therefore, the type of reactor will directly determine the various types of reactors have been successfully used for Anammox reaction, such as Sequencing Batch Reactor (SBR), Upflow Anaerobic Sludge Blanket (UASB), Rotating Biological Contactor (RBC), Membrane Brane Reactor (MBR), Fluidized Bed Reactor (FBR), Moving Bed Biofilm Reactor (MBBR), and Trickling Filter, Granular Sludge Bed Reactor (GSBR) etc.

In terms of reactors, UASB and SBR reactors are the most common in the study of anaerobic ammonia oxidation, and the UASB reactor has the advantages of large sludge concentration, high treatment load, high shock resistance, convenient gas-liquid-solid three-phase separation, and high confinement, and the nitrogen loading can reach 74.3–76.7 kg/($m^3 \cdot d$), which is much higher than that of the traditional biological denitrification technology [$\leq$ 0.5 kg/($m^3 \cdot d$)]. At the same time, the SBR reactor can realize efficient mud-water separation in the sedimentation stage, and the sludge retention capacity can reach more than 90%, and the substrate mixing is uniform in the mixing section, which can promote the Anammox bacteria to absorb the substrate for metabolism in a timely manner. However, compared with UASB, SBR has a low operating load and high automatic control requirements, which limits its application to a certain extent. In recent years, MBR has been more and more widely used in the field of wastewater, and has also been applied in the enrichment culture of Anammox, because its pore size is smaller than that of Anammox bacteria, which can effectively prevent the loss of Anammox bacteria. The selection of the reactor plays an important role in Anammox enrichment and should be based on the environmental conditions of the application site. The choice of reactor is important for Anammox enrichment.

2. Sludge properties

The nature of sludge affects the performance of the Anammox reactor, especially granular sludge, which can effectively improve the efficiency of the Anammox reactor due to its good settling performance, high sludge interception rate and high shock load resistance. In recent years, Anammox sludge has become a hot research topic because of its good settling performance, high sludge interception, high shock load resistance, and its ability to improve the treatment efficiency of Anammox reactor, which has become a hot research topic of Anammox sludge. When the hydraulic retention time was reduced to 0.3 h, the TN volumetric removal load of the EGSB reactor with Anammox granular sludge as the core was as high as 50.75 kg/(m^3 · d), which was much higher than that of the conventional biological nitrogen removal process. Meanwhile, the UASB reactor was inoculated with Anammox granular sludge as the core, which could effectively resist the toxicity of phenol and realize the simultaneous denitrification of Anammox and denitrification. Therefore, how to enhance the number of Anammox bacteria in the sludge and reduce the loss of reactor sludge is the direction of selection of inoculated sludge for the Anammox reaction.

3. Reactor packing

Selection of granular sludge can effectively reduce the loss of biomass in the Anammox reactor, but the domestication and cultivation of granular sludge is difficult, and researchers have sought new ways to reduce the loss of sludge in the Anammox reactor so as to accelerate the start-up and stable operation of the Anammox reaction. Currently, the main method to reduce the loss of Anammox bacteria is to add filler, so that Anammox bacteria attach to the surface of the filler to promote the enrichment of Anammox bacteria. Zeolite particles as Anammox growth carriers can greatly increase the biomass in the reactor and reduce biomass loss. The addition of submerged hollow fiber membrane modules to a membrane sequencing batch reactor (MSBR) was found to greatly reduce system biomass loss. Currently, nonwoven fabrics, polyethylene glycol gels, polyvinyl alcohol beads, and sponges have been successfully applied to retain Anammox bacteria, and the addition of fillers not only accelerated the startup of the Anammox reactor, but also increased the efficiency of the reaction. The addition of packing material not only accelerates the start-up of the Anammox reactor, but also increases the denitrification performance of the reactor. The addition of packing materials not only speeds up the start-up of the Anammox reactor, but also increases the denitrification performance of the reactor.

4.4.2 Influence of environmental factors

1. Light and temperature

Anammox bacteria are sensitive to light, which inhibits their activity and reduces nitrogen removal by 30%–50%. Therefore, in order to ensure the smooth running of the reaction, the Anammox test usually places the reactor in the dark or takes measures of shading to minimize the negative effect of light on its processing ability. Bacterial activities are carried out by enzymes, and enzyme activity is greatly affected by temperature. Therefore, temperature is an important factor affecting the activity of Anammox bacteria. Only in the vicinity of the optimum temperature, the bacteria will show good reaction activity, which will enhance reactor operation of the reactor. By measuring the reaction rate of Anammox at different temperatures, the researchers concluded that the optimal survival temperature of Anammox bacteria is 30–35 ℃.

2. pH value

pH value has an important effect on microbial enzyme activity, it can change the charged state of enzyme molecules and substrate molecules, affecting the binding of enzyme and substrate; too high or too low pH value will affect the stability and activity of the enzyme, so that the enzyme has been irreversibly damaged, and affects the substrate in the medium of the ionization effect. pH value has an important effect on microbial enzyme activity, it can change the charged state of enzyme molecules and substrate molecules, affecting the binding of enzyme and substrate; too high or too low pH value will affect the stability and activity of the enzyme, so that the enzyme has been irreversibly damaged, and affects the substrate in the medium of the ionization effect.

At the same time, pH also affects the concentration of NH_4^+ and NO_2^- in aqueous solution, thus affecting the substrate source of the Anammox reaction. The Anammox reaction also increases the alkalinity of the reaction system, thus increasing the pH value of the reaction system, so controlling the appropriate pH value has an important effect on the Anammox reaction. It has been found that the optimal pH value of Anammox is 7.5–8.3, and the relationship between Anammox rate and pH value can be described by the two–substrate, two–inhibitor model between pH values of 7.0 and 9.0. Anammox bacteria in different reactors had different pH adaptations to the environment, but all of them were close to neutral. The growth of Anammox bacteria in different reactors was suitable for maintaining neutral or alkaline conditions.

3. Oxygen

Anammox bacteria are anaerobic microorganisms, very sensitive to oxygen, in the process of enrichment of Anammox generally have to pass nitrogen (N_2) in the influent water to deoxidize, to ensure that the influent water dissolved oxygen is less than 0.5 mg/L. Anammox bacteria can only survive at oxygen partial pressures below 5% oxygen saturation (100% oxygen concentration in air), and their activity is inhibited once the oxygen partial pressure exceeds 18% oxygen saturation, but the inhibition is reversible. At very low oxygen (>0.04 mg/L), the Anammox reaction is inhibited, and in an aerobic environment, the nitrifying bacteria oxidize NO_2^-–N to NO_3^-–N, resulting in a lack of substrate for the Anammox bacteria, which affects the efficiency of the reaction. Typically, Anammox reactors are hermetically sealed to ensure a completely anaerobic environment and to avoid the effects of oxygen on the Anammox bacteria.

4. Electromagnetic field

Changes in electromagnetic fields can affect the growth of Anammox bacteria. It was shown that applied electric field and magnetic field could effectively increase the activity of Anammox bacteria, the start-up time of applied electric field was shortened by 1/5, and the activity of Anammox bacteria was increased by 20% when the electrode voltage was −0.05 V, while the start-up time of applied magnetic field was shortened by 1/4, and the activity of Anammox bacteria was increased by 50% when the magnetic field strength was 75 mT. The removal of nitrogen was significantly improved by either applied electric or magnetic field. Because of this, to a certain extent, the addition of an electric field of suitable intensity has a positive effect on the activity of Anammox bacterium. The Anammox metabolic response to environmental factors is shown in Figure 4–7.

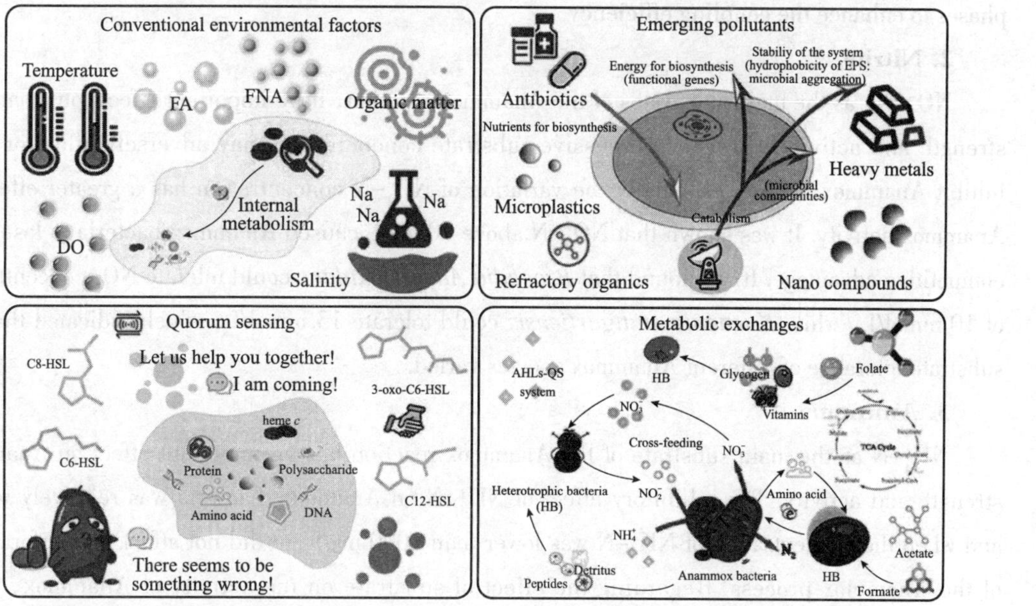

Figure 4–7 The Anammox metabolic response to environmental factors

4.4.3 Substrate effects

1. Organic matter

The presence of heterotrophic denitrifying bacteria in Anammox sludge leads to substrate competition between the two types of bacteria in the presence of organic matter. When there is an excess of organic carbon in the wastewater, the growth and reproduction of Anammox bacteria will be greatly inhibited and denitrifying bacteria will quickly take over. Theoretically, the growth rate of Anammox bacteria is about 0.066 C–mol biomass per mol ammonium consumed, which is only 1/5 of that of denitrifying bacteria, and thermodynamically the Anammox reaction is more difficult to carry out than the denitrification reaction. Studies have shown that COD above 300 mg/L will prevent or completely inhibit Anammox activity. At C/N ratios above 1, Anammox loses its competitive advantage over denitrifying bacteria for substrate.

However, it has also been shown that a certain concentration of organic matter promotes the growth of Anammox bacteria. The addition of acetate and glucose was found to increase the activity of Anammox, and denitrifying bacteria used the organic matter for nitrate reduction to produce more nitrite. The Anammox was able to operate stably in a high nitrogen and low organic carbon environment, and the TN removal rate was improved compared with that in the autotrophic system. However, when the concentration of organic matter was too high, the contribution of Anammox to TN removal decreased continuously, denitrification was intensified, and the operation of Anammox was unstable. Although it was found that Anammox and denitrification can coexist, the coexistence system is still fragile, and it is important to cultivate Anammox and denitrifying bacteria in separate phases to enhance the coupling efficiency.

2. Nitrite

NO_2^--N, as the main substrates of the Anammox reaction, have important effects on Anammox strength and activity. However, excessive substrate concentration may adversely affect or even inhibit Anammox action, especially the variation of NO_2^--N concentration has a greater effect on Anammox activity. It was shown that NO_2^--N above 70 mg/L caused Anammox bacteria to lose their competitive advantage. It was found that *Brocadia Anammoxidans* could tolerate NO_2^- concentration of 10 mmol/L, while *Kuenenia stuttgartiensis* could tolerate 13 mmol/L, which indicated that the substrate tolerance of different Anammox species varied.

3. Ammonia

NH_4^+-N as the main substrate of the Anammox reaction has an important effect on Anammox strength and activity. The inhibitory effect of NH_4^+-N on Anammox bacteria was relatively weak, and when the concentration of NH_4^+-N was lower than 1000 mg/L, it did not affect the performance of the Anammox process. Regarding the effect of substrate on the activity of Anammox, when the concentration of NO_2^--N in the influent water was higher than 280 mg/L, it would lead to the dysfunction of the reactor, whereas the performance of the reactor was still stable when the concentration of NH_4^+-N was increased to 415 mg/L, which indicated that the sensitivity of the reaction of Anammox to NO_2^--N was higher than that of NH_4^+-N. Therefore, controlling the appropriate concentration of influent substrate can improve the denitrification performance of the Anammox reaction.

4. Phosphate concentration

The presence of moderate amounts of phosphorus, an essential element for the growth of Anammox bacteria, promotes the growth of Anammox bacteria, but like the reaction substrates, excessive amounts can adversely affect the Anammox process. The activity of Anammox bacterium was reduced by the addition of 5 mmol/L phosphate (PO_4^{3-}), but it could tolerate 1 mmol/L PO_4^{3-} concentration. The presence of moderate amounts of phosphorus, an essential element for the growth of Anammox bacteria, promotes the growth of Anammox bacteria, but like the reaction substrates, excessive amounts can adversely affect the Anammox process. The activity of Anammox bacterium was reduced by the addition of 5 mmol/L phosphate (PO_4^{3-}), but it could tolerate 1 mmol/L PO_4^{3-} concentration.

A biofilter was used to treat municipal wastewater at ambient temperature, and it was found that PO_4^{3-} had a certain reversible inhibitory effect on Anammox. When the influent TP concentration is higher than 10 mg/L, there is a milky-white sediment of magnesium ammonium phosphate (MAP) formed in the biofilter, and the physical blocking effect of MAP affects the normal transfer of the substrate of Anammox reaction. This led to a significant decrease in the denitrification load, and the rapid recovery of the denitrification efficacy of the Anammox was achieved within 2 h by stopping the addition of PO_4^{3-} and lowering the pH of the influent water while backwashing.

5. Salinity

High salt content in wastewater can damage the cell membrane of Anammox bacteria and enzymes in the body of the bacteria, and inhibit the growth of the bacteria. At the same concentration of inhibition Na_2SO_4>KCl>NaCl, low concentrations of salt had an activating effect on Anammox bacteria, whereas NaCl above 50 g/L would severely inhibit the activity of Anammox bacteria.

6. Metal ions

Metal ions are essential mineral nutrients for the growth of microorganisms and have a significant impact on the growth of Anammox bacteria. Iron ions are found in cytochromes, iron-reducing proteins and iron-sulfur proteins, and the addition of iron ions not only improves the denitrification efficiency of Anammox bacteria, but also stimulates the growth of the Anammox mixture.

7. Others

Calcium and magnesium are both divalent cations that compress the bilayer and promote the aggregation of Anammox cells. Also, their adhesion to extracellular polymers accelerates sludge granulation. The Anammox system containing Mg^{2+} could form granules after 4 days of operation, while the control group took 17 days to form granules. Ca^{2+} has a high affinity for extracellular polymers. It can act as a bridge between extracellular polymers and biological cells to form a three-dimensional structure and create a suitable environment for sludge granulation. Free intracellular Ca^{2+} is often referred to as an important intracellular substrate and a key regulator of bacterial metabolism.

4.5 Nitrite type Anammox main process

In attempts to optimize the Anammox process with regards to its NRE, cost-effectiveness, and eco-friendliness, a number of PN/A reaction processes, all in accordance with the fundamental principle of the Anammox process, have been engineered/patented. The developed processes are broadly classified as either one-stage or two-stage, and include, "single reactor high activity ammonia removal over nitrite (SHARON)-Anammox", "oxygen-limited autotrophic nitrification/denitrification (OLAND)", "completely autotrophic N removal over nitrite (CANON), and deammonification process (DEMON)", "single-stage nitrogen removal using Anammox and partial nitritation (SNAP)" and partial denitrification (PD) Anammox (PD/A) (summarized in Table 4-2). Conventional Nitrification-Denitrification compared with Anammox and its variations are shown in Table 4-3.

Table 4-2 Anammox process variations

Processes	Types	Wastewater	Type of reactor	Dominant microbial community
OLAND	One-stage	Digested black water	RBC	AerAOB
CANON	One-stage	Sludge digester effluents	SBR	*Bacteroidetes*
DEMON	One-stage	Sludge digester effluents	SBR	*Proteobacteria*
SHARON-ANAMMOX	Two-stage	Sludge digester effluents	CSTR/SBR	*Planctomycetes*
SNAD	One-stage	Opto-electronic industrial	SBR	*Candidatus Kuenenia*
SNAP	One-stage	Waste brine	Complete mixing type	*Candidatus Scalindua*
SPNAD	One-stage	Mature landfill leachate	SPNAD reactor	*Candidatus Brocadia*
SPND	One-stage	ADFE	Bioreactor	*Candidatus Brocadia*
SND	One-stage	—	SBR	Heterotrophic nitrifiers + aerobic denitrifiers
SNADCO	One-stage	Digested effluent of fish wastewater	AL-MBBR	AnAOB + Heterotrophic nitrifiers

AerAOB: aerobic ammonium-oxidizing bacteria; SBR: sequencing batch reactor; RBC: rotating biological contactor; SNAP: single-stage nitrogen removal using Anammox and partial nitritation; SNADCO: simultaneous nitrification, ANAMMOX, denitrification & COD oxidation; SPNAD: step-feed partial nitrification simultaneous nitrification and denitrification; SPND: simultaneous partial nitrification and denitrification; ADFE: anaerobic digestate food waste effluent; AL-MBBR: air-lift moving bed biofilm reactor.

Table 4-3 Conventional Nitrification-Denitrification compared with Anammox and its variations

Processes	Microbes	DO/ (kg O_2/kg N)	NLR/ [kg N/ ($m^3 \cdot d$)]	Organic carbon use/ (kg COD/kg N)	Biomass production rate/ (kg/kg N)	N Removal efficiency/%
Traditional ND	Autotrophic + heterotrophic	4.6	2-8	7.6	3.2	95
Shortened ND	Autotrophic + heterotrophic	2.3	1.5	4.6	2.0	95
CANON	Autotrophic	2.1	1.2-8.9	0	0.3	90

Continued

Processes	Microbes	DO/ (kg O_2/kg N)	NLR/ [kg N/ (m^3 · d)]	Organic carbon use/ (kg COD/kg N)	Biomass production rate/ (kg/kg N)	N Removal efficiency/%
OLAND	Autotrophic	1.7	0.1	0	0.16	85
ANAMMOX	Autotrophic	0	5.1	0	0.12	95
SHARON/A	Autotrophic	1.9	0.75	0	0.3	95
SHARON	Autotrophic + heterotrophic	2.3	1.5	2.4	1	90
DEMON	Autotrophic + heterotrophic	0.3	1.16	0	—	85

NLR–nitrogen loading rate, COD–chemical oxygen demand.

4.5.1 SHARON–ANAMMOX

1. Process Principles and Control Parameters

The SHARON, a secondary treatment process, is said to have been developed during the early stages of advancing the Anammox process. Though it was originally intended to reduce the supply of organic carbon and oxygen for the conventional denitrification process, it is now used to remove NH_4^+ from wastewater through partial nitrification of NH_4^+ to NO_2^- at 25–40 ℃. It is worth noting that the process may be carried out in either a single or two separate reactor systems, with the operation of the single reactor system being more complicated. Thus, the SHARON process could generate nitrite for the Anammox process, resulting in a two–stage process coupling with Anammox. The SHARON–ANAMMOX is distinguished by its inability to retain sludge. Thus, SHARON is basically a chemostat without retention of biomass and a dilution rate greater than the maximum NOB growth rate, but less than that of AOB. The SHARON–process has traditionally been used to treat high ammonium concentrated (500 mg N/L) wastewater. The SHARON reactor can produce a nitrite/ammonium ratio of 1 : 1 under regulated oxygen supply, controlled temperature, and dilution rate, all of which are prerequisites for the Anammox process.

Owing to the presence of bicarbonate as NH_4 counter–ion in sludge liquor, it is possible to obtain the required ratio without the need for pH control. As a result, converting half of the NH_4 in the liquor leads to depletion in the water alkalinity, resulting in a pH drop and prevention of further nitrification. Physiological difference between AOB and NOB such as temperature can also be used to ensure selective AOB retention. In a 2 years experiment that employed the SHARON–ANAMMOX process to treat ammonium–rich sludge digester effluent, 53% of ammonia was successfully converted to nitrite with the ammonia conversion to N_2 gas been greater than 80%

while operating at a loading rate of 1.2 kg N/m^3 per day. Also, after 58 days of domestication and 33 days of stabilization, SHARON–ANAMMOX was able to remove 95% of ammonium at (32 ± 0.4) ℃. The feasibility of running a full-scale SHARON–ANAMMOX has thus been realized. Other operational factors with NOB inhibition capabilities have been reported to include aeration pattern, temperature, dissolved oxygen, concentration, and substrate concentration. The SHARON–ANAMMOX is shown in Figure 4–8.

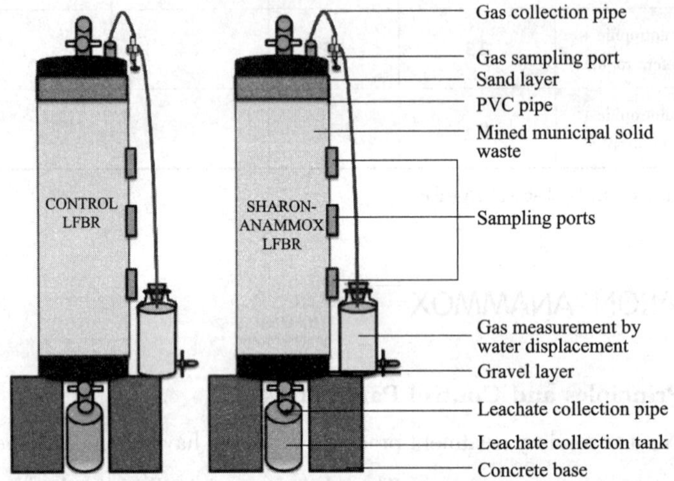

Figure 4–8 The SHARON–ANAMMOX

2. Process application

It is proposed that SHARON–ANAMMOX technology is more efficient than other autotrophic technologies because of its higher nitrogen removal rate (NRR), relatively lesser aeration requirement and lower effluent nitrate concentrations as well as lower nitrogen oxide emissions. The SHARON process, employing NO_2^- as an intermediate product is a low–cost treatment process for removing total nitrogen from highly nitrogen–concentrated wastewaters (>550 mg/L) and was the first to be commercialized and installed. The system has treated wastewater from dewatered primary sludge, waste–activated sludge (AS), sludge dryers, and incinerators. However, the number of relevant studies has substantially declined since 2008. The research focus has since shifted to CANON, as well as OLAND and other primary treatment processes.

$$2.3NH_4^+ + 2.95O_2 + 1.2CO_2 \longrightarrow 1.2CH_{1.8}O_{0.2} + N_2 + 0.3NO_3^-$$

4.5.2 CANON

1. Process principle and control parameters

CANON, just like simultaneous nitrification, ANAMMOX & denitrification (SNAD) is another example of the "single–sludge single–reactor" systems employed to treat low C/N ratio wastewaters using the principle of partial nitritation and Anammox. The wastewater in this case typically has a high NH_3 concentration, a low COD concentration, and a low oxygen demand, resulting in a cost-effective process. To successfully remove nitrogen using this system, the DO and nitrite concentrations must be controlled so as to inhibit the growth and proliferation of NOB. The CANON process facilitates the concurrent harmonious integration of aerobic ammonium oxidizing bacteria (AerAOB) and anaerobic ammonium oxidizers, because of the granular biomass configuration within the same reactor under steady microaerophilic oxygen conditions. AOB converts ammonium to nitrite while consuming oxygen, resulting in the desired anaerobic environment for AnAOB. The strong oxygen-limited condition allows the oxidation of ammonium to nitrite without the resulting nitrite being further oxidized to nitrate, thus, creating favorable environmental conditions for the occurrence of the Anammox reaction. Even though AOB and AnAOB are the predominant set of microorganisms in the CANON system as earlier stated, long-term insufficient supply of ammonia has been reported to result in the development of NOB as well.

It is reported that using the CANON process could result in an ammonium conversion rate of 0.07 kg NH_4^+ / (m^3 · d). Because of the nature of its instantaneously generated granular biomass, partial nitrification and Anammox occur concurrently in the CANON reactor. Granular biomass has been reported to enhance system operational conditions by increasing biomass settling with settling velocities ranging from 70 to 150 m/h, increasing resistance to toxic compounds as well as biomass retention, a significant improvement upon its susceptibility to shock in its early developmental stages. The AOB layers on top of an Anammox core form these granule-like biofilm structures. According to reports, 63% of energy (aeration) savings is realized relative to the conventional ND nitrogen removal setups. However, using the CANON process has been reported to result in the conversion of ~10% of influent NH_4^+-N into nitrate without subsequent conversion to N_2 gas. Thus, only 90% of influent NH_4^+-N is successfully removed.

2. Process applications

In an attempt to investigate the maximum nitrogen loading rates (NLR) for both the Anammox and CANON processes in a gas lift reactor, AnAOB removed 8.9 kg N/(m^3 reactor · d), while a combination of AOB and AnAOB removed 1.5 kg N/(m^3 reactor · d), achieving the yardstick for industrial trials or application. In addition, a 100 mg COD/L had no significant effect on the overall nitrogen removal efficiency (85%) in a CANON biofilter system containing Anammox, denitrifiers, and AOB and operating at influent ammonium concentration of 320 mg/L relative to similar bioreactor setups composed of just AOB and AnAOB. In a different study where urea was employed as the feed for the CANON system, it only took two weeks for the system to reach full capacity, broadening the scope of application to urine waste as well as other low C/N ratio wastewaters. Meihua MSG-I West (a CANON reactor), built in 2009 and located in Hebei (China), is the world's largest autotrophic nitrogen removal reactor, with a capacity of 6750 m^3. In general, the prevailing AOB population, ammonia concentration and DO concentration could be said to be the key control parameters of this process.

$$NH_4^+ + 0.85O_2 \longrightarrow 0.435N_2 + 0.13NO_3^- + 1.4H^+ + 1.3H_2O$$

4.5.3 Oxygen-Limited Autotrophic Nitrification/Denitrification (OLAND)

1. Process Principles and Control Parameters

This is another one-stage nitrite-limited system in which partial nitritation occurs in the outermost (aerobic) zones of the microbial biofilm, followed by the conversion of NO_2^- and NH_4^+ to N_2 gas in the inner anoxic zones by AnAOB, represented by Eq. OLAND is also resistant to NH_4^+ and NO_2^- shocks and does not require alkaline supplementation. OLAND, also known as a CANON variant, has advantages and mechanism similar to the SHARON-ANAMMOX and CANON processes. However, nitrifying bacteria including *Nitrosomonas eutropha* perform both aerobic and anaerobic ammonium oxidation simultaneously, as opposed to AOB and AnAOB coexisting as in the case of CANON. In addition, OLAND, unlike CANON, has a relatively short sludge retention time and a lesser need for nitrite sources. The activity of *Kuenenia stuttgartiensis*, a classic Anammox species has been reported in the OLAND system through molecular biological tests. The OLAND process is typically used to treat nitrogen-rich wastewaters (> 0.2 g N/L) including landfill leachate and sewage sludge digestate, specific industrial streams, and concentrated black water. But then again, due to the strong affinity (0.05-2.4 mg N/L) for nitrogen substrates by aerobic ammonia-oxidizing bacteria and AnAOB, it should be suitable for treating low-strength wastewaters and yield optimal removal efficiencies. The treatment of separate sludge liquor using the OLAND process resulted in a 25% reduction in the aeration requirement without any negative impact on the NLR of the AS system. The use of successive high and low loading rates in activated sludge for sludge digestion in an Austrian WWTP and nitrogen removal in its side stream resulted in an energy neutral system.

The DO concentration is regulated to ensure the maintenance a balance between the PN/A processes, thereby avoiding the potential outgrowth of AnAOB by NOB. At the moment, the OLAND process relies heavily on the rotating biological contactor (RBC) reaction setups, which is uncommon in practical engineering but is expected to become more common in the future. To successfully optimize the OLAND process, the NH_4^+ to NO_2^- oxidation rate as well as the AerAOB growth rate under anoxic conditions must be significantly improved.

$$NH_4^+ + 2O_2 + 0.83CH_3OH \longrightarrow 0.5N_2 + 0.83CO_2 + 3.17H_2O + H^+$$

2. Process application

DEMON (Deammonification process), this process was first documented when over 90% inorganic nitrogen losses was observed during the nitrification step of a RBC treating highly concentrated ammonium landfill leachate operating under microaerophilic conditions. This is typically a single reactor setup with alternating cycles of aeration and no-aeration, which is automatically regulated by the change in pH. Decreased pH results in an increase in reactor aeration and vice versa. Thus, compared to the traditional nitrogen removal processes, this process is cost effective and energy saving. According to Wett, the DEMON system consumes 1.16 kW/h per kg N as compared to the 6.5 kW/h per kg N of the conventional processes. The overall reaction mechanism is similar to the PN/A pathway and is summarized by equation below.

The efficiency of the DEMON process for the sidestream treatment of highly concentrated NH_4^+-N wastewater has been proven with numerous studies. However, if it is to be applied to the mainstream treatment of sewage, a lot of work still remains to be done, especially in its application to higher nitrogen loading rates. Nonetheless, the Changi Recycling Water Plant situated in Singapore is an exemplary full-scale setup of the DEMON process. There are other full-scale WWTP around the globe that have employed this process for so many years; Haltigen in Germany, Strass in Austria, from 2004, Glarnerland in Switzerland from 2007, and Olburgen in the Netherlands since 2006. Globally, there are over 30 full-scale DEMON process sewage treatment plants.

4.5.4 Simultaneous partial Nitrification, Anammox and Denitrification (SNAD)

1. Process Principles and Control Parameters

Unlike the CANON process which is only suitable for the treating highly concentrated ammonium-rich wastewater without COD, this process has an additional advantage of being suitable for simultaneous removal of nitrogen and COD. Here, ammonium is first partially converted to nitrite by AOB as illustrated by Eq. followed by the subsequent conversion of ammonia with nitrite to N_2 by AnAOB Eq. The overall reaction process is summarized by Eq.

$$NH_3 + 1.5O_2^{-} \longrightarrow NO_2^- + H_2O + H^+$$

$$NH_3 + 1.32NO_2^- + H^+ \longrightarrow 1.02N_2 + 0.26NO_3^- + 2H_2O$$

$$NH_3 + 0.85O_2 \longrightarrow 0.44N_2 + 0.11NO_3^- + 0.14H^+ + 1.43H_2O$$

2. Process application

The reported NH_4^+, TN, and COD removal efficiencies in the development of the SNAD process for the treatment of synthetic wastewater in a small scale, oxygen-limited, non-woven rotating biological contactor (NRBC) reactor were 79%, 70%, and 94%, respectively. In another experiment to ascertain the effects of organic nitrogen, reactor feeding mode and temperature on the SNAD process in an SBR reactor treating swine wastewater, average TN, NH_4^+-N and COD removal efficiencies of 80%, 96% and 76% respectively, were obtained, indicating their strong influence on the process. This process has also been successfully applied to remove nitrogen from other wastewater types such as; landfill leachate and opto-electric wastewater. For instance, in the treatment of landfill leachate in a partially aerated full-scale bioreactor, a 76% TN removal efficiency was achieved using the SNAD process. The SNAD process has been employed in the treatment of sewage with C/N ratios ranging from 0.4 to 4. Similarly, an average total inorganic nitrogen removal efficiency (TINRE) and NRR of 93.4% and 0.30 kg N/(m^3 · d) were achieved respectively in the investigation of the efficiency of the SNAD process for the treatment of mainstream wastewater under low DO levels (varying aeration times). The increase in aeration duration resulted in an increase in NRR while suppressing NOB activity. This could be due to the usage of AnAOB and AOB-produced nitrate through heterotrophic denitrification.

4.5.5 Partial Denitrification – Anammox (PD/A)

1. Process principle and control parameters

Partial denitrification (PD) is recommended as a sustainable NO_2^--N supply means for Anammox process. Coupled with its N_2O production minimization ability, PD/A process has started gaining attention in the treatment of mainstream wastewater. Even with low C/N ratio wastewater, the PD/A process can fully achieve enhanced nitrogen removal as well as provide additional organic material for polyphosphate accumulating organisms (PAOs) to accomplish phosphorus removal at relatively low temperature. Here, with organic carbon (mostly from sewage) as electron donor, NO_3-N is reduced to NO_2^--N which is then utilized by AnAOB in the subsequent transformation of NH_4^+-N to N_2 as illustrated in Eqs below.

$$CH_4 + 4NO_3^- \longrightarrow CO_2 + 4NO_2^- + 2H_2O$$

$$3CH_4 + 8NO_2^- + 8H^+ \longrightarrow 3CO_2 + 4N_2 + 10H_2O$$

The capacity of PD/A to treat wastewater with low C/N ratio has been demonstrated. A TINRE of 81% was achieved by a PDA biofilter process treating municipal wastewater, with Anammox removing approximately 97% of the TIN. With acetate as carbon source and a nitrate/ammonium ratio of 1 : 1, a total nitrogen removal efficiency (TNRE) of 93.6% was attained (Anammox responsible for >95% of TN). In the treatment of domestic wastewater (carbon-limited), a >95% NH_4^+-N oxidation efficiency was realized along with a significantly high nitrate/nitrite transformation ratio of 64.4% using a PD/A biofilm reactor. A PDA biofilter process treating municipal wastewater achieved a TINRE of 81%, with approximately 97% of the TIN removed by Anammox. Anammox and partial denitrification coupling nitrogen removal matrix flow direction is shown in Figure 4-9.

2. Process application

Also, the PD/A process seems to have a lot of promise in terms of aiding AnAOB proliferation and biomass retention even at low temperatures. This is exemplified by the recorded 72% TNRE of a PD/A reactor via the Anammox pathway, notwithstanding the Anammox bacteria making up

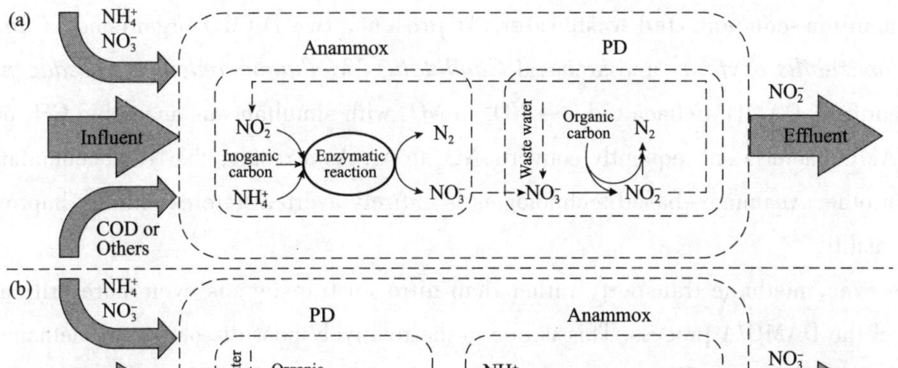

Figure 4-9 Anammox and partial denitrification coupling nitrogen removal matrix flow direction

for just 0.33% to 0.46% of the prevailing microorganisms. However, the ecological and financial viability of a large-scale PD/A operation needs further scrutiny, taking into account N_2O production and the massive power requirement (more aeration than PN/A). Furthermore, online and off-gas measurements are required in the future to offer reliable evidence and deep perspectives into the N_2O production mechanism.

$$1.32NO_3^- + 0.55CH_3COO^- + 0.088NH_4^+ \longrightarrow 1.32NO_2^- + 0.088C_5H_7NO_2 + 0.66HCO_3^- + 0.198H^+ + 0.264H_2O$$

$$NO_3^- + 0.519CH_3COO^- + 1.026NH_4^+ \longrightarrow 0.926N_2 + 0.083C_5H_7NO_2(\text{denitrifying}) + 0.062CH_2O_{0.5}N_{0.15}(\text{Anammox}) + 0.56HCO_3^- + 0.064H^+ + 2.164H_2O$$

4.5.6 Denitrifying Anaerobic Methane Oxidation/Anammox (DAMO/A)

1. Process Principles and Control Parameters

The DAMO/A process, much like PD/A process, is suitable for the treatment of nitrate and ammonium-concentrated wastewater. At present, two DAMO organisms, *Candidatus 'Methylomirabilis oxyfera'* and archaeal *Candidatus 'Methanoperedens nitroreducens'* have been identified. DAMO archaea reduces NO_3^- to NO_2^- with simultaneous anaerobic CH_4 oxidation while DAMO bacteria subsequently converts NO_2^- to N_2. As a result, the NO_3^- accumulation that occurs in other Anammox-based technologies is entirely averted here essentially improving the effluent quality.

However, methane transport, rather than nitrogen transfer, is even more critical to the success of the DAMO/A process. This is due to the relatively poor dissolution of methane as well as the distinct environmental requirements of DAMO microorganisms. The usage of a hollow fiber membrane, facilitating the attached biofilm methane consumption and the solubilizing of methane into the influent are the two proposed methane supply techniques thus far. However, the slow release of dissolved methane, particularly at low temperatures is somewhat undesirable as it may contribute to the greenhouse effect.

The DAMO/A process has yielded some appreciable treatment outcomes: a 0.7 kg N/d nitrate removal rate in a membrane biofilm reactor (MBfR) and excellent nitrogen removal (> 99%) from an anaerobic sludge digestion liquor that was partly nitrified. The DAMO/A process in a membrane bioreactor (MBR) revealed the facilitative effect of methane for the growth of both AnAOB and DAMO microorganisms while achieving a TNRE of 90% and 3 mg N/L effluent TN. This is primarily due to the utilization of NO_3-N as well as the unconsumed NO_2^--N (terminal electron acceptors) by the DAMO organisms with methane as the electron donor. In addition, the availability of multiple microorganisms aid in maintaining the stability of the reaction process. In the treatment of sidestream synthetic wastewater, a high NRR of 1.0 kg/(m^3 · d) was reported while simultaneously revealing that AnAOB and DAMO microbes are found in the inner and outer layers respectively, of layered granular sludge. With a highly efficient NRR of 16.53 kg/(m^3 · d) coupled with an allowable effluent quality (8 mg/L), the practical implementation of DAMO/A process could be said to be plausible and promising.

2. Process application

The DAMO/A process is also applicable in the two-stage processes. For instance, the 200 days operation of a PN-DAMO/A process yielded an impressive 2.5 kg/(m^3 · d) NRR. Similarly, while operating at 1 day HRT, a two-stage PN-DAMO/A recorded more than 1 kg/(m^3 · d) NRR, 0.56 kg/(m^3 · d) nitrite removal rate and 0.47 kg/(m^3 · d) ammonium removal rate without accumulating any nitrate.

Excessive energy requirement for methane sparging into the reactor, the added expenses in methane transport and methane release in actual implementations are some drawbacks to the DAMO/A process. Future studies should therefore focus on finding sustainable means of capturing methane post effluent discharge as it could negatively impact the ecological system (global warming). It has also been reported that DAMO-bacteria are susceptible to nitrite and ammonium.

$$CH_4 + 4NO_3^- \longrightarrow CO_2 + 4NO_2^- + 2H_2O$$

$$3CH_4 + 8NO_2^- + 8H^+ \longrightarrow 3CO_2 + 4N_2 + 10H_2O$$

The DAMO–Anammox is shown in Figure 4–10.

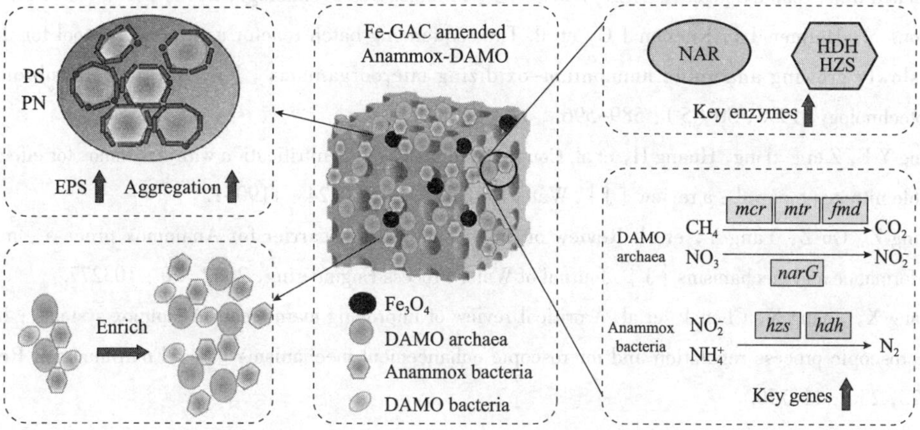

Figure 4–10 The DAMO–Anammox

Questions

[1] How was nitrite-based anaerobic ammonia oxidation discovered? What is its stoichiometric ratio relationship?

[2] What is the diversity of anaerobic ammonia-oxidizing functional bacteria of the nitrite type? What are their central metabolic mechanisms?

[3] What are the factors affecting anaerobic ammonia-oxidation of the nitrite type? What are the main areas of reactor influence?

[4] Both environmental factors and substrate concentration have significant effects on nitrite-based anaerobic ammonia oxidation, and an example of each is given to develop the mechanism of action.

[5] What are the main processes for nitrite-based anaerobic ammonia oxidation? Cite one major reaction process, describe its process principles and parameters, and discuss its process applications.

References

[1] You Q G, Wang J H, Qi G X, et al. Anammox and partial denitrification coupling: a review [J]. RSC Advances, 2020, 10 (21): 12554–12572.

[2] Yang S, Peng Y, Zhang Q, et al. Biofilm phenotypes and internal community succession determines distinct growth of Anammox bacteria in functional Anammox biofilms [J]. Bioresource Technology, 2022, 349: 126893.

[3] Chen X, Liu L, Bi Y, et al. A review of Anammox metabolic response to environmental factors: characteristics and mechanisms [J]. Environmental Research, 2023, 223: 115464.
[4] Zhang Y, Ma H, Chen R, et al. Stoichiometric variation and loading capacity of a high-loading Anammox attached film expanded bed (AAEEB) reactor [J]. Bioresource Technology, 2018, 253: 130-140.
[5] Strous M, Heijnen J J, Kuenen J G, et al. The sequencing batch reactor as a powerful tool for the study of slowly growing anaerobic ammonium-oxidizing microorganisms [J]. Applied Microbiology and Biotechnology, 1998, 50 (5): 589-596.
[6] Deng Y F, Zan F xiang, Huang H, et al. Coupling sulfur-based denitrification with Anammox for effective and stable nitrogen removal: a review [J]. Water Research, 2022, 224: 119051.
[7] Wang Z, Gu Z, Yang Y, et al. Review of biochar as a novel carrier for Anammox process: material, performance and mechanisms [J]. Journal of Water Process Engineering, 2022, 50: 103277.
[8] Zhang X, Zhang X, Chen J, et al. A critical review of improving mainstream Anammox systems: based on macroscopic process regulation and microscopic enhancement mechanisms [J]. Environmental Research, 2023, 236: 116770.

Chapter 5 Matching nitrosation for Anammox

Emphasis: Matched nitrosation reaction; Process of realization of matched nitrosation; Low temperature for matched nitrosation; Low substrate concentration for matched nitrosation; Microbial populations in matched nitrosation systems

Nut: Real-time reactor control

5.1 Matched nitrosation reaction

From equation (5-1), it can be seen that the wastewater treated by Anammox needs two kinds of substrates: NO_2^--N and NH_4^+-N, but the nitrogen in the actual wastewater is generally dominated by NH_4^+-N, and the substrate required by Anammox bacteria is low in NO_2^--N, while the matched nitrosation reaction can provide the substrate required by Anammox. The basic principle of the matched nitrosation reaction is to control the operating parameters to oxidize about 50% of the NH_4^+-N in the influent water to NO_2^--N by the ammonia-oxidizing bacteria (AOB) and inhibit the nitrite-oxidizing bacteria (NOB) from oxidizing the NO_2^--N to NO_3^--N, so as to maintain the accumulation rate of NO_2^--N and to make the $\rho(NO_2^-\text{-N})/\rho(NH_4^+\text{-N})$ of the effluent water to be 1-1.3, i.e. to achieve the matching nitrosation reaction for Anammox. The PN-Anammox type of nitrosation was achieved by matching the PN-Anammox type of nitrosation with the PN-Anamox type of nitrosation.

$$NH_4^+ + 1.32NO_2^- + 0.066HCO_3^- + 0.13H^+ \longrightarrow 1.02N_2 + 0.26NO_3^- + 0.066CH_2O_{0.5}N_{0.15} + 2.03H_2O \quad (5\text{-}1)$$

In recent years, there have been more studies on the control and influencing factors of the Anammox section in the control of the combined PN-Anammox process, and better enriched sludge has also been obtained for engineering applications. However, the front-end matched nitrosation (PN section) has become a difficulty in the engineering application of the PN-Anammox process due to its high control requirements and large fluctuation.

5.2 Process of realization of matched nitrosation

The matched nitrosation reaction is mainly achieved by selecting the reactor type, regulating the initiation mode and controlling the environmental factors, etc. The study proved that better stabilization of matched nitrosation efficacy has been obtained through the control of these parameters.

5.2.1 Selection of reactor

Different reactors will provide different microdomain environments for the growth of matched nitrosating microorganisms, which will have an important impact on microbial growth and retention of functional microorganisms. In recent years, various types of reactors have been successfully used as matched nitrosation reactions, such as sequencing batch reactor (SBR), membrane bioreactor (MBR), continuous flow reactor (CSTR), moving bed biofilm reactor (MBBR), and air-lift reactor. The initiation of matched-type nitrosation in each reactor and the process of realization are shown in Table 5-1.

Table 5-1 Initiation of each reactor to achieve matched nitrification

Reactor	Seedpod	Activation Method
MBBR	Nitrifying sludge	Gradual increase of influent NH_4^+-N initiates nitrification, and the synergistic effect of DO and pH is used to regulate matched nitrification.
SBR	Aerobic nitrification sludge	Nitrification was initiated by adjusting the hydraulic retention time and influent NH_4^+-N content (200 mg/L by mass), and the HRT was shortened to achieve matched nitrification.
SBR	Activated sludge with good nitrification	The mass concentration of influent NH_4^+-N was controlled at 160–200 mg/L, nitrification was initiated by high FA content (mass concentration >7 mg/L), and matched nitrification was achieved by low DO content (mass concentration of 0.1–0.3 mg/L).
Push-flow aerobic reactor+dual sinker	Mixed sludge of short-range and full-range nitrified sludge (3 : 1)	Reactor MLSS at ≥4 g/L mass concentration and 75% reflux volume ratio to initiate nitrification under aerobic and anoxic conditions. Adjust HRT to achieve matched nitrification.
A/O reactor	Sludge with good phosphorus removal performance and some nitrification function	Matched nitrosation by adjusting HRT.
Reaction Sedimentation Integrated Reactor	Taken from the sludge in the secondary settling tank	Nitrosation was initiated by controlling low DO content (mass concentration of 0.2–0.5 mg/L) and high FA content (mass concentration >7 mg/L) at 25 ℃, and HRT was changed to achieve matched nitrosation.

Continued

Reactor	Seedpod	Activation Method
CSTR	Activated sludge in low DO aeration tanks	After nitrosation was initiated by the synergistic effects of DO at a mass concentration of 0.3–0.5 mg/L and FA at room temperature (22–25 ℃), matched nitrosation was achieved and maintained at an anoxic, aerobic ratio time ratio of 3 : 1.

According to the reactor's sludge retention and screening effect on functional microorganisms, SBR is the preferred reactor in the control of matched nitrosation. SBR is a sequential batch, according to the intermittent aeration to run the activated sludge method, has a unique mode of operation, the water intake, reaction, sedimentation, drainage sludge discharge and idling, through the control of the drainage time can be amassed NOB out of the reactor, which is conducive to the enrichment of AOB; At the same time, the NH_4^+-N conversion ratio can be realized by aeration control, so SBR has the advantages of flexibility and controllability in matched nitrosation control.

A stable NH_4^+-N removal rate of 0.22 kg/(m³ · d) with an effluent $\rho(NO_2^-\text{-N})/\rho(NH_4^+\text{-N})$ of 1.67 was achieved by incubating matched nitrosation in an SBR for 70 consecutive days. It was found that it was difficult to achieve matched nitrosation using CSTR alone, and nitrosation was initiated after 17 d by regulating the NH_4^+-N content in the SBR first, and the nitrosation accumulation rate reached more than 90%, and then the effluent $\rho(NO_2^-\text{-N})/\rho(NH_4^+\text{-N})$ of 1.32 was achieved by running the CSTR, but the load of removal of NH_4^+-N was only 0.12 kg/(m³ · d).

However, SBR brings many limitations in application due to its low operating load and high automatic control requirements. Therefore, some modified SBRs are applied in experiments, such as sequencing batch biofilm reactor (SBBR) and modified SBR (changing operation strategy). Meanwhile, MBR, which has a high retention rate for sludge, has also been used by researchers to initiate and achieve matched nitrosation. MBR enriches microorganisms through the retention of membranes, which has the advantage of a high biomass, and has a significant advantage in terms of the retention of microorganisms for matched nitrosation. Therefore, the choice of reactor plays an important role in the enrichment of functional microorganisms for matched nitrosation, and a suitable reactor should be selected according to the environmental conditions of the application site, the characteristics of the wastewater, and the treatment objectives.

5.2.2 Alkalinity

Alkalinity is an effective means of regulating the matrix ratio of the effluent water from matched nitrosation, which is not only a source of inorganic carbon for AOB and NOB within the reaction system, but also an important buffer that prevents the pH of the reaction process from decreasing. In many studies, alkalinity is usually controlled by $NaHCO_3$ and Na_2CO_3, in the form of carbonate (e.g., HCO_3^-) in the water, according to equations (5–2) and (5–3) can be seen, HCO_3^- is the reaction substrate for ammonia oxidation, nitrosation reaction in the influent water of 1 mol of NH_4^+–N into NO_2^-–N requires the consumption of 2 mol of HCO_3^-, and matched nitrosation in the matched nitrosation, only 1 mol of HCO_3^- is required for the conversion of 1 mol of NH_4^+–N to NO_2^-–N. Therefore, it is possible to control the half alkalinity (the molar ratio of HCO_3^- and NH_4^+–N in the feed water is 1) and directly convert 50% of the feed water volume of NH_4^+–N to NO_2^-–N to realize the matched nitrosation. Alternatively, under the premise of realizing total nitrosation, $\rho(NO_2^-–N)/\rho(NH_4^+–N)$ of the effluent water is realized by reducing the amount of alkalinity and thus controlling the conversion rate of NH_4^+–N.

$$NH_4^+ + 1.5O_2 + 2HCO_3^- \longrightarrow 2H_2CO_3 + H_2O + NO_2^- \quad (5–2)$$

$$NH_4^+ + HCO_3^- + 0.75O_2 \longrightarrow 0.5NH_4^+ + 0.5NO_2^- + CO_2 + 1.5H_2O \quad (5–3)$$

The researchers studied the NH_4^+–N conversion rate from nearly 100% to 51% and finally to 32.8% when the average influent alkalinity and $\rho(NH_4^+–N)$ ratio were 10, 7.6, 6.6, 5, 3.6, 3.0, and 2.4, respectively, and the nitrosation accumulation rate had been stable at about 96%, indicating that when the alkalinity was insufficient, the NH_4^+–N conversion rate was linearly correlated with the influent alkalinity and NH_4^+–N ratio is linearly related to the alkalinity and NH_4^+–N ratio, and the matching nitrosation was realized under half alkalinity. In the SBR reactor, a low DO content (mass concentration of 0.6–0.8 mg/L) and an elevated influent NH_4^+–N load from 0.09 to 0.25 kg/(kg · d) were used to achieve short-range nitrosation, and then the semi-alkalinity was controlled to control the NH_4^+–N conversion rate at 50%–60%. In a fixed-bed biofilm reactor, the changes of effluent $\rho(NO_2^-–N)/\rho(NH_4^+–N)$ under the conditions of sufficient alkalinity and semi-alkalinity were investigated, and it was found that the removal rate of NH_4^+–N decreased from 80% to 58.6% after the alkalinity was reduced by half, and it was also verified that the matched nitrosation system could be stabilized by controlling the semi-alkalinity reaction to adapt to a certain influent NH_4^+–N shock load (Figure 5–1).

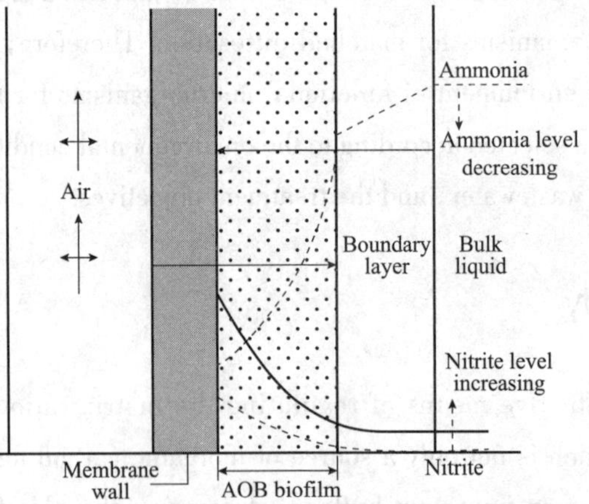

Figure 5–1 Schematic representation of concentration profile (oxygen, ammonia, and nitrite) in membrane aeration bioreactor

Controlling alkalinity in MBR under insufficient conditions [mass concentration of influent NH_4^+–N was 510 mg/L and alkalinity ($CaCO_3$ counting) was 750 mg/L] and in CSTR under insufficient conditions, about 50% of influent NH_4^+–N (mass concentration of 1.1 g/L) was converted to NO_2^-–N, which resulted in matched nitrosation, with a certain degree of stability and less influenced by substrate content and microbial activity.

5.2.3 pH

The influence of pH on matched nitrosation is mainly divided into two aspects: ① Within a certain range, the greater the pH the higher the ammonia oxidation rate, which can be used as a control parameter for nitrosation reaction time, according to the degree of change in pH of the inlet and outlet water combined with the pH "inflection point" in the nitrosation process can accurately determine the degree of oxidation of NH_4^+–N, so as to obtain the effluent target $\rho(NO_2^-\text{–N})/\rho(NH_4^+\text{–N})$ substrate ratio. ② pH will affect the reaction system FA and free nitrite. The target $\rho(NO_2^-\text{–N})/\rho(NH_4^+\text{–N})$ substrate ratio. ③ pH affects the FA and free nitrite (FNA) content of the reaction system, when the pH is high, the use of high content of FA can inhibit NOB; and when the pH is low, the FA content is low and can not reach the content of inhibition of NOB, but at this time, the FNA content is higher, and there is an inhibitory effect of FNA on NOB, which affects the realization of matched nitrosation. FNA inhibits NOB, thus affecting the realization of matched nitrosation (Figure 5–2).

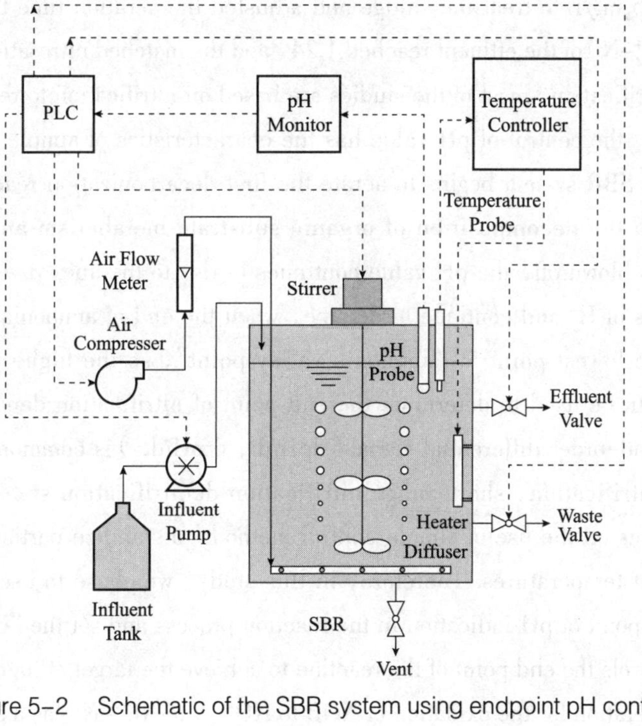

Figure 5–2 Schematic of the SBR system using endpoint pH control

The pH of the effluent $\rho(NO_2^-\text{-}N)/\rho(NH_4^+\text{-}N)$ was found to be 8.3–8.5 for the matched nitrosation under five different pH conditions (8.1, 8.3, 8.5, 8.7, 8.9), and it was found that the matching of $\rho(NO_2^-\text{-}N)/\rho(NH_4^+\text{-}N)$ could be achieved by the Anammox reaction at a pH of 8.17–8.19. In the case of the SBR treatment of leachate, it was found that the aeration was stopped at an appropriate pH and the endpoint pH was controlled at 8.17–8.19. In the SBR process, it is believed that stopping the aeration at a suitable pH point and controlling the pH at the end point between 8.17 and 8.19 can realize the $\rho(NO_2^-\text{-}N)/\rho(NH_4^+\text{-}N)$ of the effluent to match the Anammox reaction. It has been shown that FA and FNA are the real inhibitors of nitrifying bacterial activity. It was shown that the key controlling factor for successful matched nitrosation was the high FA content (mass concentration of NH_4^+-N >10 mg/L) due to high pH (~8.0), and the reactor effluent $\rho(NO_2^-\text{-}N)/\rho(NH_4^+\text{-}N)$ was 1.20 ± 0.33, which was stable for more than 250 d. In the case of the reactor, the mass concentration of NH_4^+-N ranged from 2–4 g/L, COD from 8–17 g/L, and FNA from 8 to 17 g/L. In the case of the Anammox reaction, the FA and FNA were the main inhibitors of the activity of the nitrifying bacteria. In the study of treating digestate of culture wastewater with NH_4^+-N mass concentration of 2–4 g/L and COD of 8–17 g/L, the mass ratio of NH_4^+-N converted to NO_2^--N and NO_3^--N was 58% and <5% when controlling the pH of 8.3 and maintaining the mass concentration of FA and FNA at 0.5 mg/L and 2 mg/L, respectively, and the stable treatment was maintained for 12 d. We believe that the contents of FA and FNA can be used to regulate the nitrosation of the high ammonia wastewater with matched nitrosation. of matched nitrosation. Under the condition of controlling the sludge age for about 15 d, we used FNA with a NO_2^--N mass concentration of 0.5 mg/L to treat the sludge and adjusted the aeration time to 2.5 h, so that the $\rho(NO_2^-\text{-}N)/\rho(NH_4^+\text{-}N)$ of the effluent reached 1.24, and the matched nitrosation was realized.

For partial nitrification, most of the studies are based on nitrification to realize. In the process control parameters, the control of pH value has the characteristics of simple and convenient and is widely used. the SBR system begins to aerate the first decarbonization reaction, heterotrophic microorganisms on the decomposition of organic substrate metabolism and anabolism of the formation of CO_2 is blown off, the pH value continues to rise to the apex due to the output of the nitrification process of H^+ and continue to decline, when the end of ammonia oxidation pH value appeared to be the lowest point— "ammonia valley point", so the highest and lowest points of the pH curve can be used to determine the end point of nitrification denitrification, and pH first-order or second-order differential signal (dpH/dt, d^2pH/dt^2) is commonly used to maintain the short-range nitrification, short-range nitrification denitrification stabilization. However, there are few studies on the use of simple control methods to stabilize partial nitrification under fluctuating ambient temperatures. Therefore, in this study, we chose to use the final pH value (the characteristic point of pH indication in the reaction process and set the "de-exposure point", which not only controls the end point of the reaction to achieve the target effluent, but also prevents over-exposure of aeration for the oxidation of NOB to NO_3^--N to NO_2^--N.) as a single factor indirect control parameter, through the change of pH value to judge the reaction process, so as to determine the control strategy to achieve long-term stability of partial nitrification and denitrification and decarbonization performance under ambient temperature, with a view to providing theoretical basis and technical support for the practical application of Anammox in the ambient and low-temperature environment.

5.2.4 DO content

Denitrifying bacteria are heterotrophic and anaerobic, and can only perform nitrite-nitrogen reduction in the absence of molecular oxygen and in the presence of both nitrate and nitrite ions. If the DO in the reactor is high, denitrifying bacteria will carry out aerobic respiration and inhibit the synthesis of nitrite reductase. Inhibit the synthesis of nitrite reductase. Therefore, denitrification should be controlled in anaerobic or anoxic or alternating anaerobic and aerobic conditions. In general, DO should be controlled below 0.5 mg/L.

DO content is one of the most concerned indicators in the matched nitrosation process. On the one hand, the two microorganisms involved in the reaction-AOB and NOB-are aerobic microorganisms, and AOB has a higher affinity for oxygen under the environment of low DO content, so by reducing the DO content, we can effectively control the enrichment of AOB, reduce the conversion rate of NH_4^+-N, and increase the accumulation rate of NO_2^--N, so as to make the nitrosation reaction transition to the matched nitrosation; on the other hand, the DO content can be controlled by adjusting the aeration volume and aeration time, and the aerobic aeration time is closely related to the nitrite accumulation rate. The DO content can be controlled by adjusting the aeration volume and aeration time, and the aerobic aeration time is closely related to the nitrite accumulation rate, so the use of aerobic aeration time to control and realize the reaction in real time is a kind of population-optimized control method, which is conducive to the long-term and stable operation of the matched nitrosation system.

The realization of matched-type nitrosation under low DO content environment has been widely reported. It was found that the control of DO content (mass concentration of 0.2–0.4 mg/L) was the preferred way to achieve stable matched nitrosation in a biofilter with low NH_4^+-N content in municipal wastewater at ambient temperature. Under the condition of 120 mg/L of influent NH_4^+-N in the SBR, when the mass concentration of DO was 0.5–0.7 mg/L, the NH_4^+-N removal rate was 100%, and the nitrite accumulation rate was 89%–97%, and when the DO was reduced to 0.3–0.4 mg/L, the NH_4^+-N removal rate decreased to 50%, and the nitrite accumulation rate was 97%–98%, and the effluent nitrosation rate decreased to 50%, and the effluent nitrosation rate was 97%–98%. 97%–98%, the effluent basically realized the matched nitrosation.

In the SBR, the mass concentration of influent NH_4^+-N and COD were controlled to be 40–85 mg/L and 110–320 mg/L, respectively, and the optimized aeration volume and aeration time were adjusted to be 6–12 L/(h · L) and 2–3.5 h, respectively, so that the mass concentration of DO was at 0.4 mg/L. The aeration was stopped before the ammonia valley point, so that NH_4^+-N was only partially oxidized to NO_2^--N, and $\rho(NO_2^-\text{-N})/\rho(NH_4^-\text{-N})$ was controlled to be $\rho(NO_2^-\text{-N})/\rho(NH_4^-\text{-N})$. The aeration was stopped before reaching the "ammonia valley point" so that only part of the NH_4^+-N was oxidized to NO_2^--N, thus controlling the $\rho(NO_2^-\text{-N})/\rho(NH_4^+\text{-N})$ to be in the range of 2–4 to match with the subsequently realized Anammox section.

5.2.5 HRT

Changing the HRT not only affected the load inside the reactor, but also regulated the degree of NH_4^+-N conversion during nitrosation. When the HRT was reduced from 4.5 h to 3.2 h, the microbial load in the CSTR was maintained at 1.2–2.5 g/L, the NH_4^+-N removal load was between 0.11 and 0.3 kg/(kg · d), and the effluent condition was changed from 60% NH_4^+-N conversion rate and 90% nitrite accumulation rate to the $\rho(NO_2^-\text{-N})/\rho(NH_4^+\text{-N})$ in the range of 1%–1.32%. The HRT was considered to be a prerequisite for obtaining a matched nitrosation stabilized effluent substrate ratio in CSTR. In the SBR, by reducing the HRT from 14 h to 6 h, the effluent went from 98% NH_4^+-N removal and 95% nitrite accumulation to $\rho(NO_2^-\text{-N})/\rho(NH_4^+\text{-N})$ around 1, which realized the transition from nitrosation to matched-type nitrosation; and in the study of the effect of HRT on matched-type nitrosation in the reactive precipitation integrated reactor (RPIR), after the HRT was reduced from 24 to 16 h and 12 h, the effluent had a stable substrate ratio of matched-type nitrosation. When the HRT was reduced from 24 h to 16 h and 12 h, the nitrite accumulation rate of 70% in the effluent, and the NH_4^+-N of about 0 mg/L were converted to $\rho(NO_2^-\text{-N})/\rho(NH_4^+\text{-N})$ of 1.39 and 1.46, respectively, and the matched nitrosation was realized.

5.3 Low temperature for matched nitrosation

Temperature is a key parameter in the anaerobic ammonia oxidation process. Theoretically, stable long-term nitrogen removal by autotrophic deamination is possible at temperatures ranging from 25 ℃ to 35 ℃, which favors the growth of most AOBs and AnAOBs and overrides NOB selection. Exposure to low temperatures leads to inhibition of anaerobic ammonia-oxidation activity due to direct effects of temperature on bacterial metabolism, indirect oxygen inhibition and nitrate accumulation. With a temperature range of 15–30 ℃, anaerobic ammonia-oxidizing bacteria can be stably present and active in the reactor, especially in the biofilm layer. The activity of anaerobic ammonia-oxidizing bacteria decreases with prolonged exposure at 15 ℃, and during the cold season (T<15 ℃), anaerobic ammonia-oxidizing activity decreases sharply and is almost completely inhibited. Anaerobic ammonia-oxidizing bacteria are resilient after temporary temperature inhibition and can proliferate once temperatures return above 15 ℃. At lower temperatures, nitrite and ammonium accumulation in anaerobic ammonia-oxidizing wastewater was observed due to changes in enzyme activity, growth metabolism, mass transfer and cytoplasmic mobility. It has been reported that although biofilm reactors can operate stably at 15 ℃ when treating synthetic influent, process instability occurs during prolonged operation at 11 ℃ (Figure 5-3).

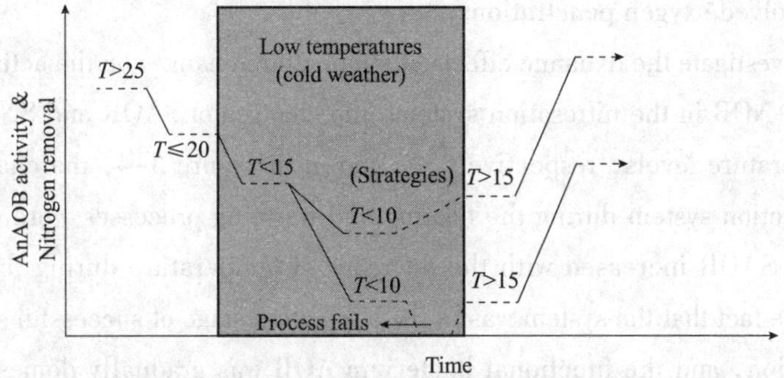

Figure 5-3 The hypothesis variations of Anammox activities and consequently nitrogen removal rates with seasons 2-2-Nitrite accumulation

Denitrifying bacteria are mostly suitable for growth at medium temperatures. In the mesophilic stage, the higher the temperature, the greater the bacterial activity. If the temperature is too low, the metabolic rate of bacteria will be slowed down and denitrification efficiency will be poor. If the temperature is too high, the metabolic enzymes of the bacteria will be denatured, or even make the cells die, thus hindering the denitrification reaction and reducing the nitrogen removal efficiency. The suitable temperature for short-range denitrification is around 25-35 ℃.

Isolation of microorganisms from low-temperature environments is most likely to compensate for the adverse effects of low temperatures. *C. Brocadia*, *C. Brocadia fulgida*, and *C. Kuenenia* have been reported to be dominant bacterial species and genera at low temperatures and under mainstream conditions. In addition, through microbial adaptation to low temperatures, the anaerobic ammonia oxidation process can operate at low temperatures without significantly altering the bacterial populations and physical properties of the biomass. It has been demonstrated that biomass acclimated to lower temperatures has higher specific anaerobic ammonia oxidation activity than those that are not acclimated, and some research scholars have reported that the anaerobic ammonia oxidation process in an SBR achieved a nitrogen removal rate of about [0.038 g N/(L·d)] over a 1000-day acclimatization period ranging from 30 to 13.2 ℃, and the ability of anaerobic ammonia oxidation to tolerate low temperatures has also been investigated. Through gradual acclimatization of anaerobic ammonia-oxidizing bacteria to low temperatures, a robust up flow Anammox sludge bed (UASB) achieved stable long-term nitrogen removal from ammonium and nitrite-modified synthetic wastewater [1.8 g N/(L·d)] and nitrite-modified pre-treated natural wastewater [1.2 g N/(L·d)] at 11 ℃. A combination of inoculation, bioaugmentation and acclimatization techniques improves the anaerobic ammonia oxidation performance of the anaerobic ammonia-oxidizing bacterial biomass and enhances nitrogen removal, and inoculation at optimal temperatures creates a robust biomass that can better tolerate low temperatures. The use of biofilms or dense particles means especially in the cold season. During the cold season, oxygen concentration increases due to the decrease in bacterial activity. Therefore, the layer structure of the pellet and biofilm system protects the AnAOB from dissolved oxygen penetration.

In order to investigate the dynamic effects of temperature changes on the activities of functional bacteria AOB and NOB in the nitrosation system, the changes of SAOR and SNOR were detected at different temperature levels, respectively. As shown in Figure 5-4, the changes of microbial activity in the reaction system during the cooling and warming processes were different, and the phenomenon that SAOR increased with the decrease of temperature during the cooling process might be due to the fact that the system was in the adaptation stage of successful startup at the early stage of the reaction, and the functional bacterium AOB was gradually domesticated under the operation and regulation strategy, and the activity was gradually increased, and the system was in the period of efficiency enhancement, and with the continuous decrease of temperature, the SAOR gradually decreased, and the magnitude of 25-30 ℃ across to 20-25 ℃ was the largest, with a 3.5-fold decrease. Theoretically, the maximum specific growth rate of AOB follows the Arrhenius equation, and the maximum specific growth rate increases by 1 ℃ for every 10 ℃ increase in temperature, but in this study, the SAOR was stabilized at about 0.074 mg N/(mg MLSS · d) after the temperature was reduced from 30-35 ℃ to 25-30 ℃, which indicated that the nitrification system was in the period of stabilization of the efficacy at this time. The reaction function bacteria were dominated by AOB. During the warming process, the SAOR increased with the increase of temperature, and the SNOR decreased with the increase of temperature, but compared with the cooling process, there were differences in the SAOR and SNOR at the same temperature level, and the SAOR and SNOR of the reaction system during the warming process were generally lower than those during the cooling process, which indicated that the low temperature had a shocking impact on the reaction system, but the small-scale increase of the SAOR at 25-30 ℃ also showed that the inhibition could be inhibited by the low temperature. However, the small range of SAOR at 25-30 ℃ also showed the reversibility of inhibition. Meanwhile, the SAOR was always higher than the SNOR during the warming and cooling process, indicating that there was no significant effect on the nitrosation system either by increasing or decreasing the temperature, and AOB was always the dominant bacterial species.

Nitrification is relatively easy to achieve at temperatures between 15 ℃ and 36 ℃, and is more difficult to achieve outside this temperature range. Stable nitrification at different temperatures shows the flexibility of the process. At the same time, the effect of temperature on the activity of AOB will definitely affect the ratio of nitrite nitrogen and ammonia nitrogen in the effluent, which requires different reactors to control the reaction temperature according to the actual situation.

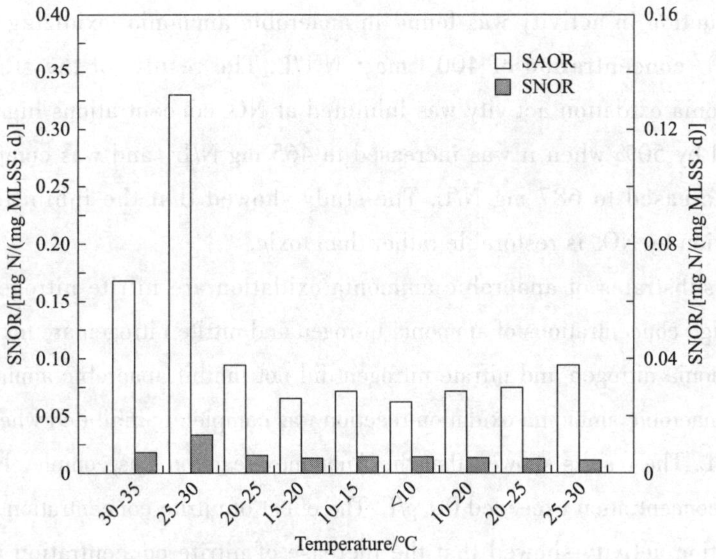

Figure 5–4 Changes of SAOR and SNOR at different temperature levels

5.4 Low substrate concentration for matched nitrosation

NH_4^+ and NO_2^- are the main reactive substrates in Anammox, but when the concentration of these substrates in the reactor is too high, it can inhibit the reaction process of Anammox. In the case of NH_4^+, it was initially reported that it did not inhibit Anammox at concentrations below 1000 mg N/L, but other studies have subsequently shown that low concentrations of NH_4^+ also inhibit anaerobic ammonia oxidation and have confirmed that this inhibition occurs primarily through free ammonia (NH_3, FA), rather than NH_4^+. It was found that no inhibition of anaerobic ammonia oxidation was observed during the gradual increase of FA concentration up to 150 mg N/L, whereas anaerobic ammonia oxidation activity suddenly decreased by 10% when FA concentration was increased to 190 mg N/L. Some scholars observed that the anaerobic ammonia oxidation activity was reduced by 50% when the FA concentration was 38 mg N/L (NH_4^+–N=770 mg/L, pH=7.8, 30 ℃), which indicated that the anaerobic ammonia oxidation activity was completely inhibited when the FA concentration was 35–40 mg N/L (NH_4^+–N=600–700 mg/L), and that the FA concentration should be lower than 20–25 mg N/L in order to maintain the stability of the anaerobic ammonia-oxidation reactor.

Anaerobic ammonia-oxidation is more sensitive to NO_2^- than NH_4^+. The threshold at which NO_2^- inhibits anaerobic ammonia-oxidation varies greatly among researchers due to different experimental operating conditions. It was observed that anaerobic ammonia-oxidation activity was completely inhibited at NO_2^- concentrations higher than 185 mg N/L, anaerobic ammonia-oxidation activity was reduced by 50% at a NO_2^- concentration of 350 mg N/L, and a 50% reduction in activity was found in anaerobic ammonia-oxidizing granular sludge system at a NO_2^- concentration of 400 (mg · N)/L. The results of the study showed that anaerobic ammonia oxidation activity was inhibited at NO_2^- concentrations higher than 244 mg N/L, decreased by 50% when it was increased to 465 mg N/L, and was completely inhibited when it was increased to 687 mg N/L. The study showed that the inhibition of anaerobic ammonia oxidation by NO_2^- is restorable rather than toxic.

The main substrates of anaerobic ammonia oxidation are nitrite nitrogen and ammonia nitrogen, and high concentrations of ammonia nitrogen and nitrite nitrogen are toxic to cells. It was shown that ammonia nitrogen and nitrate nitrogen did not inhibit anaerobic ammonia oxidation at 1 g/L, but the anaerobic ammonia oxidation reaction was completely inhibited when nitrite nitrogen exceeded 0.1 g/L. The results showed that the Anammox reaction was completely inhibited when nitrite nitrogen concentration exceeded 0.1 g/L. The effect of nitrite concentration on the anaerobic ammonia oxidation activity showed that the increase of nitrite concentration was favorable to accelerate the reaction rate of Anammox. In an upflow bioreactor, the effect of different substrate concentrations on Anammox was studied, and the critical concentration of nitrite nitrogen was found to be 280 mg/L. In an anaerobic manifold reactor, the critical inhibitory concentration of nitrite nitrogen was around 320 mg/L. Therefore, the effect of substrate concentration on anaerobic ammonia oxidation is not absolute, and the substrate tolerance concentration can be improved by changing the operation mode and reactor structure.

In systems with high-strength ammonia concentrations (e.g., waste leachate), high concentrations of substrate ammonia are only more readily achieved in low-strength systems (i.e., municipal wastewater). In a laboratory-scale moving bed biofilm reactor used for partial nitrification, nitrite accumulation increased with decreasing C/N ratio, and maximum nitrification rates were observed at C/N ratios below 6. In addition, a high nitrite accumulation rate of more than 95% was observed at a low C/N ratio of 2.33. A reduction of the C/N ratio to 0.2 had no significant effect on the oxidation of ammonia to nitrite, but a C/N of 0.3 led to competition between heterotrophic and autotrophic organisms, which adversely affected the latter.

5.5 Real-time reactor control

Real-time control, as an effective means to realize and stabilize the partial nitrosation reaction, has the advantages of flexibility, stability, and accuracy, etc. At present, real-time control is usually divided into two kinds, one is to use the on-line detection of the substrate concentration, such as COD, NH_4^+-N, NO_2^--N, and NO_3^--N, as the process control analysis, and the other is to use the numerical changes of the indirect parameters, such as pH, DO, and ORP, etc., as the reference to regulate the reaction. At this stage, in order to simplify and develop new real-time control strategies, most researchers choose indirect parameters to regulate the nitrification system: Tomlins et al. controlled the process of SBR through OGAR control system, and ended the nitrification reaction when the ORP was 400 mV, and the denitrification ended when the ORP was 150 mV; Buitron et al. controlled the process of SBR through OGAR control system, and utilized a sudden rise in DO value at the end of the nitrification reaction as an indication for control; Peng et al. showed that a sudden rise in the pH curve indicated the end of the nitrification reaction regardless of excess alkalinity. Among them, pH is the main factor affecting microbial composition and biodiversity in the reactor, and its change process and characteristic points can better indicate the reaction process, which indicates the end of denitrification at the highest point and the end of ammonia oxidation at the lowest point ("ammonia valley point"), and the pH value can be adjusted in time through these characteristic points and be used as an operating parameter. The pH value can be adjusted in time through these characteristic points, and the pH value can be used as an operating parameter, which is conducive to shortening the reaction cycle time, improving the treatment capacity and saving energy and reducing consumption at the same time, and it has become an important parameter for the regulation of partial nitrification.

Case Study:

Partial nitrification has become one of the bottlenecks in the practical application of Anammox because of the difficulty in maintaining stability due to the influence of environmental factors such as temperature etc. The SBR reactor is often used as the preferred reactor to realize partial nitrification, and its unique intermittent operation is conducive to the enrichment of the functional bacterium Ammonia-oxidizing Bacteria (AOB) and the elution of the competing bacterium Nitrite-Oxidizing Bacteria (NOB), and meanwhile, the automated real-time control of the SBR reactor provides a guarantee for the realization of partial nitrification. The SBR reactor provides a guarantee for the realization of partial nitrification, and the regulation parameters include dissolved oxygen (DO), pH, free ammonia (FA), alkalinity, and hydraulic retention time (HRT), which has strong flexibility, adaptability and accuracy. However, single-factor real-time control is difficult to maintain stable partial nitrification for a long time. In addition, temperature is one of the important factors affecting the maintenance of stable partial nitrification, the appropriate growth temperature of AOB is 20–35 ℃, but the temperature fluctuation is large in engineering applications, and the instability of the ambient temperature has a destructive effect on the activity of AOB and the effluent substrate ratio. Therefore, it is of great significance to realize the stabilization of partial nitrosation in single-factor control conditions and ambient temperature for the popularization and application of Anammox (Figure 5-5).

The SBR reactor cycle is divided into five stages: water intake, aeration, sedimentation, drainage, and sludge discharge, in which water intake is 5 min, aeration is controlled in real time, sedimentation is 40 min, drainage is 3 min, and sludge discharge is 2 min, and during the reaction

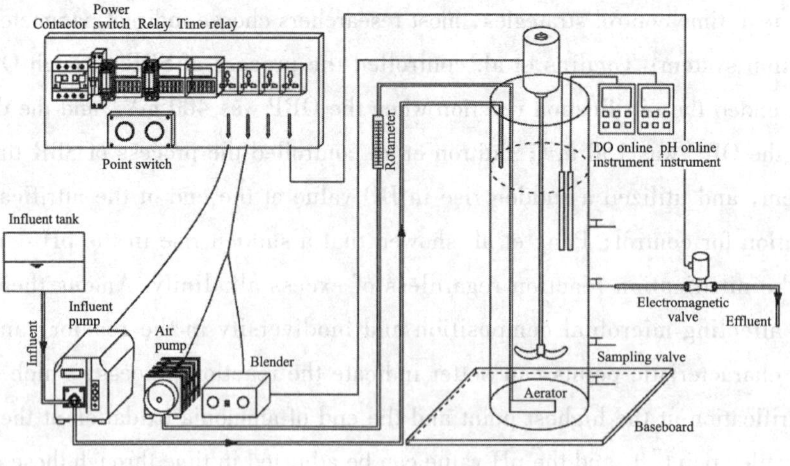

Figure 5-5 Schematic diagram of the reactor

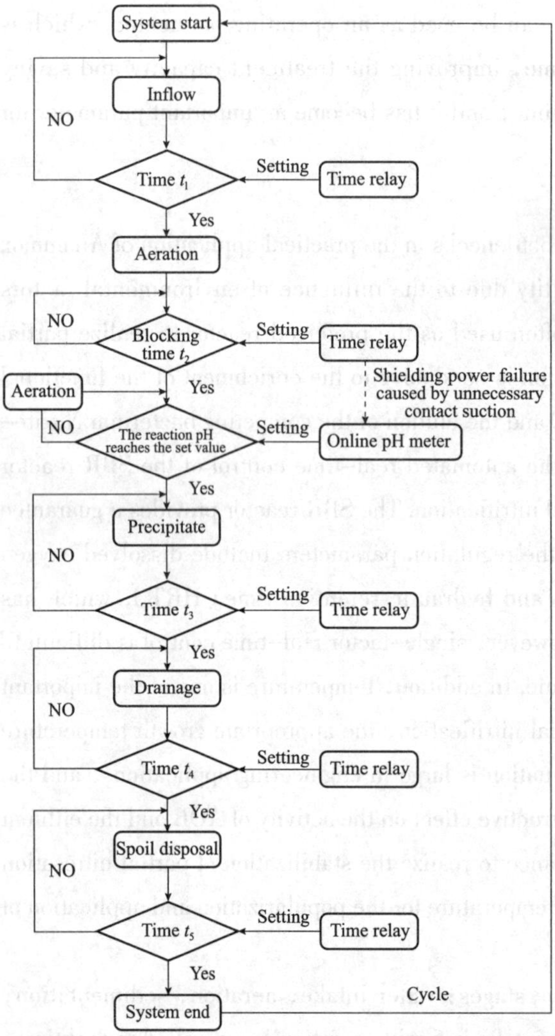

Figure 5-6 Flow chart of real-time control strategy in the SBR system

period, the system reads the pH signal value every 1 min, and the operation mode is shown in. The process, set five time variables t_1–t_5, the unit is min, the system starts, the feed pump is opened, after time t_1, the feed stops, the aeration reaction phase begins; aeration begins after first passing the shielding time t_2, shielding the online pH meter repeater connection port contact suction, preventing automatic stopping of the aeration; after arriving at the shielding time t_2, according to the pH value of the online pH meter monitoring to determine whether to Stop aeration, when the detected pH value is lower than or equal to the set pH value, the contact of the repeater connected to the online pH meter will be suctioned resulting in power failure to control the point of stopping the aeration and enter the next process; sedimentation, drainage and mud draining stages are controlled by time t_3, t_4 and t_5 respectively, and the whole set of reaction is repeated automatically after mud draining time t_5 (Figure 5-6).

5.6 Microbial populations in matched nitrosation systems

There are mainly three kinds of bacteria in the matched nitrosation system: AOB, NOB and heterotrophic bacteria (HB), and in the matched nitrosation stage, the two kinds of bacteria are mainly AOB and NOB. AOB can be roughly divided into autotrophic and heterotrophic in nature, and the former are mainly *Nitrosomonas*, *Nitrosococcus* and *Nitrosospira*, and the latter are *Arthrobacter*, *Flavobacterium* and *Thiosphaera*, while the autotrophic AOB are mainly used in wastewater biological nitrogen removal. The main NOB are *Nitrobacter*, *Nitrospina*, *Nitrococcus*, and *Nitrospira*. The current study shows that the prominent NOB species in wastewater treatment systems are *Nitrospira spp*. From the microbial point of view, matched nitrosation is not only from NH_4^+-N to NO_2^--N, but also involves a variety of enzymes and a variety of intermediates, accompanied by electron transfer and energy generation of complex biochemical reaction process, resulting in matched nitrosation reactor in the structure of the microbial community is more complex (Figure 5-7).

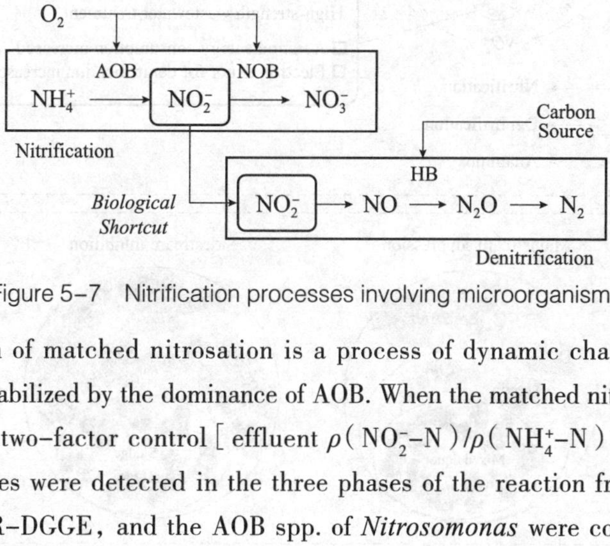

Figure 5-7 Nitrification processes involving microorganisms

The realization of matched nitrosation is a process of dynamic change of the bacterial population, and is stabilized by the dominance of AOB. When the matched nitrosation was realized by the "DO-HRT" two-factor control [effluent $\rho(NO_2^-\text{-N})/\rho(NH_4^+\text{-N})$ 1 : 1.04-1 : 1.47], the microbial changes were detected in the three phases of the reaction from pre-initiation to stabilization by PCR-DGGE, and the AOB spp. of *Nitrosomonas* were continuously enriched and new *Nitrosomonas spp.* appeared, while the AOB spp. of *Nitrosomonas* were constantly enriched, and new *Nitrosomonas* genera appeared, whereas the number of *Nitrospira* genera in NOB decreased to none. When the spatial distribution of AOB and NOB on biofilm was investigated by FISH, we found that *β-Proteobacteria* and *Nitrosomonas spp.* (AOB species) increased from $(5.1 \pm 0.1)\%$ and $(1.3 \pm 0.2)\%$, respectively, in the seeding mud to $(50.2 \pm 4.5)\%$, $(23.5 \pm 3.5)\%$, and $(23.5 \pm 3.5)\%$, respectively, in the initiation phase. and $(23.5 \pm 3.2)\%$, and finally reached $(64.1 \pm 5.0)\%$ and $(24.2 \pm 2.1)\%$ when the stable operation of matched nitrosation was achieved, while NOB species were basically maintained at 5%. In the MBR, the abundance of AerAOB in the seeded mud was only (3.8 ± 1.2)% using FISH technique, while the abundance of AerAOB increased to (31.2 ± 3.3)% when the matched Anammox inlet was realized; on the contrary, the abundance of NOB was only (1.8 ± 0.9)%.

The ratio of nitrifying bacteria to nitrifying bacteria on the biofilm was 1000 : 1 and 187.5 : 1 when the mass concentration of DO was 0.5 and 1.5 mg/L, respectively, as measured by the maximum counting method (MPN) in the MBBR with matched nitrosation; therefore, the microbial ecology of the matched nitrosation system is complex, but the regulation of the community structure of the AOB and NOB is a prerequisite for the realization of stable matched nitrosation. "NOB as the main and NOB as the auxiliary" is the prerequisite for the realization of stable matched nitrosation (Figure 5-8).

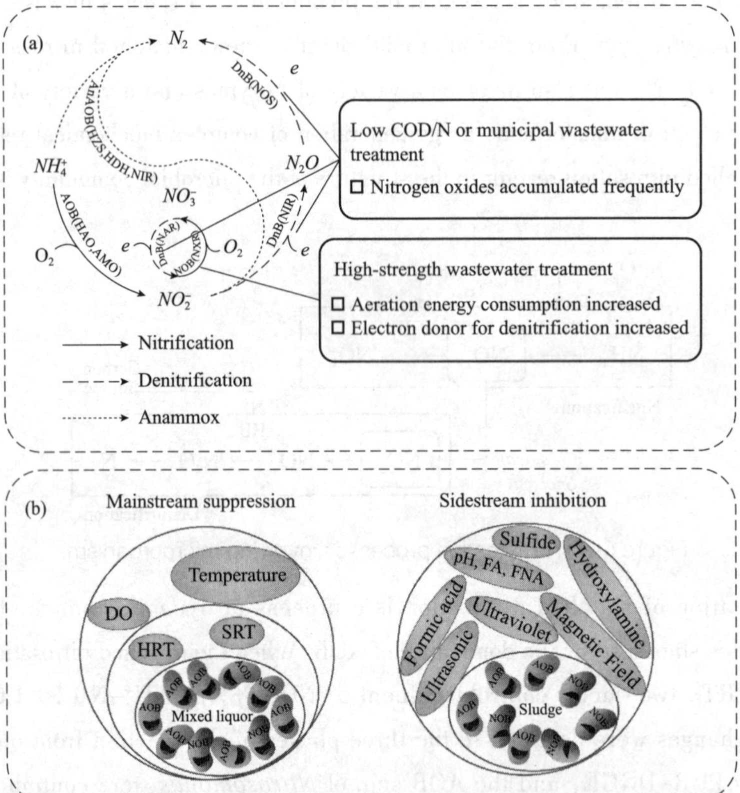

Figure 5-8 The existing problems of conventional biological nitrogen removal process (a) and the main strategies for partial nitrification through mainstream suppression and sidestream inhibition (b)

Two main kinds of nitrifiers, namely ammonia oxidizing bacteria (AOB) and nitrite oxidizing bacteria (NOB), are involved in the biological nitrification process (van Kessel et al. 2015). Ammonia monooxygenase (AMO) and hydroxylamine oxidase (HAO) in AOB catalyze the conversion of ammonia to nitrite, while nitrite oxidoreductase (NXR) in NOB catalyzes the conversion of nitrite to nitrate. In order to achieve stable partial nitrification, regulations must ensure the abundance or activity of NOB in nitrification sustained maintain very low level, while

AOB can be enriched based on the difference of their physiological characteristics. Based on the differences of growth kinetics and physiological characteristics between AOB and NOB, controlling operation conditions of nitrification in the mainstream or treating sludge physically and chemically in the sidestream could selectively suppress and then wash out NOB (Wang et al. 2020). Nevertheless, NOB has been reported to be able to gradually adapt to various selective pressures during long-term operation, resulting in the failure of partial nitrification (Ma et al. 2017a; Cao et al. 2018; Duan et al. 2019a). Obviously, the physiological and metabolic characteristics of these functional microorganisms need to be further studied.

Five genera of AOB (*Nitrosomonas*, *Nitrosospira*, *Nitrosovibrio*, *Nitrosolobus* and *Nitrosococcus*) and four genera of NOB (*Nitrobacter*, *Nitrospira*, *Nitrococcus* and *Nitrospina*) have been detected in various environments, while *Nitrosomonas*, *Nitrobacter* and *Nitrospira* are the most common nitrifiers in the biological nitrogen removal process in WWTPs. *Nitrosomonas* was the only genus of AOB involved in almost all of the partial nitrification achieved by regulating the operation conditions. *Nitrosomonas*, *Nitrosospira*, *Nitrosovibrio* and *Nitrosolobus* belong to *beta-Proteobacteria*, while *Nitrosococcus* belongs to *Gammaproteo* bacteria. Although *Nitrosomonas* converts ammonia to hydroxylamine and nitrite by ammonia monooxygenase (AMO) and hydroxylamine oxidoreductase (HAO) sequentially as other genera of AOB, it has a higher specific maximum growth rate and growth yield than others. In addition, some *Nitrosomonas* genera could use nitrite as electron acceptor to maintain growth and metabolism when the DO is deficient. These characteristics are considered to favor the enrichment of *Nitrosomonas* than other AOB genera in partial nitrification process.

Nitrosomonas is still the dominant AOB genus in the sidestream inhibition driven partial nitrification processes. This could be attributed to the initial dominant abundance of *Nitrosomonas* in activated sludge, as well as its distinctive resistance strategies under the selective pressures. The catabolism and anabolism of *Nitrosomonas* both suffered negligible influence when exposed to 16.0 mg/L FA, and half of the ammonia oxidation activity was preserved although the growth of *Nitrosomonas* was completely inhibited when exposed to more than 0.5–0.63 mg/L FNA. According to the metagenomic and metaproteomic analyses, *Nitrosomonas* maintains internal homeostasis by upregulating the expression of some enzymes and proteins involved in oxidative stress response (such as pteridine reductase and dihydrolipoyl dehydrogenase, DNA and protein repair enzymes, phage prevention proteins and iron transport proteins), energy generation (such as AMO and HAO) and nitrite reduction (nitrite reductase) when exposed to 3.64 mg/L FNA. In the presence of sulfide, *Nitrosomonas* was inferred to live in symbiosis with the sulfur oxidizing bacteria (SOB) that detoxify sulfide to elemental sulfur.

Different genera of NOB also have remarkable discrepancy in response to the physical and chemical selective pressures. *Nitrobacter* had a stronger adaptability than *Nitrospira* when exposed to FA, sulfide or hydroxylamine. Under the suppression of FNA (0.011–0.023 mg FNA–N/L), the anabolism of *Nitrobacter* was significantly inhibited, but its catabolism was not affected. Thus, rapid elimination of *Nitrobacter* from systems faces numerous challenges. *Candidatus Nitrotoga*, a novel genus of nitriteoxidizing *beta–Proteobacteria* that has not been paid more attention to previously in activated sludge, shows more resistant to FA, FNA or sulfide treatment than *Nitrobacter*. Recently, Zheng et al. systematically investigated the effects of various critical environmental factors (temperature, DO, FNA, and FA) on *Ca. Nitrotoga* in activated sludge systems, and confirmed that *Ca. Nitrotoga* had excellent tolerance to extreme environmental conditions including high levels of nitrite, FNA or FA. However, the research on the molecular mechanism is still absent and requires further study. Fortunately, the activity of *Candida Nitrotoga* as well as *Nitrospira* could be inhibited selectively and thoroughly by lowdose UVA irradiation. Compared with *Nitrosomonas*, these two genera of NOB produced more intracellular reactive oxygen species (ROS) when exposed to UVA irradiation. Previous studies have shown that *Nitrospira* and Nitrobacter lack the stress response mechanisms against ROS. Obviously, exploring a universal regulation strategy is beneficial for the simplifying of partial nitrification.

Questions

[1] What is the principle of the matched nitrosation reaction?

[2] Which reactor is better for matched nitrosation process?

[3] How temperature affects the process of matched nitrosation?

[4] Which method is used to control DO levels during nitrosation reactions?

[5] Which microbial populations are present in a matched nitrification system?

References

[1] Chen Y M. Start–up strategies for semi–nitrification in SBR reactors [J]. Journal of Xiamen University (Arts & Social Sciences), 2013, 52 (2): 232–236. (In Chinese)

[2] Feng Y, Tseng S, Hsia T, et al. 2007. Partial nitrification of ammonium rich wastewater as pretreatment for anaerobic ammonium oxidation (Anammox) using membrane aeration bioreactor [J]. Journal of Bioscience and Bioengineering, 104 (3): 182–187.

[3] Li H, Zhou S, Huang G, et al. Achieving stable partial nitritation using endpoint pH control in an SBR treating landfill leachate [J]. Process Safety and Environmental Protection, 2014, 92 (3): 199–205.

[4] Jaramillo F, Orchard M, Muñoz C, et al. Advanced strategies to improve nitrification process in sequencing batch reactors–a review [J]. Journal of Environmental Management, 2018, 218, 8 : 154–164.

[5] Javid Gholami–Shiri, Mohammad Azari, Shabnam Dehghani, et al. A technical review on the adaptability of mainstream partial nitrification and Anammox : substrate management and aeration control in cold

weather [J] . Journal of Environmental Chemical Engineering, 2021, 9 (6) : 106468.

[6] Ji P W, Liang Z L, Yongdi L, et al. A review of partial nitrification in biological nitrogen removal processes : from development to application [J] . Biodegradation, 2021, 32 : 229–249.

[7] Li H S. Processes control on partial nitritation coupled with anaerobic ammonium oxidation and its application to nitrogen removal from landfill leachate [D] . Guangdong : South China University of Technology, 2013.

[8] Zhang M, Wang Y Q, Jiang Y, et al. Progress in the study of the regulatory process of matched anaerobic ammonia oxidative nitrosation [J] . Technology of Water Treatment, 2018, 44, 5 : 7–12.

Chapter 6 Start-up of Anammox

Emphasis: The purpose and significance of quick start; Initiation factor control; Start-up process characteristics; Microbial enrichment status.

Nut: Stoichiometry ratio; Carrier selection.

6.1 The purpose and significance of quick startup

The slow growth and long doubling time of Anammox bacteria have become the bottleneck in its applications. Anammox is a process in which the bacterium Anammox uses an inorganic carbon source to convert NH_4^+ and NO_2^- directly to N_2 in an anaerobic environment, with ammonia (NH_4^+) as the electron donor and nitrite nitrogen (NO_2^-) as the electron acceptor. Compared with the traditional nitrification/denitrification process, this process has the advantages of low energy consumption, low sludge production, and no need for additional carbon source. However, the slow growth rate (maximum specific growth rate of 0.0027/h) and long doubling time (10–12 d) of Anammox bacteria, as well as their extreme sensitivity to changes in temperature, pH, and substrate concentration, have made the start-up of Anammox slow, which is a limiting factor for the large-scale application of this process. The reported start-up times for Anammox have basically exceeded 120 d. The world's first field-scale Anammox reactor in the Netherlands had a start-up time of up to 3.5 years.

Therefore, how to analyze and control the fast start-up of Anammox reactor is a critical step for the large-scale application of the Anammox process. In recent years, scholars have conducted in-depth studies on the start-up process of an Anammox reaction, which mainly involves the type of reactor, start-up mode, substrate consumption process and the difference of microbial number and species, etc., and good research progress has been achieved. The content of this chapter is intended to summarize and prospect the latest research on the start-up process of Anammox, aiming to provide a theoretical basis for the rapid start-up and engineering application of Anammox process.

6.2 Initiation factor control

6.2.1 Selection of reactor

Different reactor types will provide different microdomain environments for microbial growth, with important implications for microbial growth and retention of functional microflora. In recent years, various types of reactors such as upflow anaerobic sludge bed reactor (UASB), sequencing batch activated sludge (SBR), fluidized bed reactor, and membrane bioreactor (MBR) have been successfully applied for the initiation of Anammox, and among them, the UASB and SBR reactors are the most commonly used.

UASB reactor has the characteristics of compact process structure, high treatment load, strong shock resistance, etc., coupled with its special three-phase separator, which can effectively reduce the loss of enriched sludge, and at the same time to ensure that the reactor has a high degree of confinement, so that the Anammox bacteria are not easy to be affected by the external environmental factors, it is the preferred type of reactor to start Anammox. Some scholars initiated the Anammox reaction using a UASB reactor inoculated with sludge from a wastewater treatment plant sludge thickening tank, which was successfully started up in 119 d, and the removal rates of NH_4^+-N and NO_2^--N were 96.6% and 75.4%, respectively. Some scholars inoculated activated sludge in a UASB reactor, which was started up in 91 d, and the removal rates of NH_4^+-N and NO_2^--N reached 96.1% and 95.2%, respectively. Some scholars inoculated the UASB reactor with mature Anammox sludge and successfully started the Anammox reaction after only 14 d, and the TN removal reached 89.6% ± 0.96% after the successful startup. Due to its special hydraulic characteristics, the UASB reactor has become the first choice for Anammox startup because of its granularity, enhanced sludge settling and sludge-water separation, and prolonged sludge residence time (Figure 6-1).

The SBR process has the advantages of uniform substrate mixing in the mixing stage, efficient mud-water separation in the sedimentation stage, and sludge retention capacity of more than 90%, and Anammox initiated by the SBR reactor to promote timely substrate uptake and metabolism by Anammox bacteria. Some scholars inoculated nitrifying sludge and denitrifying sludge in SBR reactor to start Anammox reaction, respectively, after 176 d, 179 d start successfully, the average removal rate of total nitrogen reached 84.8% and 81.6%. Some other scholars used SBR reactor, after 100 d successful start Anammox reaction. Due to its own operating characteristics, the SBR reactor has a strong retention capacity for microorganisms, which can effectively reduce sludge loss and has certain advantages for the initiation of Anammox. However, compared with UASB, SBR has a lower operating load and higher automatic control requirements, which limits its application to a certain extent. Therefore, some improved SBR reactors, such as sequencing batch biofilm reactor (SBBR), have been applied

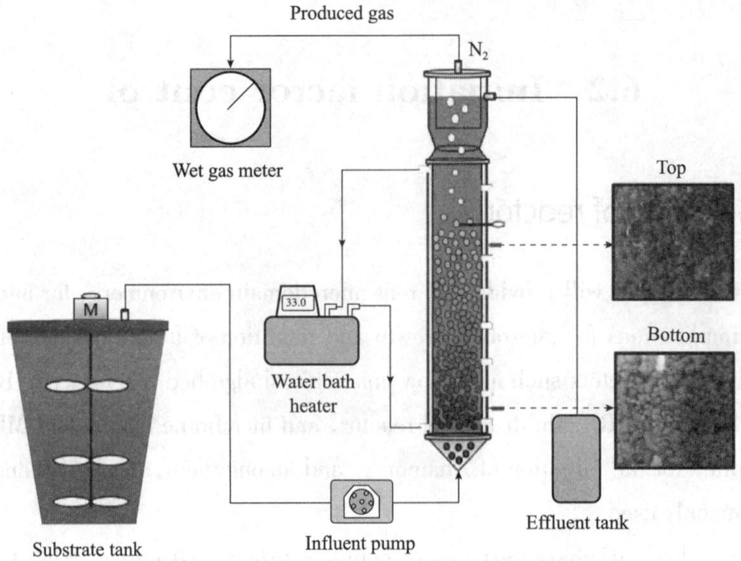

Figure 6–1 Schematic diagram of the UASB–Anammox process

to the startup of anaerobic ammonia oxidation. Chinese scholars used SBBR to start Anammox quickly, and the phenomenon of Anammox appeared from the 60th day, and after 107 d of startup, the maximum load of total nitrogen (TN) reached 0.67 kg/(m·d), and the removal rate reached 87.3%, and the relative content of Anammox bacteria reached more than 40%, which became an advantageous bacterial species in the reactor. The SBBR reactor, with built-in packing and biofilm formation, makes up for the weakness of SBR, which is difficult to withstand high loads, and significantly strengthens the impact resistance, and has low sensitivity to temperature and dissolved oxygen, so that the nitrogen removal effect is stable and highly efficient (Figure 6–2).

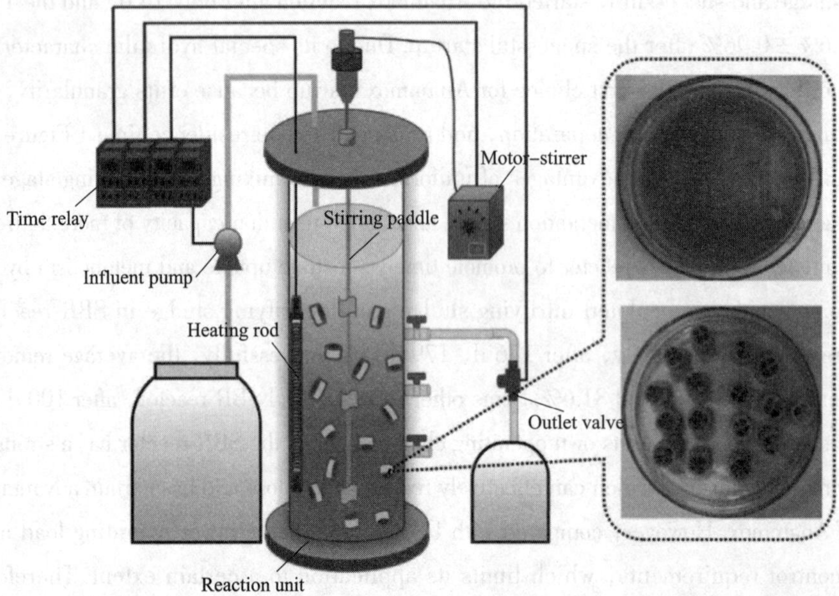

Figure 6–2 Schematic representation of the SBBR–Anammox system

Meanwhile, fluidized bed reactor, membrane bioreactor, anaerobic folded plate reactor (ABR), upflow biofilm (UBF) Anammox reactor, Anammox expanding granular sludge bed (EGSB), and biological carousel were all successfully applied to start the Anammox reaction. The characteristics and differences of each reactor application are shown in Table 6–1. From Table 6–1, the UASB, SBR, MBR, and fluidized bed reactors could successfully start the Anammox, and the TN removal rates were almost all higher than 90%, but the TN removal rate of the SBR reactor, which was started with aerobic nitrifying sludge, was only 80.5%. However, in terms of start–up time, there was a large difference in the start–up time of each type of reactor, ranging from 14 to 365 d. The sludge inoculated with the domesticated Anammox performance had a faster start–up time of only 14 d, whereas the enrichment of the Anammox from the other sludge had a longer start–up time. In conclusion, different reactors have their own advantages and disadvantages, and thus different reactor types are used in practice according to the actual situation.

Table 6–1 Various start–up reactors activated

Reactor	Reactor characteristics	Inoculate sludge	NH_4^+–N removal rate /%	NO_2^-–N removal rate /%	TN removal rate /%	Start–up time
UASB	Good sludge retention capacity, high treatment load, very strong impact resistance, high tightness	Domestication of enriched Anammox sludge	95–99.2	97.7–100	89.6 ± 0.96	14
		Landfill SBR tank sludge	90	90	90	120
		Anaerobic granular sludge	93.2	100	—	125
		Nitrified sludge	86	100	—	80
SBR	Strong sludge holding capacity, homogeneous substrate mixing, high efficiency mud–water separation; however, the operating load is low and the automatic control requirements are high	Aerobic nitrification sludge	81.2	85.7	80.5	142
		Activated sludge	79–99.9	96–99.5		365
		UASB pool sludge	79	—	98	121
		Activated sludge	78	—	98	106
		Anaerobically digested sludge	81	—	98	100
MBR	The pore size is smaller than the diameter of Anammox bacteria, effectively preventing the loss of bacteria, but with high resistance loss	Conventional activated sludge	90	90		60
		Anaerobic sludge	—	—	96	135
Fluidized bed	High biological retention, good mass transfer conditions, high load; serious wear and tear of carriers and equipment, easy to lose microbial particles	Denitrifying sludge	82	99	—	150

6.2.2 Selection of inoculated sludge

The difference of inoculated sludge has an important effect on the initiation process of Anammox and microbial enrichment process, and the inoculated sludges successfully applied to the initiation of Anammox include anaerobic sludge (e.g., denitrifying sludge, river bottom sludge, etc.) and aerobic sludge (e.g., nitrifying sludge, traditional activated sludge, etc.). Some researchers used aerobic sludge, anaerobic granular sludge and anaerobic digested sludge as different inoculated sludge to initiate Anammox, which was achieved after 61 d, 70 d and 85 d of operation, respectively. The NH_4^+–N removal rates were 82%, 92% and 91%, respectively, and the yield coefficients of Anammox (in terms of the amount of $CH_2O_{0.5}N_{0.15}$ produced by 1 mol of NH_4^+) were 0.080 mol/mol, 0.105 mol/mol, and 0.114 mol/mol, respectively, and it was concluded that the anaerobic granular sludge was the most suitable sludge source for enriching Anammox. It was concluded that anaerobic granular sludge was the most suitable sludge source for enriching Anammox bacteria. Researchers also inoculated beer wastewater treatment project in the UASB reactor, anaerobic granular sludge, anaerobic granular sludge and aerobic activated sludge mixed sludge and river sludge, respectively, after 225 d, 220 d and 250 d of operation to successfully start Anammox reaction, the removal rate of NH_4^+–N have reached more than 80%, the three start–up time and the removal rate of the substrate is not much different.

Granular sludge has good settling property, which is conducive to the enrichment of active microorganisms, and the use of granular sludge to initiate Anammox can increase the retention of Anammox bacteria and wash out other microorganisms, which can promote the initiation of Anammox, and in recent years, granular sludge has been widely used for the initiation of Anammox. Some researchers inoculated anaerobic granular sludge to start the Anammox in a SBR reactor. After 161 d of start–up operation, the sludge turned red and the granular Anammox bacteria formed NH_4^+ and NO_2^- removal rate was 60.6% and 62.5%, respectively. Other researchers inoculated anaerobic granular sludge and initiated Anammox in an SBR reactor, and by the 90th day, granular sludge with Anammox activity was successfully cultivated.

The Table 6–2 shows the initiation of the Anammox process using different reactor systems and inoculums.

Table 6–2 Startup of Anammox process with different reactor systems and inoculum

Reactor type	Type of seed sludge	pH	T/℃	Startup time/d	Main strain detected in the reactor
UASB	Aerobic granules	7.0	30	140	*Brocadia Anammoxidans*
MBR	Granular Anammox sludge	7.1–7.8	38 ± 1	About 28	Change from *Candidatus* "*Brocadia*" to *Candidatus* "*Kuenenia stuttgartiensis*".
Full–scale	Nitrifying sludge	7–8	30–40	1235	*Brocadia Anammoxidans*
SBR	Local activated sludge	7.8	35 ± 1	47	*Candidatus Brocadia Anammoxidans*
SBR	Mixture of sludges from municipal WWTPs	7.8–8.0	35 ± 1	60	*Kuenenia stuttgartiensis*
MBR	Mixed aerobic activated sludge and nitrifying	8.0	35 ± 1	19	*Candidatus Brocadia Anammoxidans* (60%)
SBR		8.0	35 ± 1	49	

Continued

Reactor type	Type of seed sludge	pH	T/℃	Startup time/d	Main strain detected in the reactor
UBF	From a full-scale wastewater treatment plant	7	35 ± 1	150	*Brocadia Anammoxidans*
UASB	Anaerobic granular sludge	7	30	100	—
Upflow column	Anaerobic activated sludge	7	35 ± 1	50	Denitrifying bacteria

6.2.3 Selection of carrier

The addition of carriers can reduce the loss of Anammox sludge, increase the residence time of Anammox bacteria in the reactor, enable rapid enrichment of Anammox bacteria, and accelerate the start-up of Anammox reactor. The carriers that have been successfully applied to the Anammox reactor include nonwoven fabric, polyethylene glycol gel, polyvinyl alcohol beads, sponges, biomass charcoal, etc. The Anammox bacteria are attached to the surface of the carriers, which promotes the enrichment of Anammox bacteria. Some researchers used nonwoven fabric as packing material, utilized a biological rotor, inoculated a small amount of Anammox mature sludge, and initiated the Anammox reaction under the conditions of 18–25 ℃, dissolved oxygen (DO) 3.6 mg/L, and 45 r/min rotational speed; the TN removal load reached 1 kg/(m^3 · d) after 90 d, which increased 20-fold compared with that of the inoculated sludge biomass. After 90 d, the TN removal load reached 1 kg/(m^3 · d), which was more than 20 times of the inoculated sludge biomass. Researchers used two types of UASB reactors with and without sponges to start Anammox, and the sponge-added reactor entered the stabilization period in the 80th day, which was 25 d less than that of the sponge-less one, and the removal rates of NH_4^+-N and NO_2^--N reached 99.8% and 99.9%, which were higher than that of the sponge-less reactor. Starting from the 2nd day of the start-up, compared with the turbid water and floating sludge in the reactor without sponge, the water was clearer in the sponge-added reactor. This indicates that the addition of sponges has an advantage in retaining sludge and facilitates the start-up and stabilization of the Anammox reactor. Other researchers used three different UASB reactors with bamboo charcoal, multi-surface hollow spheres and no filler to successfully initiate Anammox after 85 d, 97 d and 117 d, respectively, and found that the addition of filler not only accelerated the initiation of Anammox, but also facilitated the rapid propagation of Anammox bacteria.

Fillers provide a place for Anammox bacteria to attach and reproduce, which can promote the growth and reproduction of Anammox bacteria, reduce the loss of bacteria, shorten the start-up time, and have a greater application value in the initiation of Anammox. The selection of packing material is very important, it should have large surface area, low resistance, high void ratio, large surface roughness, hydrophilicity and other characteristics, and cheap and easy to obtain, high chemical stability, such as non-woven fabrics, sponges, biomass charcoal and other common porous materials can be used as the preferred packing material for the initiation of Anammox.

Discussed below are some of the biomass carriers that have been widely used in the Anammox process and their respective effects on Anammox initiation and performance.

1. Non-woven fabric

The addition of carriers to the bioreactor system enhances the attachment of biofilm, thus increasing the volumetric loading rates of the Anammox activity thereby shortening the start-up of the Anammox process. Recently, studies have indicated that non-woven materials as carriers have efficient biomass retention capability in nitrogen removal through the Anammox process. The porosity of the non-woven material aids in the attachment of biomass and prevents washouts thereby enhancing the performance of the Anammox reactor. They are also able to retain free bacteria cells and prevent membrane fouling when operated for long periods (up to 400 d).

In a study which used non-woven fabric as the biomass carrier, an efficient Anammox reactor with a super high NRR of 26 kg/(Nm^3 · d) was successfully established. In another study where a hybrid reactor (a combination of fixed-bed and FBR) was used with nonwoven fabric as the carrier, an 8% improvement was registered in the NRR. In another Anammox start-up process in an FBR where a non-woven biomass carrier was used as the fixed medium, there was an improvement in both the efficiency of biomass attachment as well as the Anammox-reactor performance. An FBR filled with non-woven ring carriers showed excellent performance with Anammox activity occurring on day 39 and the maximum activity on day 367. The total nitrogen loading and removal efficiency were 10.3 kg/(Nm^3 · d) and 90% respectively (Figure 6-3).

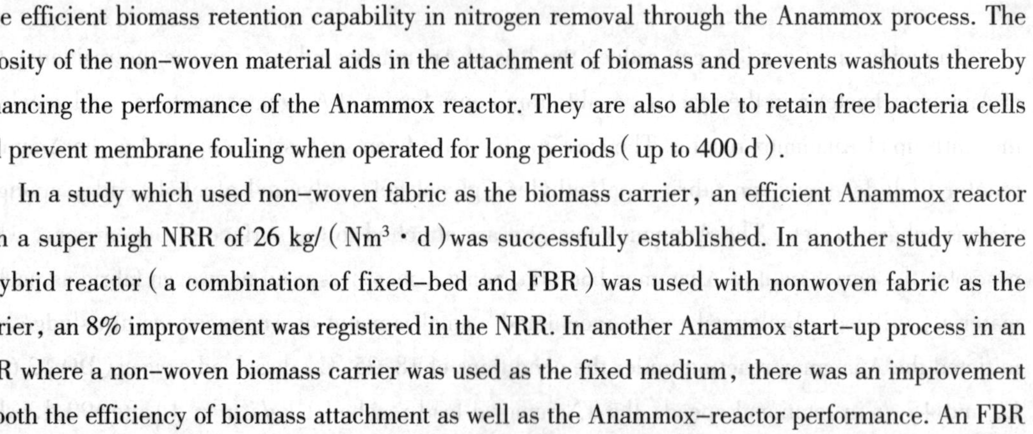

(a) (b)

Figure 6-3 Scheme of reactor configuration

(a) The non-woven ring carrier fixed bed reactor;

1—feed vessel, 2—influent pump, 3—novel carriers, 4—recycle pump, 5—heater, 6—temperature sensor, 7—heating tank, 8—sampling port, 9—effluent port;

(b) photo of novel non-woven ring carriers.

2. Biochar

These are produced by thermal decomposition of organic material, such as wood, bamboo, or leaves under a limited supply of oxygen at relatively low temperatures (>700 ℃). Biochar carriers are among the suitable carrier materials that are capable of preventing biomass washout and retaining stable immobilization of Anammox bacteria due to their high surface area, great potential cation exchange capacity (CEC), high negative surface charge and charge density. It helped mitigate nitrate leaching and has the potential of being used to influence the rates of nitrogen cycling by increasing NH_3 absorption and NH_4^+ storage.

Bamboo charcoal has been used extensively in trying to shorten the start-up of the Anammox process. Due to their large specific area, they enhance adhesion for Anammox bacteria hence supporting the growth of biofilms. They can execute this function as a result of their good NH_4^+ adsorption capacity, microporous structure that prevents oxygen from penetrating the inlet zone as well as by increasing the settling velocity of Anammox and improving Anammox retention.

Biochar addition enhanced ammonia-oxidizing bacteria abundance than control by 4-12.3 fold and promoted nitrification in dry montane forest soils. While the nitrate concentration in the effluent of the biochar amended reactor was quite low [(2.6 ± 1.7) mg/L] during the entire start-up time, that of the control fluctuated; (9.6 ± 3.0) mg/L during the expressive phase, 20.8 mg/L in the enhanced phase and then dropped to (9.1 ± 1.9) mg/L in the steady phase. The entire start-up time of the UASB reactor was dramatically shortened when amended with bamboo charcoal: a start-up of 85 days was achieved as compared with 97 days when spherical plastic was used and 117 days when no carrier was used (Figure 6-4).

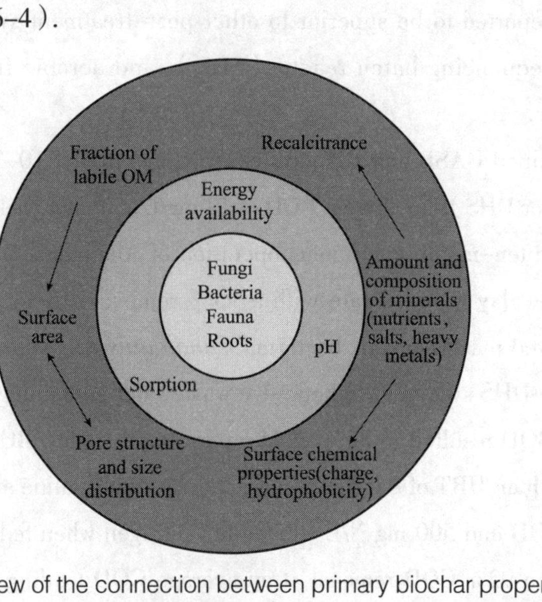

Figure 6-4 Schematic overview of the connection between primary biochar properties (outer circle), the soil processes they may influence (intermediate circle) and the soil biota (inner circle) (shorter distance give a qualitative estimate of the strength of the connection)

3. Downflow hanging sponge (DHS)

DHS is an aerobic treatment system similar to the trickling filter and uses a polyurethane foam (sponge) as the growing and supporting medium for microorganisms. Because it can act as a mobile carrier for active biomass, DHS is considered as an ideal attached growth media.

Apart from being able to reduce membrane fouling through mechanical cleaning and maintaining a balance of suspended-attached microorganisms in submerged membrane bioreactor (SMBR), DHS can also enhance dissolved organic matter and nutrient removal. Its longer SRT is a result of its high porosity which helps in entrapping the microorganisms and organic matter. It is also found to be efficient for *E. coli* and nitrogen removal. DHS systems were also found to require approximately 75% less energy as compared to the conventional activated sludge (CAS).

DHS effluent water is similar to that of an activated sludge process and satisfies adequate levels for discharge with higher nitrification efficiency. It has also been reported to be suitable for retaining the high biomass and minerals in the sponge material. It is also reported to be capable of maintaining a long sludge retention time, making it suitable for the enrichment of slow-growing bacteria. Several microorganisms including the methanogenic community from the seafloor and different types of polyphosphate accumulating organisms (PAOs) have been successfully enriched using the DHS reactor. However, part of the DHS floats on a liquid medium, resulting in the accumulation of insoluble precipitated inorganic matter on the dry area.

It is strongly recommended to use the DHS system for post-treatment of anaerobic pre-treated sewage as DHS is capable of achieving a >90% BOD removal when coupled with UASB. The DHS system has been reported to be superior to other post-treatment systems, such as the activated sludge process, sequencing batch reactor (SBR), and aerobic filter for domestic wastewater treatment processes.

When a combined UASB and DHS system was operated at 10–28 ℃ and 10.6 h HRT (8.6 h for UASB and 2 h for DHS), an effluent BOD of 12 mg/L and total Kjeldahl nitrogen content of 6 mg/L were obtained. The ten-month continuous operation of Anammox-DHS process resulted in a 1.85 kg N/m^3 sponge per day removal rate with a 95% removal efficiency. Microbial analysis in the sponge media showed that Anammox bacteria "*Candidatus Kuenenia Stugartiensis*" dominated. In a combined UASB-DHS system, synthetic dye wastewater containing 1000 mg/L of phenol as COD and 700 mg/L as BOD resulted in 80% COD removal and > 95% BOD removal operating in twelve different modes with an HRT of 8 h for each reactor. A highly saline artificial coke-plant wastewater with 1400 mg/L COD and 500 mg N/L ammonium nitrogen when fed into a combined UASB-DHS system resulted in a 92% COD removal at an average COD loading rate of 3.0 kg COD/(m^3 · d) (Figure 6-5).

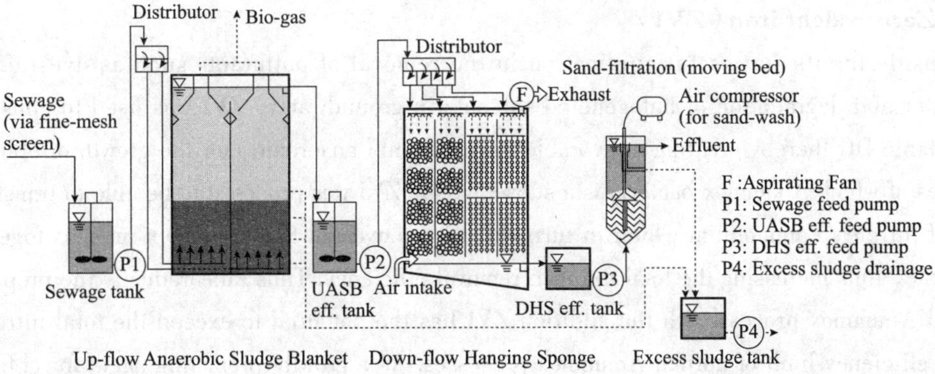

Figure 6-5 Schematic of the pilot-scale plant (DHS process is included)

4. Zeolite

Zeolites are aluminosilicate solids bearing a negatively charged honeycomb framework of micropores into which molecules may be adsorbed for environmental decontamination, and to catalyze chemical reactions. Because of the necessity for organic solvents to be minimized, they have become central to green-chemistry. They are also characterized by extensive internal porosity, very large surface areas; internally and externally, and corresponding high CECs. Currently, zeolites are being used to speed up the start-up of the Anammox process in a process called the zeolite-Anammox system.

The zeolite carrier addition could very much reduce the volatile suspended solids (VSS) of effluent just about 3 mg/L as well as increase the biomass in the SBR. In the zeolite-Anammox system, a single reactor is used to convert ammonia to nitrogen gas with a combination of nitrifying, denitrifying and bacteria. As such, they are comparatively easier, simpler and less expensive to operate. There was higher enrichment of the Anammox biomass coupled with a reduction of biomass wash-out in the effluent to values as low as 3 mg VSS/L with the addition of zeolite particles. Smaller zeolite aggregate (10 mm) achieved a greater ammonia removal fraction than the larger aggregates (20 mm) (89% vs 85%). The use of the smaller aggregates, however, resulted in the pore-clogging problems during the 15-week test period in a trickling-filter system (Figure 6-6).

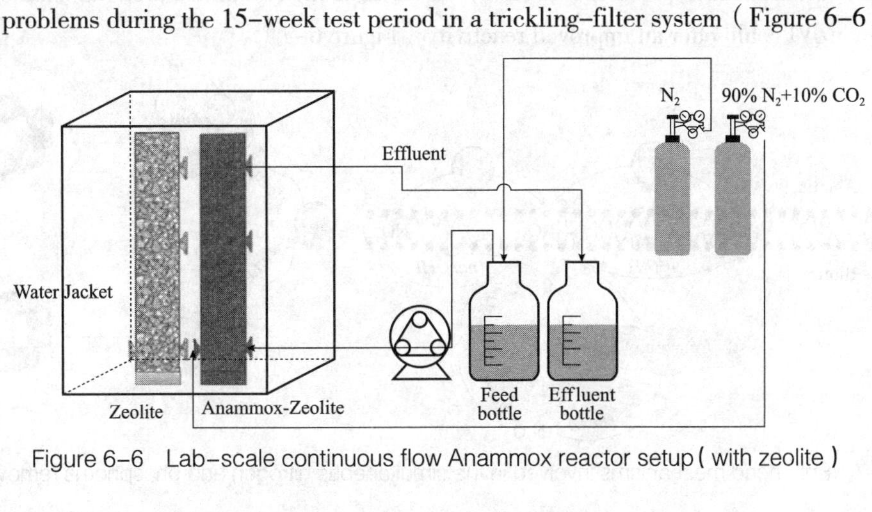

Figure 6-6 Lab-scale continuous flow Anammox reactor setup (with zeolite)

5. Zero-valent iron (ZVI)

Considering its successful application in the removal of pollutants such as dyestuffs in wastewater and degradation of halogenated solvents in groundwater, ZVI was used to consume the available DO thereby creating a favorable environment (anaerobic) for the growth of obligate anaerobes of which Anammox bacteria is a subset. Also, ZVI was proposed to be able to transform a part of nitrate to ammonium which in turn will be removed in the Anammox process together with nitrite, thus increasing the total nitrogen removal efficiency. Thus, theoretically the proposed modified Anammox process with the addition ZVI has the potential to exceed the total nitrogen removal efficiency limit of normal Anammox processes. As a growth-promoting agent for obligate anaerobes as earlier stated, ZVI also promotes the synthesis of hemoglobin thereby stimulating the formation of bacterioferritin in an anammoxosome which is essential for Anammox reactions. Iron ions generated from ZVI have been employed in anaerobic reactors to promote microbial agglomeration and secretion of the extracellular polymeric substances (EPS). This EPS possesses a vital role in the formation of multilevel porous structures on cell surfaces which is very helpful in enhancing microbial aggregation in a three-dimensional gel-like hydrated matrix through weak physiochemical interactions. This (agglomeration) helps prevent the washout of Anammox bacteria which is very beneficial for the proliferation of the slow-growing Anammox bacteria.

A 13.2% increase in total nitrogen removal was achieved while ammonium removal improved from 13.3% to 92.% when ZVI was used in the pretreatment The use of micro-ZVI (mZVI) and nano-ZVI (nZVI) in two different UASB reactors saw the start-up time of the Anammox reduced to 105 days (16.7% reduction) and 84 days (33% reduction) respectively coupled with higher nitrogen removal efficiencies.

mZVI shortened the start-up by 16.7% whereas nZVI reduced it by 33.4%. According to the performance of nZVI (42 days) on accelerating Anammox start-up was even better than the Fe-graphite electrode which had initially been used to reduce the Anammox start-up to 30 days. This could be due to the stronger oxygen-consuming ability of nZVI thereby creating a more suitable environment for Anammox proliferation. Also, due to their higher surface area but smaller average grain size, nZVI could offer an improved reactivity (Figure 6-7).

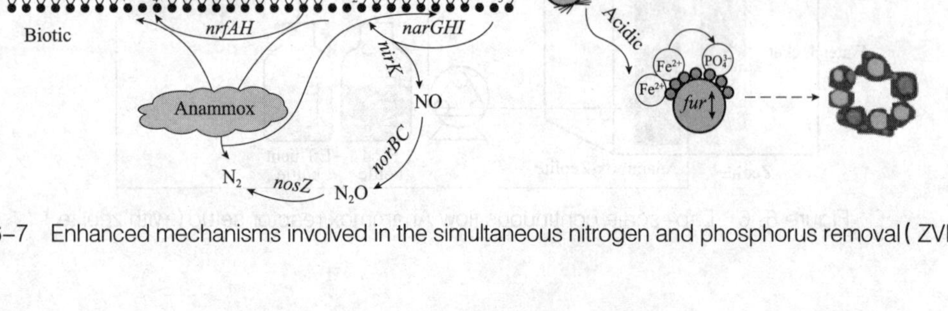

Figure 6-7 Enhanced mechanisms involved in the simultaneous nitrogen and phosphorus removal (ZVI)

6. Polyethylene glycol gel carrier

Gel entrapment technology has been successfully used to immobilize nitrification, denitrification and Anammox bacteria in biological nitrogen removal systems. Polyethylene ethylene gel carrier has excellent biomass retaining capacity. This is mostly because nitrifying bacteria immobilized in a gel carrier are easily separated from the liquid in the reactor, thus yielding the prolonged biomass retention times even with short hydraulic retention times (HRT). Relatively, gel carriers are easier to handle than sludge cultures, and hence could be simply set in a test tube after being removed from the reactor (Figure 6–8).

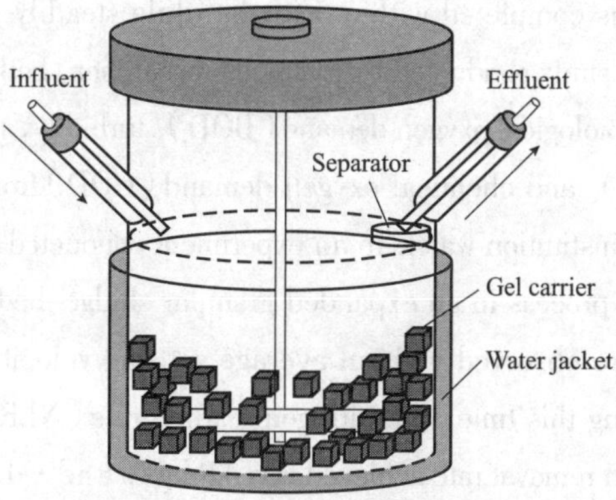

Figure 6–8 Schematic illustration of a reactor for nitrogen removal test using Anammox bacteria entrapped in gel carrier

In a sterilizing reactor with gel entrapment and 0.72 g dry suspended solids (SS) per liter reactor, a nitrogen conversion rate of 3.7 kg N/(m^3 · d) was achieved after 67 days whereas 3.4 kg N/(m^3 · d) was achieved in 25 days when 4.7 g dry SS was used. However, a maximum Anammox activity of about 11–12 kg N/(m^3 · d) with a corresponding pH increase from 8.4–8.6 regardless of the amount of the seed sludge used was achieved. Within one month of operation of a continuous feeding test using model wastewater containing a high 1,4–dioxane concentration of 400 mg/L, reactor start–up, as well as stable efficient removal of 1,4–dioxane, were achieved. More than 99% of 1,4–dioxane removal efficiency was achieved after the start–up, with the maximum 1,4–dioxane removal rate for the reactor volume achieved at 0.67 kg dioxane/(m^3 · d) under 16 h of HRT. An immobilization method where ammonium–oxidizing bacteria was entrapped in a polyethylene glycol (PEG) carrier resulted in a high nitrification rate of 2.4 kg N/(m^3 · d) at 20 ℃. In the nitrification of landfill leachate using immobilized nitrifying bacteria at 10 ℃, the reactor start–up was observed after about three months with a nitrification rate of 0.71 kg N/(m^3 · d). Though a stable nitrification activity was maintained for more than a year, complete nitrification was never achieved with 25–27 mg/L ammonium nitrogen remaining in the effluent water.

7. Activated Carbon

Activated carbon (AC) has been found to be good porous materials and effective in the adsorption of solutes from aqueous solutions. The chemical nature of AC greatly influences its adsorptive, catalytic, electrochemical and other useful properties. Activated carbon (AC) has been used in different states of applications after its discovery as a strong and reliable adsorbent. For example, it has been used extensively to remove dye from industrial wastewater. Granular activated carbon (GAC) can act as a biofilm carrier by taking on the structural function of filamentous growth and can also stimulate the Anammox bacteria growth. GAC is also suitable for retention of bacteria due to its porous microstructure and a specific gravity close to that of water.

In one study in which the sludge was seeded with GAC, Anammox activity was achieved in approximately 25 weeks. Comparatively, the formation of an Anammox granular bed in this study with the GAC seed was completed within 5 weeks while steadily increasing the NLR to 6 kg N/(m^3 · d). According to a study conducted by a famous researcher, both GACC and GACP could be used to remove high ratios of biological oxygen demand (BOD), turbidity, total dissolved solids (TDS), total suspended solids (TSS), and chemical oxygen demand (COD) from wastewater using coconut and palm shell GAC to treat institution water. In an experiment conducted using GAC to treat synthetic wastewater by the Anammox process in an expanded granular sludge-bed (EGSB) reactor, excellent retention characteristics were observed with an average settling velocity of 200 m/h over a startup period of only 38 days. During this time, the nitrogen loading rate (NLR) was increased 0.5 to 6 kg N/(m^3 · d) while the nitrogen removal rate achieved was 4.8 kg N/(m^3 · d) (Figure 6-9).

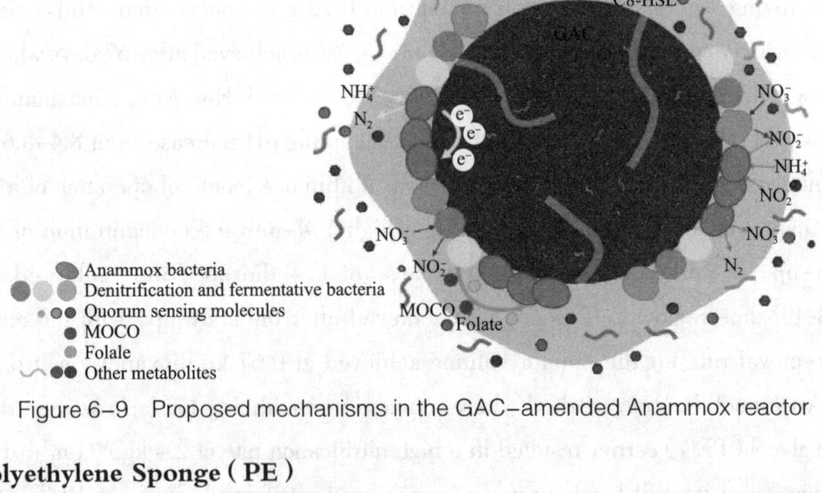

Figure 6-9 Proposed mechanisms in the GAC-amended Anammox reactor

8. Polyethylene Sponge (PE)

When PE was used in an up-flow fixed-bed Anammox column reactor, they exhibited good treatment performance for autotrophic nitrogen removal from synthetic wastewater. In estimating the Anammox reaction state, the ratio between the removal of NH_4^+-N and NO_2^--N is a very key criterion. A ratio of 1.26 : 0.21 which is similar to 1.31 : 0.26 (the standard Anammox reaction state ratio) was obtained for the experiment using PE sponge in an up-flow fixed-bed Anammox column reactor. The reactor with 8 strips of PE sponge showed a relatively better capacity for adapting than that with 6 strips, presumably due to the large surface area of the former (Figure 6-10).

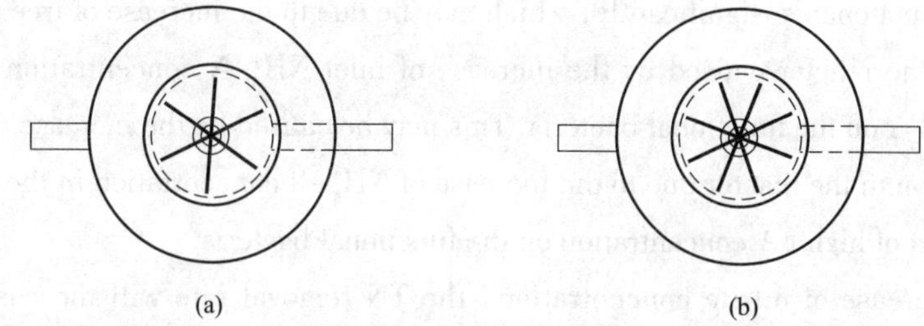

Figure 6-10 PE sponge carriers with 6(a) and 8 strips(b)

Therefore, to ensure a quick and robust start-up of the Anammox system, the selection of the proper carrier among other operational strategies is very key. Without consideration to reactor type, carrier amendment seems to be the most feasible of all the approaches employed in optimizing the performance of the Anammox process (increment in both Anammox population and acceleration of the start-up time) as evidenced by some of the data provided (Table 6-3).

Table 6-3 Performance of different carriers with different reactor configurations

Carrier material	Reactor type	Start-up/d	N-removal performance	Maximum NLR/ [kg N/(m^3 · d)]	Highest NRR/ [kg N/(m^3 · d)]
Bamboo charcoal	UASB	85 days	98%	—	—
Non-woven membrane	Anammox reactor	8 months	90.9-NH_4^+	1.263	1.048
Non-woven material	Biofilm reactor	—	60% ± 6.3%	—	~0.041
Non-woven fabric	Hybrid reactor	38 days	>70%	8.9	6.6
Non-woven fabric rings	FBR	39 days	90%	10.3	26
Polyethene sponge	Up-flow fixed-bed reactor	—	90%	8.4	7.6
Polyethene sponge strips	Upflow column reactor	Within 56 days	>85	4 ± 0.1	3.6
Sponge material	DHS		95%	5.96	2.27
Submerged hollow fibre membrane	MBR	Within 80 days	—	—	0.71
Non-woven membrane	Fixed bed biofilm	50 days	—	—	—

6.2.4 Start-up load

The control of reactor start-up load will also directly affect the start of the Anammox reaction, the initial influent substrate concentration is generally controlled at 40-200 mg/L. Some scholars found that when the influent NO_2^--N concentration is controlled at 64-80 mg/L, the NH_4^+-N volumetric loading has a great influence on the oxidative removal rate of NH_4^+-N, while it has little influence on the NO_2^--N reduction removal rate. When the NH_4^+-N load was 7.2 g/(m^3 · d) or less, the difference between the inlet and outlet water NH_4^+-N concentration increased with the load, but when the load exceeded the value, the difference between the inlet and outlet water NH_4^+–N concentration did not change significantly, which may be due to the increase of free ammonia (FA) concentration in the reactor caused by the increase of inlet NH_4^+–N concentration, and the toxic effect of the high FA on the functional bacteria. This may be caused by the increase of free ammonia (FA) concentration in the reactor due to the increase of NH_4^+–N concentration in the influent water, and the toxic effect of high FA concentration on the functional bacteria.

With the increase of nitrite concentration, the TN removal rate will increase, but the TN removal rate by Anammox bacteria started to decrease after a certain value. Some scholars studied the effect of nitrite concentration on Anammox reaction. When the NO_2^-–N concentration was greater than 84.4 mg/L, the NO_2^-–N reduction step in Anammox system would be inhibited, and the TN removal rate began to decrease. It has been reported that when the concentration of NO_2^-–N is more than 100 mg/L, the Anammox reaction will be inhibited. Therefore, it is recommended to control the concentration of nitrite in the influent water less than 100 mg/L during the startup process; however, the inhibition mechanism of Anammox bacteria by high NO_2^-–N concentration is not clear, and needs to be further investigated.

Therefore, high concentrations of NH_4^+–N and NO_2^-–N both inhibited Anammox bacteria, but the degree of inhibition was different. It is reported that NH_4^+–N concentration more than 1000 mg/L does not inhibit Anammox bacteria, while NO_2^-–N has a stronger inhibitory effect on Anammox bacteria, which plays a crucial role in the initiation of Anammox. It is recommended to control the concentration of NO_2^-–N in the influent water less than 100 mg/L during the start–up process.

6.2.5 Low temperature start–up control

The optimal growth temperature of Anammox bacteria is about 35 ℃, and most scholars have used water bath to control the growth at high temperature, which has good effect on the growth of Anammox bacteria. However, maintaining a higher temperature requires a large amount of energy consumption, which is difficult to realize in practical applications. Therefore, how to realize the initiation of Anammox at low and medium temperatures has become a hot research topic in recent years.

In general, the growth of Anammox bacteria at low temperature is slow, the start–up time is long, and the removal rate of nitrogen is relatively low. The temperature of the reaction system has an important effect on the growth and activity of Anammox bacteria. Some researchers used three different low–temperature domestication strategies to initiate the Anammox reaction, corresponding to A_1 (17.2 ℃ activated sludge placed in a biochemical incubator at 16 ℃), A_2 (31 ℃ biofilm placed in a biochemical incubator at 16 ℃) and A_3 (31 ℃ biofilm placed in a biochemical incubator at 31 ℃), with a gradient of –3 ℃/12 d (3 ℃ per 12 d). –3 ℃/12 d (every 12 d) as a gradient of gradual cooling down to 16 ℃, the three reactors showed different Anammox activities after 62 d, 56 d, and 70 d of domestication, respectively, with the activities of $A_3 > A_1 > A_2$ in the order of $A_3 > A_2$, indicating that the adverse effect of low temperature on Anammox activity can be reduced by the change of the cooling strategy. Researchers inoculated freshwater sediments with porous polyester nonwoven non-woven fabric, enriched with Anammox bacteria at low temperatures of 25 ℃ and 18 ℃, the nitrogen removal reached 0.07 to 0.26 kg/ (m^3 · d), indicating that the addition of carrier materials can promote the initiation of Anammox under low and medium temperatures to a certain extent.

6.3 Start-up process characteristics

Different choices of reactor, packing, inoculated sludge, loading and temperature have important effects on the start-up process of Anammox, which provides a good reference value for the rapid start-up of Anammox. At the same time, the researchers also investigated the substrate consumption pattern during the start-up process of Anammox, in order to clarify the characteristics of nutrient dynamics during the start-up process.

6.3.1 Start-up stage

In the initiation process of Anammox, the substrate consumption has obvious characteristics, and the initiation process of Anammox is generally divided into 3 stages, namely activity stagnation, activity enhancement and stable operation. Some scholars also divided the initiation process of Anammox into 4 stages, i.e., hydrolysis period, activity stagnation period, activity enhancement period and stable operation period.

1. Hysteresis stage

The hysteresis stage, also called the bacterial autolysis stage. Due to the autolysis of bacteria, the NH_4^+-N concentration of the effluent water is greater than that of the influent water. At the same time, the organic matter produced by the lysis of bacteria can be used as an electron donor to reduce nitrite, showing a certain degree of denitrification, resulting in a NO_2^--N concentration of effluent water is significantly smaller than that of influent water. With the initiation of time, the cell lysogeny gradually weakened, the production of organic matter gradually reduced, the lack of organic substrate heterotrophic bacteria, and the inorganic substrate for the competition of autotrophic bacteria gradually lost the advantage of denitrification, denitrification weakened, as represented by Anammox bacterium, autotrophic bacteria to provide space for the survival of the effluent NH_4^+-N concentration is gradually close to the influent water concentration.

2. Activity enhancement stage

In the stage of activity improvement, the activity of Anammox bacteria gradually increased, and Anammox gradually replaced denitrification and cell lysis as the dominant reaction. The removal rate of NH_4^+-N in the reactor gradually increases, the concentration of NO_2^--N in the effluent also shows a gradual decline, while the concentration of NO_3^--N in the effluent gradually increases. In the late stage, NH_4^+-N and NO_2^--N decreased greatly in proportion, and Anammox was basically realized.

3. Stable operation stage

Stable operation stage, after the activity improvement stage, the reactor enters the stable stage, and the Anammox reaction becomes the dominant reaction of the reactor. Keeping the operation conditions unchanged, the nitrogen removal efficiency is high and stable, the TN removal rate is above 85%, and the average removal rate of NH_4^+-N and NO_2^--N generally reaches above 95%.

6.3.2 Stoichiometry ratio

Currently, the total stoichiometric chemical formula for the catabolic synthesis of Anammox bacteria, which is generally accepted by scholars, is shown in Equation (6–1).

$$NH_4^+ + 1.32NO_2^- + 0.066HCO_3^- + 0.13H^+ \longrightarrow 1.02N_2 + 0.26NO_3^- + 0.066CH_2O_{0.5}N_{0.15} + 2.03H_2O \quad (6\text{–}1)$$

From equation (6–1), the theoretical ratio of NH_4^+-N to NO_2^--N consumption and NO_3^--N production in Anammox reaction is 1 : 1.32 : 0.26.

However, in the actual initiation process, the substrate consumption and product generation do not strictly conform to the theoretical values, and the differences in environmental conditions will directly affect the stoichiometric ratio values of the reaction. Some researchers inoculated the activated sludge of SBR tank of landfill to start Anammox, and the ratio of NH_4^+-N removal, NO_2^--N removal and NO_3^--N generation in stable operation was 1 : 1.44 : 0.26, which was close to the theoretical value. Researchers also inoculated oxidation ditch return sludge to start Anammox reaction, in the stable stage, the ratio between NH_4^+-N and NO_2^--N consumption and NO_3^--N production was 1 : (1–1.8) : (0.13–0.26), where the ratio between NH_4^+-N consumption and NO_2^--N consumption was mainly concentrated between 1 : (1.4–1.5), and NO_2^--N consumption was high. Other researchers inoculated municipal wastewater treatment plant sludge thickener sludge to initiate Anammox, and the ratio between NH_4^+-N and NO_2^--N consumption and NO_3^--N production during the steady state operation period was 1 : 1.54 : 0.3, again with a high consumption of NO_2^--N, the ratio between NO_2^--N consumption and NO_3^--N production is

1 : 1.54 : 0.3. In general, the NO_2^--N consumption was higher than the theoretical value, probably because the feed water contained a small amount of COD and organic matter produced by bacterial autolysis, resulting in the existence of a certain intensity of denitrification in the reactor, coupled with other autotrophic denitrifying bacteria in the sludge, which provided the conditions for the removal of NO_2^--N. Moreover, some DO in the influent water also provides a way for NO_2^--N to NO_3^--N conversion. However, it has also been shown that the ratio of NO_2^--N and NH_4^+-N consumption during startup is lower than the theoretical value. There are studies that show that inoculated aerobic nitrifying sludge and anaerobic granular sludge to start Anammox, and the ratios of NO_2^--N and NH_4^+-N consumption during stable operation were 1.12 and 1.15, respectively, which were lower than the theoretical value of 1.32, which could be due to the high pH values (8.02 and 7.94) of the mixture in the starting reactor.

6.4 Microbial enrichment status

6.4.1 Trends and extent of microbial enrichment

During the start-up process of Anammox, the uniformity of the microbial community gradually deteriorated, and the relative abundance of the Anammox bacterial group increased continuously, forming a relatively single, stable and mature biological community. At the beginning of the initiation period, the microbial community diversity changed little, and when entering the period of increased activity of Anammox, the microbial community diversity decreased rapidly, and the relative abundance of Anammox bacilli increased continuously, and the dominant colony was formed rapidly. Typical Anammox cells contain a large amount of cytochrome C and appear red in color. Therefore, when the sludge or biofilm in the reactor shows a large area of red color and the substrate is removed stably and efficiently, it can be judged that the Anammox has been successfully initiated.

Generally speaking, the content of Anammox bacteria in the activity retardation stage was low and the proportion of total bacteria increased slowly because the high organic matter content in the water caused by bacterial autolysis at the beginning of the start-up period inhibited the growth and reproduction of Anammox bacteria while promoting the growth of denitrifying bacteria. Entering the activity improvement stage, the activity of Anammox bacteria was improved, but the relative proportion of Anammox in this stage was still low due to the large total bacterial base. And into the stable operation stage, Anammox bacteria gradually matured, other microorganisms died due to substrate limitation, and the relative proportion of Anammox bacteria increased rapidly. Some scholars found that the proportion of Anammox bacteria in the total bacteria increased gradually during the initiation process, from the initial 5.7% to 15.8% and 17.3%, and the proportion of Anammox bacteria reached 23.3% in the 123 day of the stabilization phase. Other scholars found that at 229 d (activity stagnation), 304 d (increased activity), and 487 d (stable operation), the percentage of Anammox bacteria was 1.8%, 3.7%, and 9.0%, respectively, with the greatest increase occurring in the increased activity stage.

6.4.2 Enrichment population categories

Anammox belongs to the order Planktonomycetes, which has 9 genera, among which 5 genera are classified as Anammox. There are 9 species of Anammox bacteria, and the common Anammox bacteria in the process of wastewater denitrification are *Candidatus Brocadia Anammoxidans* and *Candidatus Kuenenia stuttgartiensis*. Research shows initiated Anammox using a nonwoven biological rotor, and the dominant bacterium was *Candidatus Brocadia Anammoxidans* after the successful completion of the process. Other studies have shown that compared the differences between SBR and MBR in initiating Anammox, and found that the dominant Anammox bacteria in both reactors were *Candidatus Brocadia Anammoxidans*. Another study used UASB to initiate Anammox and analyzed the bacterial community structure in the reactors by clone library technique, and found that the Anammox bacteria belonged to the genus *Candidatus Kuenenia*. Experts successfully initiated Anammox using a biofilter, and the Anammox bacteria in the filter were *Candidatus Kuenenia stuttgartiensis*, which had a wide range of adaptability to temperature and substrate concentration. Since each genus has its own unique survival characteristics, there is usually only one Anammox dominant genus in a particular environment.

The use of different enrichment culture environments can have a significant impact on the dominant Anammox species after successful initiation. A foreign scholar suggested that *Candidatus Kuenenia stuttgartiensis* has a higher affinity for nitrite substrates, and that there is a tendency for the dominant Anammox bacteria to evolve from *Candidatus* During the enrichment process, there was a tendency for the dominant Anammox bacteria to evolve from *Candidatus Brocadia Anammoxidans* to *Candidatus Kuenenia stuttgartiensis*. A Chinese researcher used SBBR reactor to initiate a Anammox, the initiation process used industrial wastewater, after the successful initiation of the Anammox dominant strain of *Candidatus Anammoxoglobus* propionicus, the dominant bacteria is more special, may be contained in the wastewater propionate, resulting in organic carbon sources on the dominant bacterial coercive effect as well as dominant populations of differences. The dominant bacteria were more specific, probably containing propionate in the wastewater, causing the stress effect of the organic carbon source on the dominant bacteria and differences in the dominant population. These differences may be related to the reactor type, inoculated sludge, substrate concentration, operation mode and other environmental factors, but for the influence law of each environmental factor is still lack of research, and need to be further deepened. In particular, the emergence of new molecular biology tools has provided a technical guarantee for the improvement of functional genes and metabolic pathways of Anammox bacteria, which indirectly lays a microbiological foundation for realizing the rapid startup of Anammox.

6.5 Conclusions and prospects

The start-up of Anammox is a rate-limiting step for the promotion and application of Anammox-based wastewater treatment processes in practical engineering. In this paper, the latest research on the start-up process of Anammox is summarized in terms of reactor type, start-up mode, substrate consumption and microbial enrichment, which provides a certain theoretical basis for the application of Anammox as the main wastewater treatment process. However, the rapid initiation of Anammox and the rapid proliferation of Anammox bacteria are still in the initial stage in China, and the following work is still needed.

(1) Improvement of functional genes and optimization of physiological and biochemical properties of Anammox bacteria using new molecular biological methods.

(2) To carry out in-depth research on the specific growth environments of different Anammox bacteria, and to realize the mode of cultivating different bacteria in different environments.

(3) To establish a mathematical model for the growth of Anammox bacteria, and to provide a quantitative calculation basis for the initiation of Anammox.

Questions

[1] What are the factors influencing the rapid initiation of Anammox?

[2] What are the stages of initiation of Anammox?

[3] Brief description of microbial enrichment during initiation of Anammox.

[4] In the context of what you have learnt, talk about other factors that influence how quickly or slowly Anammox is initiated.

[5] What are the effects of adding carriers on the initiation of Anammox with examples?

[6] In the context of recent research, tell us what else you know about the initiation of Anammox and your vision for the future.

References

[1] Gao M, Dang H Y, Zou X, et al. Deciphering the role of granular activated carbon (GAC) in Anammox: effects on microbial succession and communication [J]. Water Research, 2023, 02 (233): 119753.

[2] Zhang L, Wang Y P, Hao S W, et al. Anammox-synchronous zero-valent iron oxidation promoting synergistic nitrogen and phosphorus removal from wastewater [J]. Bioresource Technology, 2022, 347: 126365.

[3] Yapsakli K, Aktan C K, Mertoglu B. Anammox-zeolite system acting as buffer to achieve stable effluent nitrogen values. Biodegradation, 2017, 28: 69-79.

[4] Chen X, Liu L, Bi Y, et al. The recovery of a sequencing biofilm batch reactor—Anammox system: performance, metabolic characteristics, and microbial community analysis. Sustainability, 2023, 15 (13): 10454.

[5] Ma H Y, Niu Q G, Zhang Y L, et al. Substrate inhibition and concentration control in an UASB-Anammox process. Bioresource Technology, 2017, 238: 263-272.

[6] Adams M, Xie J, Xie J, et al. The effect of carrier addition on Anammox start-up and microbial community: a review. Rev Environ Sci Biotechnol, 2020, 19: 355-368.

[7] Isaka K, Date Y, Sumino T, Tsuneda S. Ammonium removal performance of anaerobic ammonium-oxidizing bacteria immobilized in polyethylene glycol gel carrier: Anammox bacteria immobilized in gel carrier. Appl Microbiol Biotechnol, 2007, 76 (6): 1457-1465.

[8] Isaka K, Yoshie S, Sumino T, Inamori Y, Tsuneda S. Nitrification of landfill leachate using immobilized nitrifying bacteria at low temperatures. Biochem Eng J, 2007, 37 (1): 49-55.

[9] Isaka K, Udagawa M, Kimura Y, Sei K, Ike M. Biological wastewater treatment of 1, 4-dioxane using polyethylene glycol gel carriers entrapping *Afipia* sp. D1. J Biosci Bioeng, 2016, 121 (2): 203-208.

[10] Lehmann J, Rillig M C, Thies J, et al. Biochar effects on soil biota: a review. Soil Biol Biochem, 2011, 43 (9): 1812-1836.

[11] Johannes Lehmann, Matthias C. Rillig, Janice Thies, et al. Biochar effects on soil biota - A review. Soil Biology and Biochemistry. 2011, 43 (9): 1812-1836.

[12] Tanaka H, Takahashi M, Yoneyama Y, et al. Energy saving system with high effluent quality for municipal sewage treatment by UASB-DHS. Water Sci Technol, 2012, 66(6): 1186-1194.

[13] Ibrahim M, Yusof N, Mohd Yusoff MZ. Enrichment of Anaerobic Ammonium Oxidation (Anammox) bacteria for short start-up of the Anammox process: a review. Desalination Water Treat, 2016, 57 (30): 13958-13978.

[14] Meng F, Chae S R, Drews A, et al. Recent advances in membrane bioreactors (MBRs): membrane fouling and membrane material. Water Res, 2009, 43 (6): 1489-1512.

[15] Fujii T, Sugino H, Rouse J D, Furukawa K. Characterization of the microbial community in an anaerobic ammonium-oxidizing biofilm cultured on a nonwoven biomass carrier. J Biosci Bioeng, 2002, 94 (5): 412-418.

[16] Wang T, Zhang H, Yang F, et al. Start-up and long-term operation of the Anammox process in a fixed bed reactor (FBR) filled with novel non-woven ring carriers. Chemosphere, 2013, 91 (5): 669-675.

[17] Chen C J, Huang X, Xiao L C, et al. Improving Anammox start-up with bamboo charcoal. Chemosphere,

2012, 89 (10): 1224–1229.

[18] Zhang L, Yang J, Ma Y, et al. Treatment capability of an up-flow Anammox column reactor using polyethylene sponge strips as biomass carrier. J Biosci Bioeng, 2010, 110 (1): 72–78.

[19] Awaluddin N, Akiyoshi O. Downflow hanging sponge (DHS) reactor for wastewater treatment—a short review. In: MATEC web of conferences, 2019, 280: 05004.

[20] Tandukar M, Uemura S, Machdar I, et al. A low-cost municipal sewage treatment system with a combination of UASB and the "fourth-generation" downflow hanging sponge reactors. Water Sci Technol, 2005, 52 (1-2): 323–329.

[21] Tanaka H, Takahashi M, Yoneyama Y, et al. Energy saving system with high effluent quality for municipal sewage treatment by UASB-DHS. Water Sci Technol, 2012, 66 (6): 1186–1194.

[22] Grismer M E, Collison R S. The zeolite-Anammox treatment process for nitrogen removal from wastewater: a review. Water, 2017.

Chapter 7 Promoters and inhibitors of Anammox

Emphasis: Anaerobic ammonium oxidation (Anammox); Nitrogen removal; Microbial enrichment; Exogenous substance addition; Anammox inhibition; Strategies to control the inhibition.

Nut: Promoters and inhibitors of Anammox mechanism.

The excessive use of fertilizers in agriculture and the discharge of industrial and domestic wastewater release excess nitrogen into water bodies, deteriorating water quality, endangering human health, and impeding sustainable development. Nitrogen removal is an essential method to improve water quality and address eutrophication. Conventional biological nitrogen removal is the most commonly used nitrogen removal technology, usually consisting of nitrification and denitrification processes. The mechanism of nitrogen removal is the conversion of organic nitrogen and ammonia into nitrate (NO_3^-) using nitrifying bacteria under aerobic conditions, followed by the reduction of NO_3^- to nitrogen gas (N_2) using denitrifying bacteria under anoxic conditions. But it also has limitations, the aerobic process requires a large amount of aeration to maintain a high dissolved oxygen (DO) level; the denitrification process requires a sufficient carbon source to provide electrons. In addition, the activated sludge process inevitably produces large amounts of residual sludge. These limitations result in complex processes, substantial energy consumption, high operating costs and weak shock resistance of the traditional biological nitrogen removal technology. Therefore, an effective and economical wastewater nitrogen removal process is urgently needed to solve the water pollution problem.

Compared to the conventional nitrification–denitrification process, the anaerobic and autotrophic characteristics of AnAOB can reduce aeration by 60%, conserve 100% of organic carbon, and reduce sludge production by 90%. Nitrogen removal rate of 81% along with 68% reduction in greenhouse gases (GHGs) emissions was successfully achieved by Anammox. The application of partial nitrification and Anammox (PN–Anammox) reduced mainstream total inorganic nitrogen in effluent by 4.4 mg/L, saving \$3.5 million in operational costs annually due to the reduction of organics addition.

However, the slow growth rate, long doubling time and high sensitivity to various environments of AnAOB do not satisfy the needs for large–scale applications. In addition, NH_4^+

and nitrite (NO_2^-) are substrates and inhibitors in Anammox reactions. Lack and oversupply of nutrient salts negatively affect the stability of Anammox reactors. These constraints suggest the need to optimize the Anammox process for large-scale applications. Since the discovery of Anammox in the early 1990s, many outcomes have been achieved, including discovering new AnAOB species, enabling high nitrogen removal performance in lab-scale reactors, and optimizing the characteristics, structure, packing pattern, and flotation mechanisms of Anammox granules. Currently, pilot and full-scale Anammox reactors have been constructed to treat different types of wastewater.

7.1 Exogenous additives for improving the Anammox process

Under anaerobic or anoxic conditions, the anaerobic ammonium oxidation (Anammox) process gets converted ammonia (NH_4^+-N) and nitrite (NO_2^--N) to nitrogen (N_2) and nitrate (NO_3^--N). Compared to traditional nitrogen removal processes, the anaerobic and autotrophic characteristic features of Anammox bacteria (AnAOB) allow it to reduce oxygen demand by 62.5% and sludge production by 90%, while conserving 100% of organic carbon and reducing greenhouse gas (GHG) emissions by 68%. Existing research indicates the viability of Anammox for treating domestic wastewater with high NH_4^+-N, including anaerobic digestion liquid, landfill leachate, and livestock wastewater. However, AnAOB's growth rates are exceptionally low. mostly around 50 mg/L in low NH_4^+-N concentration, and a 15–30 d doubling time, which prohibits rapid start-up. For instance, two years were taken to initiate the first full-scale Anammox reactor. It took about 90 days of operation to achieve relatively high simultaneous NH_4^+-N and NO_3^--N removal . Though inoculated with AnAOB granular sludge, the UASB Anammox reactor still required 120 days to successfully start up. In addition, the Anammox process's stability is threatened by high organic concentration, low temperature and extremely high/low substrate concentration, under which the activity of Anammox is easily inhibited, the relative number of functional bacteria decreases, and likewise, the nitrogen removal rate is reduced. In recent years, how to increase the growth and activity of AnAOB and make the Anammox process more stable have become a research focal point in the field of autotrophic nitrogen removal.

The influences of exogenous additives on the Anammox process have attracted widespread attention. By adding metals, organic matter, inorganic matter, and reaction intermediates are added to improve the biomass and activity of AnAOB in the reactor, maintain long-term stable operation of the Anammox system. The corresponding promotion mechanism is also explained. It is done to solve the problems of Anammox's slow start-up and the difficulty of quick growth and aggregation of AnAOB, which will help the technology move forward and be used more.

7.1.1 Metals addition

The predominant effects of metal ions on Anammox sludge are the activation of enzymes and the production of AnAOB. Metals are essential to AnAOB as the trace elements. Some metal ions may become potential electron donors/receptors to promote the reaction. Not only the type of metal ions, but also the appropriate concentration needs to be considered. Addition of the correct concentration of metal can stimulate key enzymes in Anammox biomass cells, which can improve the efficiency of the reactor.

7.1.1.1 Iron material

1. Zero–valent iron

Zero–valent iron (ZVI) is an important element for the growth of AnAOB. It has a low price and poisoning, and the use of it will not result in secondary environmental contamination. ZVI is frequently employed for wastewater treatment due to its high reducibility (E° = 0.44 V), increasing the amount of viable bacteria in the microbial community by enhancing physiological functions and stimulating activity, which is advantageous for the growth and spread of AnAOB. At the same time, the corrosion of ZVI is accompanied by consumption of the dissolved oxygen in water, thus, providing a suitable living environment for AnAOB. And the Fe^{2+} released by ZVI enhances microbial agglomeration through neutralization and compression of electric double layer, which promotes the formation of sludge particles. The possible mechanism of ZVI promoting Anammox is shown in Figure 7–1.

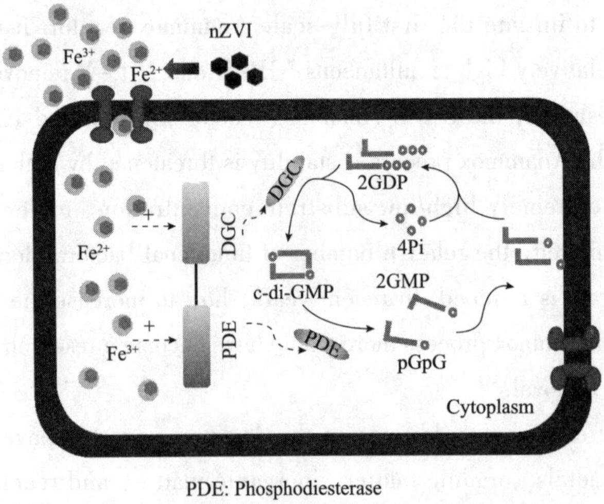

Figure 7–1 The possible mechanism of Fe promoting Anammox

It was reported that in a reactor with 50 mg/L nano zero–valent iron (nZVI), the removal rate of NH_4^+–N reached 67.85%, while in the reactor without nZVI, 58.38% was recorded. Notable is the fact that the former significantly increased *Ca. Brocadia* from 42.1 to 52.5%. Anammox–synchronous ZVI process shows a better effect of nitrogen removal. In addition, ZVI also contributes to Anammox quick startup. A decrease in the time of Anammox startup from 96 to 87 d was observed at concentrations of 1000 mg/L ZVI, meanwhile the growth rate of AnAOB increased by 20%, and the biomass of Anammox increased by 54%.

2. Iron ion

Iron ion promotes the formation of heme c and exists in cells in the form of ferritin. In addition, AnAOB contains a substantial number of dense iron particles, which may be associated with iron respiration. Consequently, the participation of Fe may be a viable strategy for enhancing the activity of AnAOB.

Appropriate ion addition can boost AnAOB growth, increasing the reactor's nitrogen removal efficiency (Table 7–1). In a study involving the enhancement of marine Anammox bacteria (MAB) with Fe^{2+}, maximum ammonia removal efficiency (ARE) and nitrite removal efficiency (NRE) were observed at 4 mg/L. TN removal efficiency increased by 7.77% as Fe^{2+} concentration rose to 4.5 mg/L. *Ca. Jettenia* increased from 2.18% to 3.17% in the middle of the reactor. Appropriate quantities of Fe^{3+} can significantly improve the SAA. The addition of 3.68 mg/L Fe^{3+} resultedin a 533.2% improvement in SAA. Moreover, the addition of Fe^{3+} increased the average nitrogen removal efficiency and maximum nitrogen removal rate (NRR) from 64.7% and 4.1 kg N/(m^3 · d) in the control tests to 67.4% and 4.9 kg N/(m^3 · d), respectively. In addition, with increasing particle size, ARE and NRE as well as SAA increased The addition of Fe^{3+}–minerals increased granule formationassociated cellular processes and pathways, which consequently increased the values of particle size by 144%. Under nitrogen shock loading, the addition of Fe^{3+} could still stimulate the growth of MAB. *Ca. Scalindua* was the dominant genus in the reactor containing Fe^{3+}, accounting for 27.70%. It was significantly higher than control (17.07%).

3. Iron–based material

As an insufficiently crystalline Fe^{3+} mineral, ferrihydrite is a highly reactive material, connecting with numerous essential nutrients. It may also induce a novel microbial Fe^{3+} reduction coupled with Anammox (Feammox), providing an additional route for nitrogen removal. Ferrihydrite facilitated the formation of densely clustered and intricate microbial topologies, according to network–based analyses, thereby enhancing resistance to environmental disturbance, and effectively increasing the intervention of cellular processes and pathways. Notably, 5 mmol ferrihydrite increased nitrogen removal by 4.1%. Ferrihydrite also promoted *Planctomycetes* AnAOB in the reactor, resulting in an 11% increase. In addition, nano–scale magnetite increased the nitrogen removal substantially. The maximum SAA of the nanomagnetite–modified group was 33.3% higher. The results showed that nano–magnetite quickly adhered to the granules, which may have hastened the release of Fe^{2+} or Fe^{3+} to enhance the efficiency of Anammox. Nonetheless, the form of iron influences the procedure. When $FeCl_3$ or ferric ammonium citrate was the iron source, there was virtually no reaction . The main cause is that $Fe(OH)_3$ precipitate formed by Fe^{2+} and

Table 7-1 Effect of ZVI, Fe^{2+} and Fe^{3+} additives on the Anammox process

Metal	Abundance	N/(mg/L)	EPS	Result
ZVI (100 mg/L)		NH_4^+: 80; NO_2^-: 80	84.8 mg/g VSS (PS/PN of 0.26)	ARE and NRE reached 88.19% and 88.73%, better than control (58.71% and 50.38%)
ZVI (750 mg/L)		NH_4^+: 236; NO_2^-: 246		SAA reached 0.24 kg N/(kg VSS · d), better than control [0.20 kg N/(kg VSS · d)]
ZVI (4 g/L)	18.13% (*Ca. Brocadia*)	NH_4^+: 208; NO_2^-: 213		NRE was enhanced from 81.67 to 86.02%
ZVI (1000 mg/L)		NH_4^+: 100; NO_2^-: 120		Start-up time was reduced from 96 to 87 d; Anammox growth rate was increased by 20%; Anammox quantity was increased by 54%
nZVI (50 mg/L)	67.4% (*Ca. Brocadia*)	NH_4^+: 70.23; NO_2^-: 84.45	143.96 mg/g sludge (PN); 134.00 mg/g sludge (PS)	ARE and NRE reached 67.85% and 82.44%, better than control (58.38% and 67.94%)
Fe^{2+} (3 mg/L)	3.17% (*Ca. Jettenia*)	NH_4^+: 157.5; NO_2^-: 262.5		TN removal efficiency was increased by 6.89%
Fe^{2+} (5 mg/L)	28.4% ± 2.6% (*Ca. Brocadia*)	NH_4^+: 28.54; NO_2^-: 37.68	189.4 ± 22.3 mg/(g VSS)	SAA reached 10.49 mg TN/(g VSS · h), better than control [7.82 mg TN/(g VSS · h)]
Fe^{2+} (4 mg/L)		NH_4^+: 80; NO_2^-: 105.6		ARE and NRE reached 100% and 89%, better than control (52.67% and 51.70%)
Fe^{2+} (1 mg/L)	11.30% (*Ca. Kuenenia*)	NH_4^+: 236; NO_2^-: 246		NRR was increased to 0.58 kg/(m^3 · d), with a ratio increase of 15.3%
Fe^{2+} (5 mg/L)	13.89% (*Ca. Kuenenia*)	NH_4^+: 50; NO_2^-: 60	46.8 mg/g SS (protein); 4.3 mg/g SS (polysaccharide)	NRE was increased from 82.6 to 91.3%
Fe^{2+} (1 mg/L)	11.04% (*Ca. Kuenenia*)	NH_4^+: 236; NO_2^-: 246	71.41 mg/g SS	NRE was increased by 62.31%
Fe^{2+} (1 mmol/L)	29.3% (*Ca. Brocadia*)	NH_4^+: 40; NO_2^-: 50	43.39 mg/g VSS (PS); 110.4 mg/g VSS (PN)	NRR was improved from 76.6 to 81.5 g N/(m^3 · d)
Fe^{3+} (6 mg/L)		NH_4^+: 85; NO_2^-: 110		SAA was increased by 33.41%
Fe^{3+} (3.68 mg/L)		NH_4^+: 100; NO_2^-: 100		Maximum SAA was increased by 533.2%; NRR was increased by 20%
Fe^{3+} (0.14 mmol/L)	3.85% (*Ca. Brocadia*)	NH_4^+: 35.6; NO_2^-: 60.3		NRR was 1.94 times of the control
Fe^{3+} (2 mg/L)		NH_4^+: 100; NO_2^-: 120		NRE was increased by 3.09%
Fe^{3+} (5 mg/L)		NH_4^+: 90; NO_2^-: 117		NRE was increased by 5.32%

Fe^{3+} inhibits the mobilization of substances and blocks metabolism. Thus, polymeric ferric sulfate (PFS) was chosen to couple Anammox and Feammox. Fe^{3+} dissolution was slower from PFS than from iron salt. So, less $Fe(OH)_3$ precipitated, which could coat microbe cells. When the ratio of NO_2^--N to NH_4^+-N in the influent was 0.90, both were nearly consumed. TN removal efficiency reached 95.60%, exceeding Anammox's theoretical value (88.79%). PFS made extracellular polymeric substances (EPS) drop and deformable, which facilitated substance transfer. In the meantime there was no observable precipitation on the cell surface.

Iron-carbon micro-electrolysis (ICME) is now used as a new technology in treating wastewater due to its low price, operational convenience and high treatment efficacy. With the Fe-C particles addition, ARE and NRE was 83.90% and 88.84%, respectively. Moreover, TN removal load was increased by 8.4%. In another research, the combined system's average TN removal efficiency was increased by 39.63%. *Ca. Kuenenia* and *Ca. Brocadia* are the predominant Anammox genera, comprising 1.01% and 0.15%, respectively.

7.1.1.2 Mn

Mn^{2+} is a necessary trace element for the metabolism and development of AnAOB. It is also a component of many intracellular enzymes and coenzymes. Mn^{2+} was reported to be an important constituent of a form of superoxide dismutase. In Anammox reaction system, Mn^{2+} can inhibit the growth of autotrophic and heterotrophic bacteria, which inadvertently fosters the proliferation of AnAOB.

The addition of 0.05 mmol Mn^{2+} increased ARE and NRE by 8%. Similarly, adding 0.05 mmol Mn^{2+} increased maximal SAA by 6.25%, and a good nitrogen removal effect was obtained in the treatment of saline wastewater with rich nitrogen. The fact that the NRR was double that of the Anammox process without MnO_2 also demonstrates the beneficial effect of MnO_2 on the Anammox reaction. Mn^{2+} simultaneously stimulates AnAOB activity and affects the distribution of EPS. With Mn addition, the prevalence of AnAOB increased from 2.68% to 4.50%. In addition, the PN/PS ratio, which correlates strongly with sludge setting-up, decreased significantly from 1.33 to 1.23. The decreasing PN/PS ratio in the Mntreated group indicated that Mn could accelerate the setting-up of the sludge in nitritation-Anammox.

7.1.1.3 Ni

Nickel metal shares properties and functions with iron in the process of cell growth. Additionally, it can be used to synthesize proteins required by cells, accelerate cell growth, and stimulate metabolic processes. When 1 mg/L of Ni^{2+} was added to a solution, ARE and NRE was increased by 8.75% and 5.84%, respectively. It demonstrates that the addition of 1 mg/L Ni^{2+} stimulates the activity of AnAOB and was advantageous for the removal of NH_4^+-N and NO_2^--N. With 1.47 mg Ni^{2+} addition, NRR and SAA were increased by 14.6% and 57.9%, respectively. The supplementation of 0.2 mg/L Ni^{2+} resulted in a 2-fold increase in heme c concentration compared to the initial level, which increased the biomass of AnAOB . The low concentrations of Ni^{2+} (2–5 mg/L) could effectively stimulate the synthesis of dehydrogenase protein, which characterized Anammox organisms' capacity to metabolize organic matter.

The short–term inhibition of Ni produced tolerance to Ni after longterm acclimation of Anammox system, which was relieved, and the composition of microbial community was also affected. Previous study showed that ARE and TN removal rates in effluent experienced a rise and then a fall under the concentration of 5 mg/L Ni^{2+}, which showed that AnAOB had a certain Ni^{2+} tolerance threshold after a period of domestication. Ni^{2+} is reported to inhibit the activity of AnAOB significantly in the short–term experiment, while in the long–term experiment, the tolerance of AnAOB increased with the decrease of Ni^{2+} dosage. The nitrogen removal efficiency of Anammox system reached (93 ± 0.03) % in the long–term operation of 10 mg/L Ni^{2+}, thus, still maintaining excellent performance. However, Ca. Kuenenia decreased by 38.07%, which revealed that the addition of Ni^{2+} had a considerable influence on the composition of the functional microbial community.

7.1.1.4 K

The Anammox process is vulnerable to the effects of wastewater containing inorganic salts. Long–term salt stress on the reactor not only decreases nitrogen removal performance, but also undermines the sludge's sedimentation capacity. While treating industrial wastewater with high NH_4^+–N and salinity stress through Anammox, long–term domestication of K^+ impeded the formation of Na^+ upon its addition. Alleviate the pressure of Na^+ and K^+ to preserve cell osmotic pressure in high salt environments. Thereby reducing energy consumption, increasing enzyme activity, and improving nitrogen removal efficiency. Therefore, the Anammox process for treating wastewater with a high salinity can be enhanced with the exogenous addition of K^+. Under medium temperature (30 ℃) and high salinity (30 g/L), 5 mmol/L K^+ promoted TN removal by 86.22%, 10.88% more than 0 mmol/L K^+. NRR increased by 10.87%, from 0.83 to 0.95 kg/ (m^3 · d). Another study investigated the performance of biological nitrogen removal when treating salty wastewater. 8 mmol/L K^+ addition achieved the best effect, the average ARE and NRE were increased from 60.4% and 56.8% to 89.24% and 84.87% respectively, while AnAOB activity reached its maximum.

7.1.1.5 Ca

Divalent cations, essential nutrients for bacterial metabolism and growth, reactivate Anammox. Ca^{2+} could enhance the Anammox reaction by accelerating the expression of *Ca. Brocadia* and the manufacturing of diguanylate cyclase, which serves to produce c–di–GMP. C–di–GMP has been reported to stimulate syntrophy of AnAOB, thus speeding up the start of the Anammox reactor. Intracellular c–di–GMP was discovered to be a key regulator of Anammox biomass, particularly under adverse conditions. Furthermore, Ca^{2+} might promote the secretion of EPS and enhance granulation. The addition of Ca^{2+}, which has antagonistic effect on Na^+, can also reduce the inhibition of salinity. Ca^{2+} supplementation at the optimized concentration of 200 mg/L was reported to increase the SAA by up to 78% of its initial activity. It also increased the amount of Na^+ that sequestered in the soluble EPS, which appears to alleviate salinity stress. In addition, A strong correlation between external Ca^{2+} concentration during feeding and steady-state intracellular free Ca^{2+}, which favors Anammox consortium restoration. The recovery rate of Anammox activity and the percentage of living bacteria increased by 78% with increasing Ca^{2+}. At the same time, steady-state intracellular free Ca^{2+} in the flora also increased.

7.1.2 Organic matter addition

In the presence of low concentrations of chemical oxygen demand (COD), Anammox would combine with denitrification to form the simultaneous nitritation Anammox and denitrification (SNAD) process, thereby enhancing the nitrogen removal efficiency. In addition to supplying SNAD at low COD concentrations, it has been discovered that the addition of some common organic substances can directly stimulate the growth of AnAOB.

7.1.2.1 SNAD process

In Anammox process, elevated levels of biodegradable organic matter are always considered to pose inhibitive effect due to the advancement of heterotrophic bacteria (HB) giving rise to competitive NO_2^- utilization between the latter and AnAOB. SNAD process makes different microbial communities coexist in one reactor, such as ammonia oxidizing bacteria (AOB), AnAOB and denitrifying bacteria, which has the potential of initiating denitrification while removing organic matter at the same time (Figure 7-2). It can remove nitrogen with an efficiency of 70%-90%, offering a low cost and energy consumption alternative for treating high and low nitrogen streams. Reported that NH_4^+-N, TN, and COD removal efficiencies in the development of the SNAD process treating synthetic wastewater were 79%, 70%, and 94%, respectively. Long-term SNAD process at NH_4^+-N and organic loading rates of 360 and 180 g/(m^3 · d), resulted in TN and COD removal efficiencies greater than 88% and 90%, respectively. The SNAD process has been employed in the treatment of sewage with C/N ratios ranging from 0.4 to 4.

7.1.2.2 Direct promotion of Anammox activity by organic compounds addition

Specific organic compounds such as FA and humic acid (HA) can be used to promote the metabolic activity of AnAOB. Regarding the short term effect, exogenous 5 mmol FA increased the activity of AnAOB by 15.5%. Moreover, the efficiency of total inorganic nitrogen removal in FA amended reactor was 58.6%, better than the control (42.1%), the Anammox activity increased by 2.69-fold, respectively, *Ca. Kuenenia* increased continuously from 9.74% to 19.8%. Earlier research confirmed that the addition of FA managed to make *Ca. Kuenenia* more competitive.

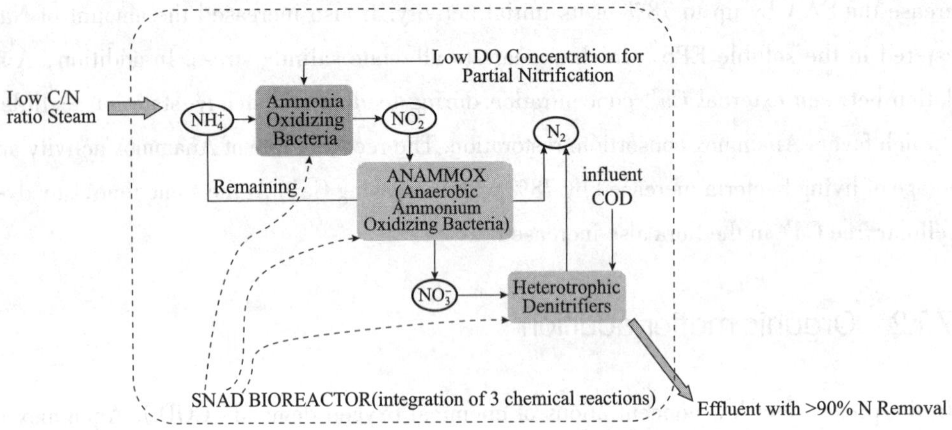

Figure 7-2 General schematic of SNAD process

When the concentration of COD (provided by HA) increased to 684.39 mg/L, NRE increased fractionally from 84.1% to 95.3%, and NRR also increased from 1.21 to 1.61 kg N/ (m^3 · d) compared to the performance of the blank HA. After being fed 500 mg/L of COD (supplied by HA), the relative abundance of *Planctomycetes* increased. These results suggested that HA could increase the functional bacteria, thereby promoting the growth of AnAOB. FA and HA are the prevailing components of humic substances, which is involved in the biochemical reaction of bacteria. Specifically, they have the ability to transfer electrons. Further research revealed that quinones are the predominant redox molecules in humic substances. Components of the bacterial respiratory chain, microbial respiratory quinones play a crucial role in electron transfer during microbial respiration. Through the addition of 0.05–0.8 mmol quinoid redox mediators, the activity of key AnAOB enzymes increased by a factor of 0.6–3. Anthraquinone-2, 6-disulfonate (AQDS) was reported to have increased nitrogen removal by 18.3% during Feammox (Figure 7-3). AQDS accelerated electron transfer between NH_4^+-N and Fe^{3+}, thus further enhancing Feammox production.

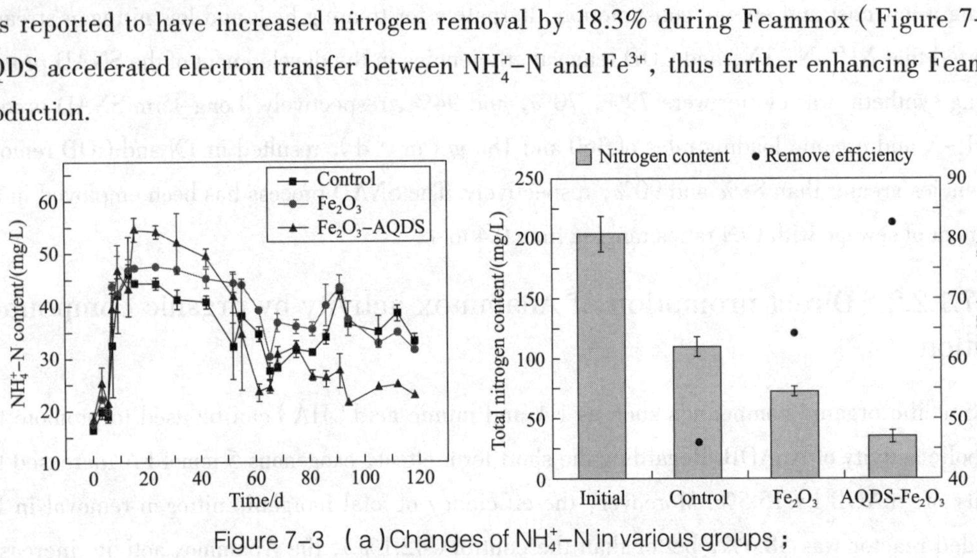

Figure 7-3 (a) Changes of NH_4^+-N in various groups ;
(b) total nitrogen content in various groups following the experiment

7.1.3 Inorganic matter addition

7.1.3.1 Biochar

Biochar is a carbon-rich substance derived from the pyrolysis of organic matter in an oxygen-depleted environment. Due to its large surface area, high charge density, and low ecotoxicity risk, biochar was suggested as a possible carrier material in the Anammox process. Under the same conditions, a combined Anammox and bamboo biochar system reduced startup time from 97 to 85 d with spherical plastic and 117 d without a carrier. The addition of 5% coconut biochar decreased the start-up time from 69 to 46 d with an average ARE of 96%.

As a carrier, biochar immobilized bacteria in the process of Anammox to obtain higher biological retention, which made AnAOB more resistant and flexible to environmental changes and disturbances, and the nitrogen removal process was more stable (Figure 7-4).

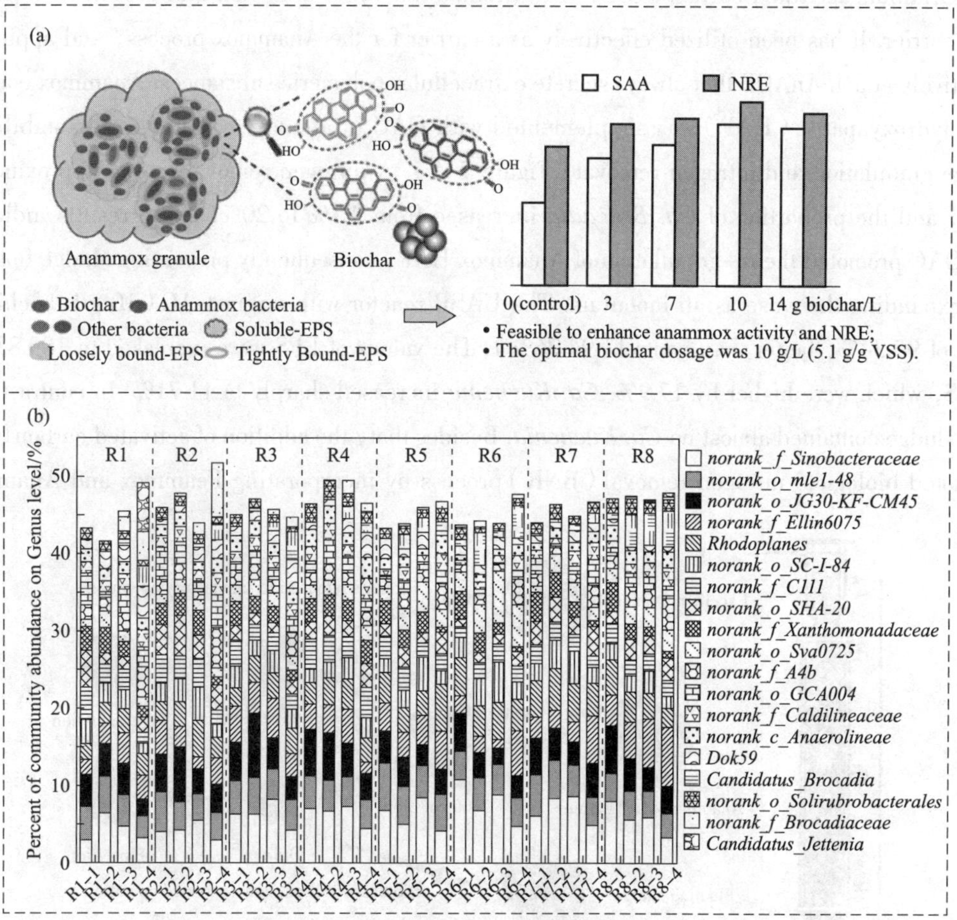

Figure 7-4 (a) The viability and mechanism of redox-active biochar in enhancing Anammox performance and AnAOB; (b) Influence of biochar to microbial community

R1: reactor without biochar addition.

According to a research, biochar amendment caused a 4–12.3-fold increase in the abundance of the AnAOB and enhanced nitrification. The addition of different kinds of biochar such as bamboo, jujube, and peach biochar improved performance with maximum ARE and average TN removal efficiency of 84.28% and 88.39%, which were considerably greater than those of the control (NH_4^+-N: 77.00%; TN: 83.81%). *Planctomycetes* in jujube biochar-enhanced bioreactors increased by a maximum of 54.5%, and their relative abundance increased from 4.6% to 7.1%. Biochar can also reduce the inhibition of Anammox by COD and NO_2^--N. Under the inhibition of 150 mg/L COD, the addition ofbamboo charcoal increased the average ARE by 5.3%. The NO_2^--N concentration of 600 mg N/L showed a serious inhibitory effect, and the total NRE increased by 231%, 149% and 51.0% respectively after adding three distinct kinds of pyrolytic biochar.

7.1.3.2 Activated carbon

Granular activated carbon (GAC) can promote bacterial adhesion and growth as a porous inert carrier. It has been utilized effectively as a carrier for the Anammox process, and applied to selectively attach AnAOB that always secrete extracellular polymeric substances. Anammox coupled with hydroxyapatite (HAP) was supplemented with GAC in a study to maintain the stability of sludge granulation and nitrogen removal (Figure 7–5). A nitrogen removal rate of approximately 88%, and the proportion of *Ca. Brocadia* increased from 2.1% to 20%. These results indicated that GAC promoted the re-granulation of Anammox HAP, consequently preventing sludge flotation and expanding their use as an inoculum. The UASB reactor with virgin GAC (10 g/L) achieved NRE of 95.34%, better than control (94.26%). The values of EPS were increased to 109.83 mg/g VSS, which were higher by 17.9%. *Ca. Kuenenia* increased sharply to 11.71%. In contrast, the seed sludge contained almost no *Ca. Kuenenia*. Besides that, the addition of activated carbon (AC) promoted biological nitrogen removal (BNR) process by incorporating Feammox and Anammox.

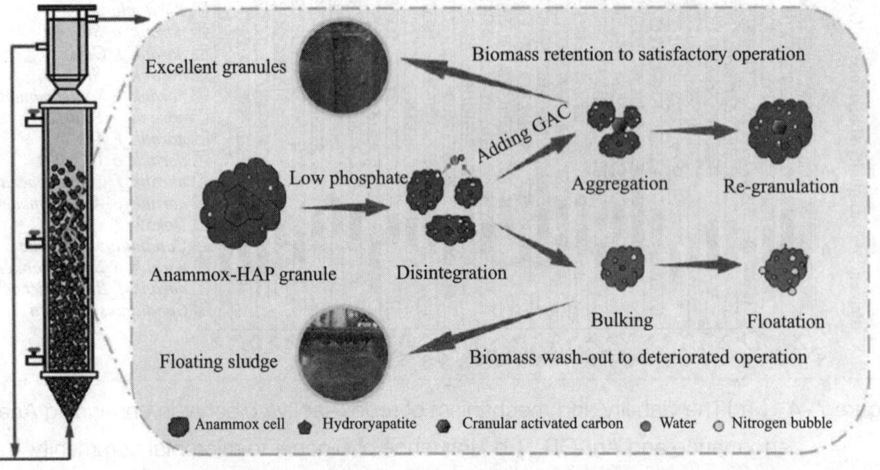

Figure 7–5 Re-granulation of Anammox–hydroxyapatite granules is promoted by GAC

As an electron shuttle, AC was added to an incorporated BNR system composed of Feammox and Anammox processes. With a nitrogen removal rate of 0.46 kg TN/(m³ · d), the TN removal efficiency reached approximately 82.9%. *Ca. Jettenia* (2.10%) and *Ca. Brocadia* (1.18%) were the most abundant.

7.1.4 Intermediates addition

7.1.4.1 Hydroxylamine

Hydroxylamine (NH_2OH) is one of the most crucial intermediate materials of Anammox, which plays an important role in the utilization and conversion of Anammox reactor substrate (Figure 7-6). Further, NH_2OH is an exemplification of reducing agent. It was reported that NH_2OH enhanced Anammox activity recovery following Cr^{6+} inhibition. Extracellular Cr^{6+} was converted to Cr^{3+}. Cr^{3+} sank due to low solubility and mobility. Specifically adding NH_2OH lowered Cr^{6+} in the solution, which decreased the inhibiting impact on the Anammox process. The oxidation of NH_2OH to nitric oxide (NO) in AnAOB produces a great reduction equivalent. At the same time, a large amount of reduction equivalent or electrons produced by the oxidation of NH_2OH may in turn stimulate the reduction of NO_2^- to NO, which promotes the nitrogen removal process. NH_2OH drastically reduced and consistently suppressed nitrite oxidizing bacteria (NOB), whereas the number of AOB remained stable, hence promoting the start and stability of nitrification.

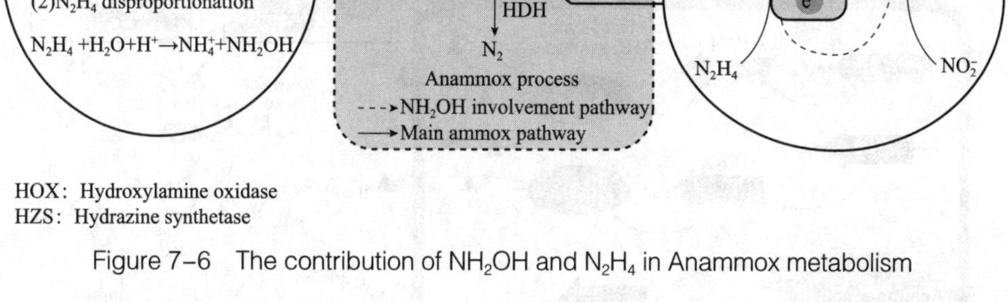

Figure 7-6 The contribution of NH_2OH and N_2H_4 in Anammox metabolism

7.1.4.2 Hydrazine

The supplemented hydrazine (N_2H_4) can be used by AnAOB to generate additional energy, which helps to accelerate the growth of Anammox biomass while also promoting the removal rate of TN. The electrons removed from the reaction of supplemental N_2H_4 with HDH, which replaced electrons from NO_2^--N oxidation to NO_3^--N, refilled AnAOB anabolism and reduced NO_3^--N production (Figure 7-6). According to a study, N_2H_4 had a considerable impact in promoting AnAOB (with an increase of 31.1%) and inhibiting NOB (with a decrease of 78.8%).

7.1.5 N-acyl-homoserine lactones addition

AnAOB is a Gram-negative bacteria with a quorum sensing (QS) system capable of secreting N-acyl-homoserine lactones (AHLs) self-inducers. QS is a common communication mechanism among microorganisms. Microorganisms sense the cell density around them by secreting self-inducers, thus regulating gene expression, such as expression of virulence factors and bioluminescence.

Furthermore, researchers have revealed that the AHLs-regulated QS in the Anammox system participates in gene expression, bacterial activity, and metabolic functions (Figure 7-7). It has been revealed that both C6-HSL and C8-HSL AHLs can be extracted from the supernatant and mud phase of Anammox mixture. It is also speculated that these two self-inducers have positive effects on AnAOB, enhancing the bacterial special growth rate by 28% and 32%, respectively, genes involved in nitrogen removal (*hzo*, *hzsB*, *nirS*, *and ccsB*) were expressed more positive in the C6-HSL and C8-HSL groups. The modification of C6-HSL or C8-HL increased the genes involved in Anammox. Some data also indicated that the addition of particular types of AHLs may selectively promote the activity and growth of AOB to provide a competitive advantage over NOB in nitrifying systems, hence contributing to a constant accumulation of nitrite.

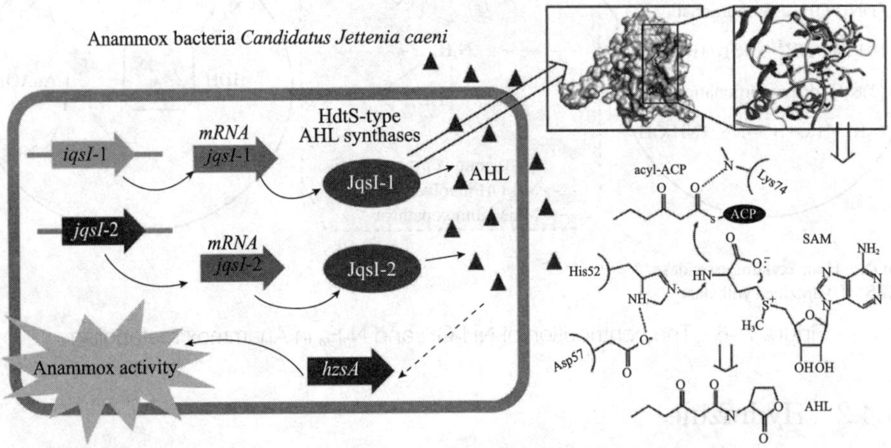

Figure 7-7 A model of Anammox gene expression induced by AHLs

7.2 Exogenous substances for inhibiting the Anammox process

Anammox is a promising novel biological nitrogen elimination process with attractive, application prospects. However, the Anammox process is inhibited by many factors, which prevent improvements on the process as well as the application of the Anammox process. A variety of inhibitory substances, such as substrates (ammonia and nitrite), organic matter (nontoxic organic matter and toxic organic matter), salts, heavy metals, sulfide, are commonly present in the practical applications. As a result of differences in operating conditions, experimental methods, physical structures of sludge (flocculent sludge, biofilm or granular sludge) and microbial populations involved, the effect of inhibitors varies widely. Temperature and pH adjustments, dissolved oxygen and oxidation–reduction potential controls, substrate concentration and load controls, sludge acclimatization and addition of auxiliary agents can significantly improve the performance of Anammox process under inhibitory conditions.

7.2.1 Different inhibitory factors of Anammox process

7.2.1.1 Inhibition of organics

Organic matter reported to inhibit the Anammox process is divided into biodegradable organic matter (glucose, formate, acetate, etc.) and nonbiodegradable organic matter (aldehydes, phenolics, antibiotics, etc.). Different organic matter had different inhibition threshold concentrations on the Anammox process, such as those of glucose (1 mmol/L), formate (0.5–3 mmol/L), sucrose (700 mg/L) and acetate (50 mmol/L). With high concentrations of biodegradable organic matter, the proliferation of heterotrophic bacteria (HB) in the Anammox system led to competition with AnAOB for the utilization of NO_2^-. High–concentration nonbiodegradable organics also affected the Anammox process, and NRE was significantly reduced. It decreased from 96.5% to 47% in the Anammox system under the stress of 300 mg/L phenol. The structure of the microbial community also changed, and organisms that were able to degrade phenol became dominant. Additionally, the high tolerance of AnAOB to organic matter could be achieved through long–term acclimatization. Previous works included the conclusion that AnAOB oxidizes volatile acids and then converts them through the acetyl–CoA pathway, leading to products synthesized from all biomass precursor metabolites through the acetyl–CoA pathway. In addition, NO_3^- is reduced to NH_4^+ by volatile fatty acid (VFA)–mediated dissimilation of nitrate nitrogen and then used in an Anammox pathway to produce N_2. Under starvation conditions, VFAs were the nutrients for AnAOB, and they combined and accumulated in the microbial cytoplasm. The storage of glycogen in AnAOB began when AnAOB were close to their zero-growth stage. The Anammox process is not effective in treating wastewater with high concentrations of organic matter. However, in long-term operation, PN/A systems maintained robust nitrogen removal when treating wastewater containing organics and high NH_4^+ concentrations. Antibiotics are also widely applied organic matter and are described in detail in the following chapter.

7.2.1.2 Inhibition of substrate

AnAOB use NO_2^- as an electron acceptor to oxidize NH_4^+ to N_2 under anoxic conditions. However, high concentrations of NO_2^- and NH_4^+ could inhibit Anammox activity. In addition, the high substrate concentration also affected the particle sizes of Anammox sludge and the components of EPS. The particle sizes of Anammox sludge and the mass transfer efficiencies significantly decreased as the influent NH_4^+ concentration decreased from 300 to 50 mg/L, and the nitrogen loading rate (NLR) of the Anammox system decreased from 7.42 ± 0.74 to 1.24 ± 0.03 kg N/($m^3 \cdot d$).

NO_2^-, NH_4^+, free nitrous acid (FNA) and free ammonia (FA) have been reported to inhibit Anammox activity at high concentrations. It was reported that the Anammox process was inhibited by nitrite with a concentration of 100 mg/L. In addition, the half-maximal inhibitory concentration (IC50) of nitrite in the Anammox process was 50 mg (N/L). The Anammox inhibition threshold for NO_2^- was lower than that of NH_4^+ and NO_3^-, and such inhibition was reversible. Reversible inhibition refers to a reversible decrease in catabolic activity during exposure, while toxicity or irreversible inhibition refers to damage to microorganisms that results in an irreversible decrease in microbial activity. Anammox floating sludge was more easily inhibited by a high concentration of NO_2^- compared to Anammox granular sludge because the outer layers of granules were used as shells to protect the corer. At low pH, the inhibition of NO_2^- on AnAOB was more severe than that at high pH. A decrease in SAA by 50% was observed at concentrations of 38 mg NH_3/L, while 100 mg NH_3/L caused inhibition of SAA by 80%. It was reported that NH_4^+ inhibition of the Anammox process was essentially caused by FA, while NO_2^- inhibition was mainly related to FNA. FA and FNA change the intracellular pH, neutralize the transmembrane potential, and even cause cell lysis. It was reported that FA inhibited the Anammox process at concentrations of 1.7–2.0 mg NH_3/L. The presence of FA could change the structure of the microbial community, destroy the EPS, and even kill living cells. FNA generated via acidification also had negative effects on bacterial metabolism and further influenced the performance of the Anammox process.

7.2.1.3 Inhibition of metals

Metals are frequently detected in wastewater. As trace elements, metals are essential to AnAOB. However, manymetal ions affect the Anammox process, such as Zn^{2+}, Pb^{4+}, Cd^{2+}, Cr^{3+},

Cu^{2+}, Co^{2+}, Fe^{2+}, Ni^{2+}, Mn^{2+} and Hg^{2+}, all of which promote AnAOB activity at low concentrations and inhibit it at high concentrations.

Cd^{2+} showed both acute and persistent inhibition of the Anammox process because it crossed the cell membrane and entered the cell, damaged the enzyme structure and reduced the amount of cytochrome c in the cell. A previous study showed that the IC50 of Cd^{2+} was 5.43 mg/L for short–term exposure, while the longterm inhibition threshold was 20 mg/L. Hg^{2+}also reduced nitrogen removal performance and the activity of the Anammox process. The IC50 of Hg^{2+} was 60.39 mg/L, and the nitrogen removal rate from sludge was 6.61 mg N/(g VSS h). The abundance of the Hg resistance gene was positively correlated with Hg^{2+} concentration. At low concentrations, Co^{2+} improved the nitrogen removal performance of the Anammox process and increased the contents of key enzymes in the cell. However, it inhibited the activity of the Anammox process at high concentrations. Previous studies reported that after Cr^{6+} entered the cell, it destroyed the structure of the enzyme in the cell, blocked the functional groups, and reduced the nitrogen removal performance of the microorganisms. Moreover, the low concentrations of Fe^{3+} (0.08 and 0.14 mmol/L) accelerated the granulation of Anammox sludge, and the NLRs of the Anammox system were 993.0 and 1105.8 kg N/(m^3 · d), respectively. Additionally, an Fe^{2+}–binding chelator promoted the synthesis of cytochrome c.

Zn^{2+} and Cu^{2+} have been particularly studied because of their higher inhibitory effects and bioavailabilities for AnAOB. It was reported that Zn^{2+} had antibacterial properties and absorbed UV radiation. Zn^{2+} is an essential micronutrient for vital cofactors of metalloproteinases and certain enzymes. High concentrations of Zn^{2+} Accumulate in cells and bind with proteins (PNs) involved in electron transport systems to reduce Anammox activity. Previous studies indicated that the IC50 of Zn^{2+} was 6.9 mg/L in short–term tests, and the longterm inhibition threshold of Zn^{2+} on the Anammox process was 3.39 mg/L. The IC50 of Cu^{2+} was determined to be 74 mg/L when the concentrations of FA were between 48.4 and 95.0 mg/L in the Anammox system. However, the IC50 of Cu^{2+} was also reported to be 1.9–30 mg/L when the concentration of influent substrate was 280 mg/L. Such a difference suggested that substrate conditions probably influenced the inhibitory effects of various pollutants. In the past several years, studies have been conducted not only on the effect of a single variable of metals on the Anammox process but also on the combined effects of metals and other pollutants, such as metals and antibiotics. A previous study showed that Cu^{2+} bioavailability and inhibition were decreased when it was bound with antibiotics.

As Figure7–8 shows, metals exhibit adverse effects on AnAOB by destroying the cell structure and the intracellular and extracellular components, such as by affecting the synthesis of nitrite reductase and damaging DNA. Microorganisms have several mechanisms for metal resistance, such as extracellular precipitation, intracellular chelation, methylation, reduction, and increasing selectivity at the membrane. Zn^{2+} promotes the production of free radicals, such as reactive oxygen species, and causes oxidative stress, DNA damage and oxidative decomposition of lipids, which further inhibits Anammox activity through cytomembrane dissociation. Similarly, Cu^{2+} catalyzes the generation of hydroxyl free radicals, promotes stress throughout the redox cycle and destroys the cell membrane. Cu^{2+} combined with antibiotics to form complexes and increase the amount of Cu^{2+} adsorbed.

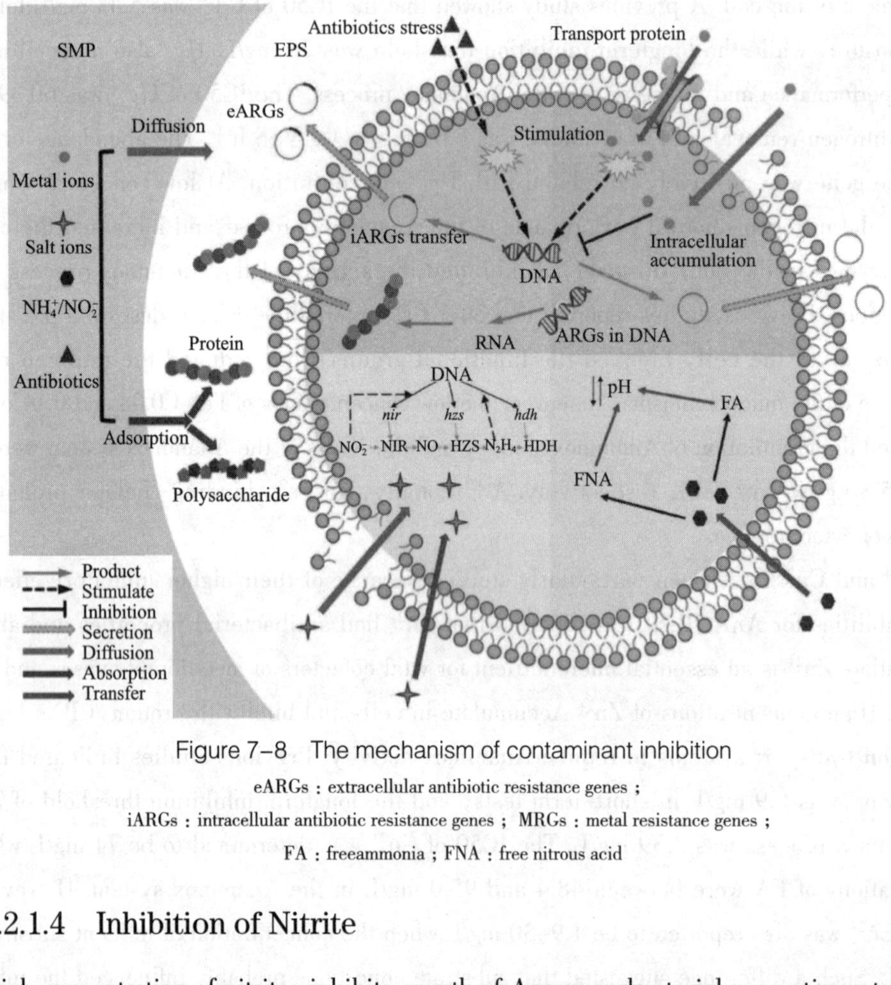

Figure 7–8 The mechanism of contaminant inhibition

eARGs : extracellular antibiotic resistance genes ;
iARGs : intracellular antibiotic resistance genes ; MRGs : metal resistance genes ;
FA : freeammonia ; FNA : free nitrous acid

7.2.1.4 Inhibition of Nitrite

High concentration of nitrites inhibits growth of Anammox bacterial consortium at a broad-range as per reported literature. Nitrite acts as one of the substrates during the Anammox process (Figure 7–9) but if nitrite concentration exceeds the threshold value than it shows inhibitory impacts on Anammox process, since nitrite is an electron acceptor in the process and converted by Anammox bacteria. The ammonia is converted to nitrite by ammonia nitrifiers while nitrite oxidizing bacteria converts nitrite into nitrogen, also the denitrification takes place from nitrite and this conversion is mediated by Anammox bacteria as given in Figure 7–9. The threshold values of nitrite inhibition remained at a minimal point than the inhibition threshold values of ammonium i.e. the Anammox process performance is more affected by nitrite than ammonium. Many studies showed that for constancy of Anammox process nitrite concentration is very critical as it inhibits the process depending on time of exposure. However, these studies differed on the nitrite inhibition threshold values, with reported threshold values ranged from 5–280 mg N/L nitrite. It has been also reported that in continuous feeding mode and maintaining a low concentration of nitrite does not inhibit Anammox activity which is evident from point of view of nitrite inhibition.

Other studies also reported higher inhibition threshold values during Anammox process for nitrite. The entire deactivation of Anammox process occurs at higher nitrite concentration (>185 mg N/L) while keeping the constant initial ammonium concentration of 42 mg N/L. The Anammox activity reduced by half of total activity at high nitrite concentration (350 mg N/L). The variation in the findings of these studies was mainly due to different operating conditions used by different researchers where the ammonium concentration of 42 mg N/L used for Anammox activity.

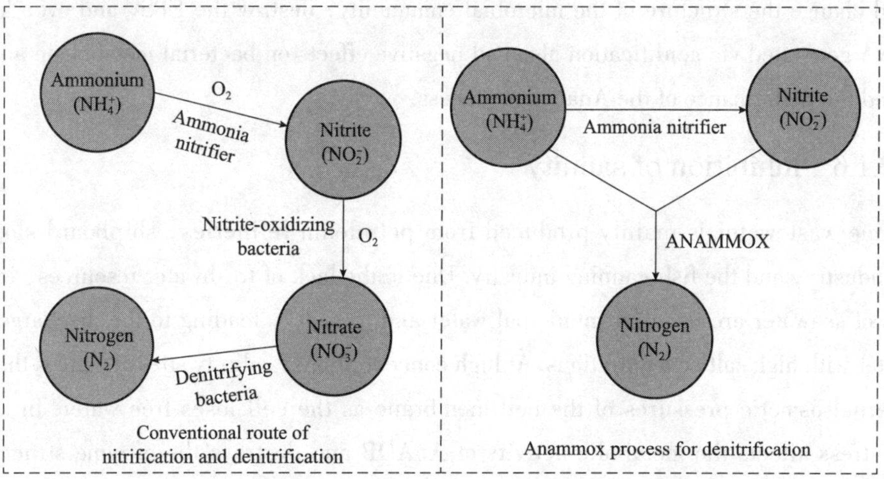

Figure 7–9 Process pathway comparison of conventional route and Anammox process for denitrification.

7.2.1.5 Inhibition of substrate

AnAOB use NO_2^- as an electron acceptor to oxidize NH_4^+ to N_2 under anoxic conditions. However, high concentrations of NO_2^- and NH_4^+ could inhibit Anammox activity. In addition, the high substrate concentration also affected the particle sizes of Anammox sludge and the components of EPS. The particle sizes of Anammox sludge and the mass transfer efficiencies significantly decreased as the influent NH_4^+ concentration decreased from 300 to 50 mg/L, and the nitrogen loading rate (NLR) of the Anammox system decreased from 7.42 ± 0.74 to 1.24 ± 0.03 kg N/($m^3 \cdot d$).

NO_2^-, NH_4^+, free nitrous acid (FNA) and free ammonia (FA) have been reported to inhibit Anammox activity at high concentrations. It was reported that the Anammox process was inhibited by nitrite with a concentration of 100 mg/L. In addition, the half-maximal inhibitory concentration (IC50) of nitrite in the Anammox process was 50 mg (N/L). The Anammox inhibition threshold for NO_2^- was lower than that of NH_4^+ and NO_3^-, and such inhibition was reversible. Reversible inhibition refers to a reversible decrease in catabolic activity during exposure, while toxicity or irreversible inhibition refers to damage to microorganisms that results in an irreversible decrease in microbial activity. Anammox floating sludge was more easily inhibited by a high concentration of NO_2^- compared to Anammox granular sludge because the outer layers of granules were used as shells to protect the corer. At low pH, the inhibition of NO_2^- on AnAOB was more severe than that at high pH. A decrease in SAA by 50% was observed at concentrations of 38 mg NH_3/L, while 100 mg NH_3/L caused inhibition of SAA by 80%. It was reported that NH_4^+ inhibition of the Anammox process was essentially caused by FA, while NO_2^- inhibition was mainly related to FNA. FA and FNA change the intracellular pH, neutralize the transmembrane potential, and even cause cell lysis. It was reported that FA inhibited the Anammox process at concentrations of 1.7–2.0 mg NH_3/L. The presence of FA could change the structure of the microbial community, destroy the EPS, and even kill living cells. FNA generated via acidification also had negative effects on bacterial metabolism and further influenced the performance of the Anammox process.

7.2.1.6 Inhibition of salinity

Saline wastewater is mainly produced from petroleum refineries, shipboard slopes, the leather industry, and the fish canning industry. Due to the lack of freshwater resources, increasing amounts of seawater are used for municipal water and industry, leading to the discharge of more wastewater with high salt concentrations. At high concentrations, salinity stress changes the internal and external osmotic pressures of the cell membrane as the cell loses free water. In addition, salinity stress affects the metabolic activity of AnAOB and destroys the enzyme structure. Due to the selective pressures of salinity, the microbial community changes, and halophilic bacteria are selected. In the population shift strategy, *Ca. Kuenenia* was the dominant genus at low salt concentrations. With an increase in salt concentration *Ca. Kuenenia* showed a NO_2^- half-saturation constant lower than those of other AnAOB genera. In addition, physical and chemical parameters related to AnAOB changed. Therefore, the Anammox process has broad prospects in the treatment of high-salt wastewater.

7.2.1.7 Inhibition of toxic

Toxic inhibition occurs due to permanent poisoning of Anammox bacteria or enzyme inactivation, and the inhibition is often not reversible. Several toxic inhibitors are phenols, alcohols, aldehydes and antibiotics. These toxic inhibitors have single as well as combined effects on Anammox.

1. Antibiotics inhibition

Antibiotics are present in abundant in wastewater and due to their antimicrobial nature Anammox bacteria and denitrifying bacteria are inhibited or killed. The antibiotics such as polymyxin B, penicillin G-Na, chloramphenicol, and kanamycin sulfate cause Anammox inhibition. The toxicity order of antibiotics is as follows: polymyxin B >chloramphenicol> penicillin

G–Na> kanamycin sulfate. The IC50 inhibitory concentrations of these antibiotics are 5114.4 mg/L for polymyxin B, 409.9 mg/L for penicillin G–Na, 5254.1 mg/L for kanamycin sulfate and 10.2 mg/L for chloramphenicol. Moreover, the immediate poisonousness of multi–component antibiotics demonstrates synergism. During coupling of denitrification and Anammox process, the impacts of erythromycin shows transient inhibition at low concentration, but the Anammox activity can be restored within short span of time. However, on higher concentrations the inhibitory effects are irreversible.

Some studies have also discussed combined effects of multiple antibiotics on Anammox consortia. The acute toxicity of Anammox is also due to combined effects of various antibiotics including chloramphenicol + penicillin G–Na, chloramphenicol + polymyxin B sulfate, kanamycin sulfate + chloramphenicol, kanamycin sulfate + polymyxin B sulfate, kanamycin sulfate + penicillin G–Na, and polymyxin B sulfate + penicillin G–Na. Furthermore, the addition of azithromycin, trimethoprim, norfloxacin and sulfamethoxazole had inhibitory impact on Anammox performance, microbial community and substrate composition of the completely autotrophic nitrogen removal over nitrite process.

2. Inhibition by aldehydes, alcohols and phenols

Methanol and ethanol are main toxic alcoholic groups for Anammox activity. During inhibition mechanism, the methanol is converted into aldehyde groups by an enzyme methanol dehydrogenase, and ethanol to acetyl–coenzyme before getting oxidized to acetaldehyde (C_2H_4O). On adding methanol, there is no accumulation of nitrite in bioreactor that reduces denitrification, and the coupling system is inhibited. At 3–4 mmol methanol concentration no activity of Anammox was noted and the inhibition was permanent. At 400 mg/L and 1000 mg/L of intermittent phenol concentrations the activity of Anammox is reduced by 56% and 76%, respectively. The inhibition of AOB and change in microbial community during Anammox process reduced the Anammox activity. Benzene, benzoic acid, toluene and phenol have IC50 value of 9.21 mg/L, 3833.12 mg/L, 35.28 mg/L and 124.55 mg/L, respectively.

7.2.1.8 Inhibition of sulfide

Sulfide is one of the common inorganic inhibitors of Anammox activity. Sulfate (SO) is often reduced to hydrogen sulfide (HS) during anaerobic digestion, which reduces the Anammox activity. The concentration of sulfide that exert inhibitory effect varies in different studies (Table 7–2) due to the differences in the seed sludge and the operating conditions.

The Anammox microbe mostly used to study sulfide inhibition is *Ca. Kuenenia stuttgartiensis*. The short time exposure of hydrogen sulfide affects the Anammox bacterial sensitivity enriched from WWTPs and concluded that the IC50 (half maximal inhibitory concentration) value was in the low range. The critical inhibitory concentration of sulfide seems to be even lower with low concentrations of sulfide in the environment completely inhibiting Anammox activity in sediments.

The correct molecular process of sulfide toxicity is not known yet, however, it has shown that hydrogen sulfide inhibits cytochromes and other heme-proteins reversibly. Thus, the process sensitivity of Anammox because of sulfide is due to the fact that Anammox bacteria harbor an exceptionally high number of heme-containing proteins. Even the inhibitory effect of hydrogen sulfide on proteins having heme such as the mitochondrial cytochrome oxidase is reversible that showed no increase in nitrite-reducing activity. Some studies described that sulfide in influent decreased the granule diameter and damaged the cell.

Additionally, the prolonged consequences of sulfide were also studied on Anammox activity. The instant drop in Anammox activity was observed due to accumulation of nitrite at the loading rate of approximately 70 μmol/d. Further, in 7 days interval the optical density was dropped from 0.32 to 0.18 due to cell lysis. The inhibition threshold levels of sulfide also showed short term effects at fixed initial substrate level and short-term effects follow non-competitive inhibition pattern.

Table 7-2 Reported data on the sulfide inhibition of Anammox process.

Seeding sludge	Anammox species	Operational mode	Sulfide Concentration/(mmol/L)	Effect
Anammox bacteria	*Ca. Kuenenia stuttgartiens*	Batch	>5	100%
Anammox bacteria	*Ca. Kuenenia stuttgartiens*	Batch	2	60%
FBR sludge		Batch	1 or 5	>100%
FBR sludge		Continuous	2	20%
FBR sludge		Continuous	2	60%

7.2.1.9 Inhibition of emerging pollutants

With accelerating urbanization, more concern has been directed to the side effects of emerging pollutants, such as NPs, microplastics and antibiotics. These pollutants exist widely in the environment, in locations ranging from natural water to sewage treatment plants. Emerging pollutants have also been reported to be toxic to AnAOB.

1. Inhibitory effect of microplastics

Microplastics are defined as plastic particles <5 mm, which were produced by breaking down large pieces of plastic. At high concentrations, microplastics affect the nitrogen removal performance of the Anammox process and the structure of the microbial community. Two types of microplastics exist in the environment: biodegradable and nonbiodegradable. Biodegradable microplastics consist of fibers and proteins, such as polylactic acid, polybutenic acid, and polyhydroxybutyrate. Polyvinyl chloride and polyethylene are nonbiodegradable. Microplastics affected the structural components of EPS and inhibited the aggregation of AnAOB. In microorganisms, the cell membrane plays an important role in maintaining the stability of the cell. Nanoscale microplastics bound to the cell membrane and had acute inhibitory effects on key enzymes, metabolic intermediates and metabolic end products. Nanoscale microplastics destroy cells and damage material exchange channels and DNA. Microplastics have a large specific surface area and absorb other pollutants, such as polycyclic aromatic hydrocarbons (PAHs), polychlorinated biphenyls (PCBs), metals and antibiotics. The harm from these pollutants mainly came from their desorption. Microplastics are vectors for pollutants such as antibiotics, heavy metals and persistent organic pollutants. These pollutants affected AnAOB through desorption from the surfaces of microplastics.

2. Inhibitory effect of NPs

Due to the needs of industry and life, NPs are widely used in cosmetics, coatings, electronic products, energy materials, and catalytic materials. The main NPs include nZVI, graphene materials (GO), TiO_2 NPs, CuO NPs, ZnO NPs, Fe_3O_4 NPs, carbon nanotubes (CNTs), and Ag NPs. However, it is inevitable that some of these NPs will be discharged into the environment and flow into sewage treatment plants. Not all NPs have negative effects on the Anammox process. NPs can have positive or negative effects on the Anammox process, depending on their concentration. At low concentrations, NPs promoted the nitrogen removal performance of the Anammox process and the diversity of the microbial community. Previous studies reported that nZVI promoted the recovery of Anammox activity after inhibition by dissolved oxygen (DO). Other studies showed that graphene nanocrystals (GNs) and γ-Fe_2O_3 NPs increased the nitrogen removal performance of the Anammox process and increased the activity of hydrazine dehydrogenase (HDH) because GNs shortened the electron transport distance and promoted the process. As an electron shuttle, γ-Fe_2O_3 NPs travel between different enzymes to promote electron transport. Electron transfer is the most important process in the reaction, and GNs and NPs facilitate electron transfer, thereby fundamentally facilitating the Anammox reaction. At high concentrations, NPs inhibit the Anammox process. Two examples are given below to illustrate this issue.

ZnO NPs are widely used in pigments, coatings and cosmetics, and they increase the generation of intracellular reactive oxygen species, leading to cell leakage and reduced survivability of AnAOB. ZnO NPs inhibited the activity of the Anammox process, and the inhibitory effect increased with increasing ZnO NP concentration. ZnO NPs absorb a variety of elements, including As, Al, Mo, Hg, Pb, Cu, Ni and Cd, and change their biological availability by changing their morphologies in the medium. A previous study reported that the IC50 of ZnO NPs was 84.7 mg/L and that the cumulative reversible inhibition dose was 100 mg/L. Ag NPs adsorbed metals, such as Cu^{2+} and Zn^{2+}. The adsorbed metals then desorbed into the environment, increased the concentration of metals in the water environment and increased their detrimental effects on the Anammox process. TiO_2 NPs have a strong capacity for absorption of Cd. GO promotes oxidative stress, and it absorbs organic pollutants and metals. A study has shown that Cu NPs interact with EPS by specifically binding to groups containing tyrosine or tryptophan through C–O–C and –OH moieties.

7.2.2 Strategies to control the inhibition

7.2.2.1 Temperature

The ideal range of temperature for Anammox process is 30–40 ℃. An irreversible fall in Anammox activity takes place at a temperature higher than 45 ℃. And when temperature falls from 30 to 20 ℃ or 10 ℃ the Anammox activity reduces by approximately ten-folds. At 37 ℃, NRR of 11.50 kg N/($m^3 \cdot d$) was observed, while at 20–22 ℃ the NRR was 8.10 kg N/($m^3 \cdot d$). The system for Anammox process could be effectively operated at 18 ℃ but at this temperature nitrite accumulation takes place and Anammox system stability at 15 ℃ was completely lost. It is reported that activity of Anammox may possibly be observed at a temperature around 20 ℃ and through acclimatization the Anammox microbes can acclimatise to lower temperature conditions. In natural systems, the Anammox bacteria have ability to tolerate high temperature that ranges from 2–85 ℃. These reported results gave the possibility that the process of Anammox can be acclimatized and subsequently can be employed to treat various nitrogen rich wastewaters at varied temperatures. At optimal temperature range, the ability of Anammox to remove was still suboptimal even by acclimatization such as nitrogen removal rates were relatively lower at temperature between 6.3 and 25 ℃. But the temperature variation during Anammox process affects the concentrations of un-ionized ammonia and un-ionized nitrous acid. Therefore, to control the Anammox inhibition it is especially important to control the temperature.

At 30–40 ℃, the growth and performance of Anammox microbes is, preventing other adverse effects such as inhibition effects. Few studies reported that high temperature over 35 ℃ cause substrate inhibition due to availability of free NH_4^+, high NO_2^- concentration and other toxic metals. Some studies also reported that temperature range between 40–45 ℃ can be detrimental to AnAOB, causing bacterial cell lysis due to release of cytochrome c after temperature shock and consequently reducing Anammox activity. Recently, there is growing interest to use the Anammox technology at high latitudes with low temperature range of 10–20 ℃. At these lower temperature ranges the low temperature Anammox bacteria are involved such as *Ca. Brocadia* fulgida and *Ca. Brocadia* sinica. Another Anammox bacterial species *Ca. Kuenenia stuttgartiensis* was also observed in a continuous bioreactor operating at temperature less than 20 ℃. Even though these strains can acclimate to lower temperature ranges (10–12 ℃), but the optimal activity is observed at temperatures between 18–23 ℃ with high SAA and high growth rate.

7.2.2.2 pH control

The pH change in the culture can influence the Anammox process greatly due to the accumulation of toxic compounds that is accountable for inhibition of Anammox activity. To achieve high NRR over a prolonged period it is necessary to maintain constant pH. The pH range of 7–8 is optimum for Anammox species and seems ideal range for high performance of Anammox system. At pH 7–8 eliminates the inhibition of Anammox process by un-ionized nitrous acid and un-ionized ammonia. The free ammonia concentration decreased due to low pH (less than 7) but un-ionized nitrous acid concentration increases, while at high pH free ammonia increases and concentration of free nitrous acid decreases. So, to prevent un-ionized ammonia and un-ionized nitrous acid inhibition the pH is usually adjusted to a neutral pH.

7.2.2.3 Oxidation-reduction potential

ORP reflects DO concentration, organic substrate, organism activity and some toxic compounds in reactor systems with some operational conditions. An ORP value between −50 to +50 is beneficial for denitrification when Anammox process is coupled with partial denitrification. It was reported that the ORP has been widely used as a control parameter for online monitoring and control of nitrification and denitrification.

7.2.2.4 DO

Dissolved oxygen (DO) is also one of the important parameter defining the performance and efficiency of Anammox activity. It was also reported that Anammox performance was repressed reversibly by DO at low levels of <2% air saturation. Researchers observed that a reversible inhibition of the Anammox process occurred at low oxygen concentrations (<1% air saturation), but an irreversible inhibition happened at higher oxygen concentrations (>18% air saturation). Therefore, the DO should be strictly controlled in Anammox systems for avoiding oxygen inhibition.

7.2.2.5 Nitrogen loading rate and substrate concentration

Among the major causes of inhibition of Anammox activity the substrate concentration is a key factor but substrate concentration lower than inhibition threshold value do not hinder the activity of Anammox. To control the substrate inhibition, it is required to have lower concentration throughout the Anammox process, but controlling the substrate concentration during start up and maintaining its low quantity leads to substrate limitation as substrate is important for growth of Anammox biomass.

Therefore, this problem needs to be resolved by applying the effluent recirculation strategy while treating nitrogen rich wastewaters with low substrate concentration. Effluent recirculation is advantageous in nitrogen removal rate with other optimal operational conditions. Various discharge recirculation ratios established on different substrate concentrations are also reported. Using recirculation technique, the NRR went up to 34.50 kg N/ (m^3 · d) with 4.3–7.4 day required by AnAOB to double, which is less than 11 days as reported previously. Researchers tested recirculation ratio (the amount of wastewater that flows through the advanced pre-treatment component divided by the volume of wastewater that is sent to the final treatment and dispersal component) of 0.5 in a laboratory scale using Anammox UASB reactor having granular sludge, and (74.3 ± 6.7) kg N/(m^3 · d) value of NRR was obtained at high concentration of influent substrate of 360 mg/L NO. The ratio of influent NH_4^+ and NO_2^- was critical determining parameter for high efficiency of nitrogen removal where 26 kg N/(m^3 · d) of NRR at high ratio of influent NH_4^+ and NO_2^- was achieved. The Anammox bacterial growth is a key parameter in determining the ability of nitrogen elimination that is highly dependent on influent nitrogen. Also, at higher nitrogen loading rate the efficiency of Anammox system is particularly sensitive to varied operational parameters that reduces the ability of resistance to unfavorable factors. The influent loads repeatedly lead to system instability. Therefore, to have better performance the nitrogen loading rate values less than the highest tolerable value must be regulated to prevent uncertainty of the Anammox system. It was also reported that pH would not impede the NRR when process was operated at a low level and moderate load, though at high nitrogen loading rate at pH range from 7.5–8.1 the NRR reduce by 61%.

7.2.2.6 Sludge/biofilm acclimatization

AnAOB exist in various kinds of adverse environments, such as coke-oven wastewater sludge, high-organic-load and methanol methanogenesis sludge. AnAOB adapt to inhibition factors by sludge acclimatization. Such as Anammox bacteria acclimatized to antibiotic rich pharmaceutical wastewater, high salts water and phenolic wastewaters. The inhibition could be relieved by adding up the supporting agents during adaptation that also facilitates sludge culture for example the addition of hydroxylamine or hydrazine is a relief from nitrite inhibition.

Anammox biomass in reactor systems have many categories: free cell or flocculent sludge, granular sludge and biofilm. Because AnAOB have a trend to grow in aggregates, granule and biofilm systems have advantage to the growth of Anammox. The granule and biofilm systems are often used to enrich the AnAOB.

7.2.2.7 Using fresh Anammox sludge during the process

As per studies, AnAOB is active at high cell concentration of. NRR is also directly proportional to the bacterial population and addition of fresh Anammox sludge to the system increased the sludge concentration and reduces the loading rate. This puts the growth factors in the system that relives the inhibition. For example, addition of small amount of granular sludge during the Anammox process in the system while treating pharmaceutical wastewater, the nitrogen loading rate was increased to 9.4 kg/(m^3 · d).

Questions

[1] What is the effect of exogenous organic matter on anaerobic Anammox?

[2] What is the effect of inorganic carriers such as biochar on Anammox?

[3] Which exogenous substances can promotion of nitrogen removal and microbial enrichment on Anammox? Please list three and brief explanations.

[4] What is the effect of inhibitors on Anammox process?

[5] Is the effects of the traditional and emerging pollutants on Anammox process are different? Please briefly explain.

References

[1] Ma M, Cao Q, Mabruk A, et al. Promotion of nitrogen removal and microbial enrichment on Anammox by exogenous substance addition: a critical review [J]. Journal of Water Process Engineering, 2022, 49: 103096.

[2] Huang D Q, Fu J J, Li Z Y, et al. Inhibition of wastewater pollutants on the Anammox process: a review [J]. Science of The Total Environment, 2022, 803: 150009.

[3] Talan A, Tyagi R D, Drogui P. Critical review on insight into the impacts of different inhibitors and performance inhibition of Anammox process with control strategies [J]. Environmental Technology & Innovation, 2021, 23: 101553.

[4] Aklujkar M, Coppi M V, Leang C, et al. Proteins involved in electron transfer to Fe (Ⅲ) and Mn (Ⅳ) oxides by Geobacter sulfurreducens and Geobacter uraniireducens [J]. Microbiology, 2013, 159 (Pt_3): 515-535.

[5] Gu W, Wang L, Liu Y, et al. Anammox bacteria enrichment and denitrification in moving bed biofilm reactors packed with different buoyant carriers: Performances and mechanisms [J]. Science of the Total Environment, 2020, 719: 137277.

[6] Yu C, Li L S, Jiang C K, et al. Insights into Anammox activity inhibition under trivalent and hexavalent chromium stresses [J]. Biochemical Engineering Journal, 2019, 147: 118-125.

[7] De Marchi L, Pretti C, Gabriel B, et al. An overview of graphene materials: properties, applications and toxicity on aquatic environments [J]. Science of the Total Environment, 2018, 631: 1440-1456.

[8] Malovanyy A, Plaza E, Trela J, et al. Ammonium removal by partial nitritation and Anammox processes from wastewater with increased salinity [J]. Environmental Technology, 2015, 36 (5): 595-604.

[9] Liu Y, Xu W, Bao L, et al. Cell surface characterization and trace metal adsorptive properties of anaerobic ammonium-oxidizing (Anammox) consortia [J]. Chemosphere, 2019, 221: 11-20.

[10] Shi Z J, Hu H Y, Shen Y Y, et al. Long-term effects of oxytetracycline (OTC) on the granule-based Anammox: process performance and occurrence of antibiotic resistance genes [J]. Biochemical Engineering Journal, 2017, 127: 110-118.

[11] Tang C J, Zheng P, Wang C H, et al. Performance of high-loaded ANAMMOX UASB reactors containing granular sludge [J]. Water Research, 2011, 45 (1): 135–144.
[12] He S, Chen Y, Qin M, et al. Effects of temperature on Anammox performance and community structure [J]. Bioresource Technology, 2018, 260: 186–195.
[13] Kimura Y, Isaka K, Kazama F, et al. Effects of nitrite inhibition on anaerobic ammonium oxidation [J]. Applied Microbiology and Biotechnology, 2010, 86: 359–365.

Chapter 8 Coupling of Anammox and denitrification

Emphasis: Mechanism of SAD; Start-up methods of SAD reactor; Control conditions for reactor start-up; Factors influencing SAD reaction efficiency; Community structure and location of microorganisms.

Nut: Mechanism of SAD

8.1 Reaction mechanism of simultaneous Anammox and denitrification process

It is found that in the environment where organic carbon sources exist, Anammox and denitrification reactions can be synchronized and show certain mutual promotion effects. This provides a basis for simultaneous Anammox and denitrification (SAD) in true wastewater treatment. However, to truly understand the coupling reaction of Anammox and denitrification, it must be studied from its mechanism. The mechanism of the coupling reaction is mainly explained by chemometric theory and microbiological theory.

8.1.1 Stoichiometry in simultaneous Anammox and denitrification

Wastewater denitrification describes the use of NO_3^- or NO_2^- ions by denitrifiers to degrade carbonaceous biological oxygen demand (cBOD). Most denitrifiers are facultative anaerobicheterotrophs that transfer redox equivalents from the oxidation of a carbon source to an N-oxide under anaerobic conditions. The modular organization of denitrification respiratory systems utilizing NO_3^-, NO_2^-, NO and N_2O is shown in Figure 8-1. In addition, the overall energy yielding (catabolism or dissimilation) and cell synthesis (anabolism or assimilation) reactions of denitrification in the presence of acetic acid are shown as Eqs. (8-1) and (8-2), respectively. The hydroxyl ion (OH^-) and some of the carbon dioxide (CO_2) produced during denitrification are returned in the system alkalinity [Eq. (8-2)].

$$1.6NO_3^- + CH_3COOH \longrightarrow 0.8N_2 + 2CO_2 + 1.2H_2O + 1.6OH^- \ (\Delta G^\circ = -843kJ/M) \quad (8\text{-}1)$$

$$1.2NO_3^- + CH_3COOH + 0.1NH_4^+ \longrightarrow 0.1C_5H_7O_2N + 0.6N_2 + 1.5CO_2 + 1.1OH^- + 1.3H_2O \quad (8\text{-}2)$$

Denitrifying bacteria degrade cBOD in the absence of free molecular oxygen to obtain energy for cellular activity and carbon for cellular synthesis under a redox potential range from +50 to −50 mV. Most denitrifiers reduce NO_3^- via NO_2^- to molecular nitrogen without accumulation of intermediates.

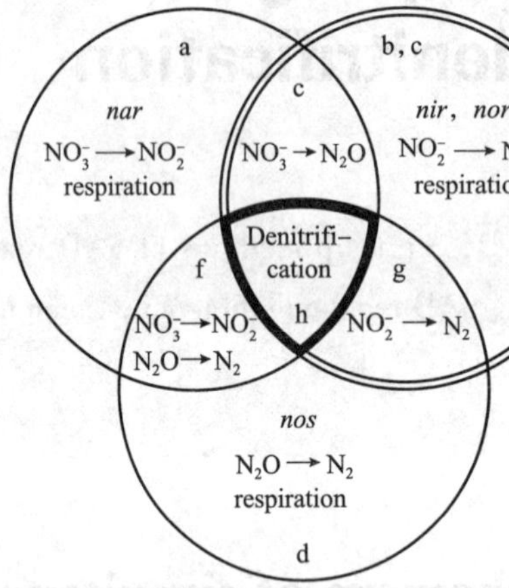

Figure 8–1 Modular organization of denitrification respiratory systems utilizing NO_3^-, NO_2^-, NO and N_2O

In Anammox, ammonium is oxidized to N_2 strictly under anoxic condition using nitrite as the electron acceptor [Eq. (8–3)]. Anammox organisms utilize CO_2 as the source of carbon and NO_2^- as the electron acceptor for ammonium oxidation. Concurrently, NO_2^- is used as the electron donor for the reduction of CO_2. Hydroxylamine (NH_2OH) and hydrazine (N_2H_4) were identified as the metabolites of Anammox process. The overall Anammox reaction is specified in Eq. (8–4) and the schematic demonstration of the Anammox process is shown in Figure 8–2. The $^{15}N^-$ studies have shown that one N–atom of the produced N_2 originate from NO_2^- and the other from ammonium. However, the growth of Anammox organism is reversibly inhibited even by oxygen concentrations below 0.5% air saturation.

$$NH_4^+ + NO_2^- \longrightarrow N_2 + 2H_2O \quad (\Delta G = -297 \text{ kJ/mol}) \tag{8–3}$$

$$NH_4^+ + 1.32NO_2^- + 0.066HCO_3^- + 0.13H^+ \longrightarrow 1.02N_2 + 0.26NO_3^- + 2.03H_2O + 0.066CH_2O_{0.5}N_{0.15} \tag{8–4}$$

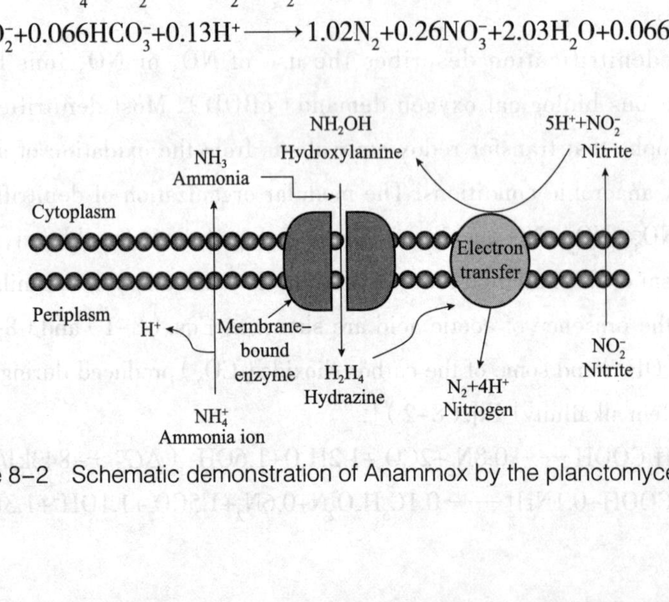

Figure 8–2 Schematic demonstration of Anammox by the planctomycetales

Comparing Eqs. (8–1) and (8–3), the standard free energy (ΔG) of denitrification reaction is higher than the Anammox reaction; therefore, the denitrifcation reaction is thermodynamically more feasible compared to Anammox reaction.

A similar coupling between Anammox and denitrification was observed in the marine environments. The experimental observations of the various physicochemical data and the theoretical calculations using the stoichiometric relationship of Anammox (i.e. 1 NH_4^+ requires 1.32 NO_2^-) and denitrification (i.e. quantity of organic carbon required for the conversion of NO_3^- to NO_2^-) can demonstrate the contributions of denitrification and Anammox in nitrogen removal. Alternatively, $^{15}N_2$ labelling experiments can be used to demonstrate the N_2 production associated with the above two processes.

In addition, the overall characteristics of the Anammox process are quite remarkable, especially considering the high toxicity of the intermediate N_2H_4, which is used as a rocket fuel and constitute an intermediate in the production of explosives and pesticides. Although Anammox catabolism takes place within the ladderanes, in some instances, N_2H_4 and NH_2OH are added to speed up the Anammox process. Under such circumstances, the tolerance of denitrifiers to N_2H_4 is another issue in the success of simultaneous carbon and nitrogen removal.

8.1.2 Microbiology in simultaneous Anammox and denitrification

This theory explains the coupling of Anammox and denitrification from the perspective of microbiology, considering the physiological characteristics and mutual relationship of various microorganisms. There are currently the following two views: one view is that Anammox and denitrification reactions are carried out in different microorganisms. Under certain conditions, these two reactions affect and promote each other to achieve the coupling; Another view is that AnAOB and denitrifiers have metabolic diversity, the same bacteria can show denitrification ability and anaerobic ammonia oxidation ability at the same time. The two reactions can be carried out in the same microorganism to achieve coupling under certain conditions.

Denitrifiers are microorganisms that reduce NO_3^- and NO_2^- to N_2. Four enzymes are involved in a complete denitrification system, i.e. reduction of NO_3^- to N_2. The reduction of NO_3^- to NO_2^- is catalyzed by the enzyme nitrate reductase (Nar). This is a membrane–bound molybdenum–iron–sulphur protein that is found in denitrifiers as well as in other dissimilatory nitrate reducing organisms. Both the synthesis and activity of nitrate reductase are inhibited by oxygen. The second enzyme in this pathway is nitrite reductase (Nir), which catalyzes the conversion of NO_2^- to N_2O. Nitrite reductase is unique to denitrifying organisms, which is found in the periplasm. Nitric oxide reductase (Nor), a membrane–bound protein, is the third enzyme in the pathway, catalyzing the conversion of N_2O to NO. Nitrous oxide reductase (Nos), a periplasmic copper–containing protein, is the last enzyme in the pathway and converts NO to N_2. Both the synthesis and activity of all four denitrification enzymes are controlled by oxygen. Nitrous oxide reductase is the most sensitive denitrification enzyme and it is inhibited by DO concentrations less than 0.2 mg/L. However, some denitrifiers lack key enzyme systems to denitrify completely, and the lack of these enzyme systems can allow the production and accumulation of free intermediates. Organisms with the capability of denitrification belong to a variety of groups and encompass a wide range of physiological traits.

Many genera of denitrifying bacteria can use NO_3^- and NO_2^- to degrade cBOD, some genera such as *Enterobacter* and *Escherichia can* use only NO_3^-. On the other hand, some genera of denitrifying bacteria including *Thiosphaera pantotropha* or *Paracoccus* denitrificans, *Magnetospirillum magnetotacticum* and *Pseudomonas stutzeri* SU2 can denitrify under aerobic or microaerophilic conditions. In addition, *Nitrosomonas*-like microorganisms including *Bacillus* cereus, *Bacillus* subtilis and *Bacillus* licheniformis, nitrify and denitrify simultaneously even under fully oxic or anoxic condition with N_2 as main final product. *Nitrosomonas eutropha* is an obligate lithoautotrophic nitrifying bacterium and also a denitrifying organism that uses hydrogen as the electron donor and nitrite as the electron acceptor. The denitrification activity of N. eutropha could be stimulated by adding gaseous nitrogen oxide under anaerobic conditions. Mostly, the denitrifying nitrifiers are detected in a bioreactor operating with the coupling of aerobic and anaerobic ammonia oxidation, for example, SNAP.

Anammox species express various unusual lipids that contain ladderanes, i.e., lipids build from concatenated cyclobutane rings, which form a molecular ladder. These ladderane lipids surround the anammoxosome, a special compartment of the Anammox cell, in which the anaerobic oxidation of ammonium to N_2 is taking place.

As the core of Anammox process, AnAOB have been divided into six genera: *Candidatus Kuenenia*, *Candidatus Brocadia*, *Candidatus Jettenia*, *Candidatus Anammoxoglobus*, *Candidatus Scalindua*, and *Candidatus Anammoximicrobium*. A large spectrum of other unidentified Anammox species or subspecies are listed in today's databases. These lineages all have '*Candidatus*' status, as they do not exist in pure cultures. This highlights the significance of understanding the universal interaction between AnAOB and denitrifiers. It is generally believed that AnAOB cannot use organic matter, and the presence of organic matter will inhibit the activity of anaerobic ammonia bacteria. However, the latest studies have found that the activity of AnAOB is even enhanced in the presence of organic matter.

Heterotrophic bacteria are affiliated with the phyla Chlorobi, Bacteroidetes, Chloroflexi and *Proteobacteria* and comprise a large fraction of the microbial community in Anammox bioreactors. There is a universal interaction between OHO and AnAOB, demonstrated by the high phylogenetic similarity shared across different wastewater treatment systems.

The symbiosis of AnAOB and denitrifiers is common in granular sludge, suspended sludge, and biofilm. The premise behind Anammox coupled with denitrification is stably supplying nitrite for AnAOB. In one Anammox granulation process, the bacterial diversity gradually increased, accompanied by an increase in particle sizes as the internal microenvironment changes; some denitrifiers co-occurring with AnAOB continued their metabolic activities; during this process, the carbon source for denitrification was mainly from the degradation of the granule core. In an anoxic biofilm system, NO_3^- reductase (Nar) levels were significantly higher than NO_2^- reductase (Nir) and Thauera contained significantly more Nar than Nir. The discrepancy between Nar and Nir is assumed to be the driver for high NO_2^- accumulation. Moreover, another recent study using genome-resolved metagenomics supported the hypothesis of a "nitrite loop" in a full-scale partial-nitritation Anammox (PNA) system. In this case, some denitrifiers in granular sludge reduced the excessive NO_3^- formed by NOB and AnAOB as electron acceptors producing NO_2^- for AnAOB.

8.2 Research on the coupling of Anammox and denitrification

Although AnAOB and denitrifiers have some similar physiological needs and growth environments, due to the differences in physiological characteristics between them, it is necessary to pay attention to the start-up strategy of the coupling reactor and investigate the various factors affecting the coupling process in order to achieve the best coupling effect of the two in a system.

8.2.1 The coupling reaction of Anammox and denitrification

The coupling of Anammox and denitrification refers to the process in which the anaerobic ammonia oxidation reaction removes ammonium and nitrite under certain conditions, and the generated nitrate is removed by denitrification. At the same time, the organic carbon is consumed to reduce the inhibition on Anammox, achieving simultaneous nitrogen and carbon removal (Figure 8-3).

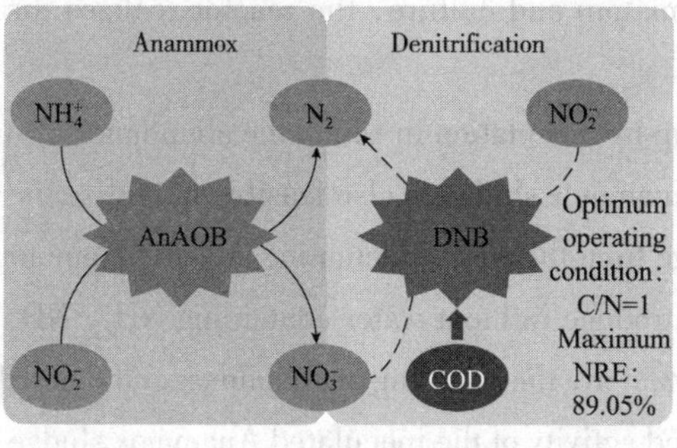

Figure 8-3 Flow diagram of SAD process

At the beginning of the discovery of Anammox, researchers conducted experiments on the addition of organic compounds to the Anammox mixture taken from the membrane biological fluidized bed, and proved that the addition of acetate and glucose can improve the Anammox activity. It was later found that this was due to the presence of denitrifying bacteria in the mixture using organic matter for nitrate reduction to produce more anaerobic ammonia substrate (nitrite). This proves that Anammox and denitrification can coexist. Since then, some scholars have found that Anammox and denitrification reactions can coexist and be coupled in the same system under the environmental conditions of organic matter. In the study of Anammox of aquaculture wastewater, it was found that the concentration of methanol to completely inhibit the anaerobic ammonia oxidation activity was 5.6 mmol/L, which was higher than the complete inhibitory concentration of methanol on purified anaerobic ammonia-oxidizing bacteria (1 mmol/L). Subsequently, the existence of Anammox and denitrification in the aquaculture water was proved by stoichiometry, and the existence of AnAOB and denitrifiers in the same system was verified by molecular biology methods such as FISH technology. Thus, the above phenomenon is explained: the rapid oxidation of methanol by a large number of heterotrophic bacteria (including denitrifying bacteria) alleviates the inhibitory effect of excessive concentration of methanol on AnAOB to a certain extent. The development of follow-up studies and the success of engineering examples show that the coupling reaction of Anammox and denitrification is not only feasible, but also has broad prospects and great practical application value for the treatment of nitrogen-containing organic wastewater.

8.2.2 Start-up of coupled Anammox and denitrification reactor

1. Start-up strategy

The start-up of the coupling reactor of Anammox and denitrification is the basis for the study of Anammox-denitrification process. At present, the most widely used start-up method called "in-situ startup" is to start the Anammox reactor with artificial water distribution containing NH_4^+ and NO_2^- according to the stoichiometric ratio of about 1 : 1.32, and enrich and cultivate AnAOB. When the reactor is successfully started, a certain amount of organic matter is added to the influent water, and after a period of acclimation and culture, the reactor realizes the coupling of Anammox and denitrification.

In addition, the startup by inoculation in which the abundance of AnAOB can be improved in a short time by inoculating Anammox sludge is also feasible to realize the coupling. The main method is to first take out the sludge from the well-functioning anaerobic ammonia reactor, inoculate it into a new reactor, and then introduce influent water containing NH_4^+, NO_2^- and organic matter into the reactor. After regular acclimation, the coupling of Anammox and denitrification is initiated. There is no doubt that the amount and activity of the inoculated Anammox sludge determine the startup time, and the startup time significantly limits the startup mode of inoculation from enriched sludge from pilot-scale reactor to the full-scale reactor.

There is no essential difference between the two initiation methods. The main reason for this is that AnAOB have a long multiplication time and are sensitive to environmental conditions such as the inhibition of chemical oxygen demand (COD), while denitrification is relatively easy to initiate and relatively insensitive to changes in environmental conditions. Therefore, Anammox is usually started first to speed up the start-up of the coupling reactor.

In the start-up process, attention should also be paid to the following three aspects: ① appropriate reactor and HRT; ② suitable environmental conditions, such as DO and pH; ③ appropriate substrate concentrations, such as NH_4^+, NO_2^-, COD and C/N.

2. Analysis of the start-up process

The process of starting Anammox reactor into a coupled reactor can be divided into three stages: lag stage, adaptive stage and stable stage. The characteristics of each stage are shown in Table 8-1. The purpose of starting the coupled reactor is to achieve simultaneous removal of carbon and nitrogen. At present, the main indicators to judge whether the coupling reaction of anaerobic ammonia oxidation and denitrification is successfully started are to investigate the synergistic removal effect of ammonium, nitrite, nitrate and organic matter in the reactor and the removal ratio of each pollutant.

Table 8-1 Characteristics of different phases on Anammox-denitrification coupling reactor

Stage	Characteristics
Lag stage	The removal of COD is mainly based on the adsorption of anaerobic ammonia sludge. The removal effect of ammonium decreased slightly. The reactor removes nitrite at a very high level. The removal rate of TN in the reactor was lower than that without the addition of organic matter.
Adaptive stage	In this stage, the bacteria in the reactor were domesticated to enrich the denitrifiers, and the denitrifiers were rapidly enriched in the later stage, and the removal rate of COD was gradually improved. The removal effect of the reactor on ammonium first decreased sharply, and then gradually increased. The removal of nitrous nitrogen from the reactor first decreased, then increased, and finally stabilized. The removal rate of TN in the reactor also decreased first and then gradually.
Stable stage	The removal rate of COD in the reactor reached the highest at a stable level. The removal rate of ammonium increased and eventually stabilized. The removal effect of nitrite nitrogen was not much different from that of the previous stage. The removal rate of TN gradually increased and reached the maximum removal rate.

8.2.3 Factors influencing the coupling of Anammox and denitrification

8.2.3.1 environmental factors

1. pH

The pH range suitable for the growth of AnAOB is 6.7-8.3, and 6-9 for the growth of denitrifiers. Theoretically, 6.7-8.3 is the suitable pH range for the coupling of Anammox and denitrification. However, It is found that Anammox can be carried out in the initial pH value of 6.02-9.00 under the presence of organic carbon source, and the optimal pH value is 8.00. Currently, the removal rates of COD, NH_4^+-N and NO_2^--N reach the ideal state. If the pH is too low (below 5.06) or too high (above 9.00), both denitrification and Anammox are stopped. It means that compared to denitrification, Anammox had a wider physiological pH range in the coupling reactor. The denitrification activity was suppressed in acidic and alkaline environment, and it was more sensitive to the latter. At present, the pH value used by various scholars to start the coupling reaction of Anammox and denitrification is between 6–8.

2. DO

Dissolved oxygen (DO) is an important factor influencing the microbial activity of denitrification and Anammox processes. In the oxygen-regulation experiments, increasing the DO concentration was observed to have a negative effect on both denitrification and Anammox activities. The dissolved oxygen that initiates the coupling reaction should be controlled at 0.5 mg/L or less. Nevertheless, the Anammox process had a stronger tolerance to higher DO concentration compared to the denitrification process. The high tolerance towards oxygen observed in the Anammox process could be explained as follows: during the incubation in a non-stirred exetainer, microorganisms and particulates can settle down from the suspension and create a local hypoxic environment where Anammox process was active at the bottom of the vials. The injection of air-saturated water contained relatively high oxygen concentration did not change the hypoxic condition in the bottom. These findings provide the possibility to couple it with other wastewater treatment process. At present, CO_2 or argon deoxygenation is mainly used to avoid the influence of oxygen on the coupling reaction.

3. temperature

Temperature can have an important effect on AnAOB and denitrifiers. AnAOB are suitable for survival in mesothermal environment, and have anaerobic ammonia oxidation activity when the temperature is higher than 11 ℃ or lower than 45 ℃. The optimal temperature for Anammox is 30–35 ℃, and the suitable temperature for denitrifiers is about 20–35 ℃. In the current study, it shows that the increase in temperature stimulated both denitrification and Anammox activities in the range of 10–35 ℃, indicating there was more cooperation between denitrification and Anammox rather than antagonistic competition with the increase of temperature within the ambient range. Improper temperature may lead to alterations in genetic expression during denitrification, affecting the rate of substrate depletion and the generation time of denitrifying bacteria. Therefore, the temperature of the coupling reaction of Anammox and denitrification should be controlled between 30–35 ℃ to exert its best effect.

4. reactor type

The reactor type is another important factor influencing the coupling of Anammox and denitrification. It directly affects the residence time of sludge solids and the upflow rate, which in turn affects AnAOB enrichment, process start-up speed, and operational stability. Different types of bioreactors, including moving bed biofilm reactor, biofilter, sequencing batch reactor, and UASB reactor, can successfully start the Anammox process.

8.2.3.2 Substrate factors

1. Organic matter

The right concentration and type of organic matter can improve the nitrogen removal efficiency of the process with Anammox as the main body, but higher concentrations of organic matter will inhibit the activity of AnAOB. However, the organic matter content in the actual wastewater fluctuates greatly (generally between 100–1000 mg/L), and how to use Anammox as the main process in the actual wastewater treatment has become a difficulty.

Organic matter with different concentrations that have different effects on the Anammox reaction are generally in competition with AnAOB in the reaction, and they compete for the electron acceptor nitrite. The presence of organic matter will cause the proliferation of heterotrophic bacteria, which will allow the denitrification to proceed, and the corresponding AnAOB will be reduced due to the lack of substance. When the reaction is inhibited for a short time, Anammox can be quickly resumed by creating suitable environmental conditions for the growth of AnAOB (such as maintaining pH at 6.7–8.5, temperature at 20–43 ℃, and regulating the reaction mass ratio of COD : NH_4^+–N : NO_2^-–N to 1.6 : 1 : 1.92, etc.). However, when the organic matter inhibits the operation of the reactor and the activity of AnAOB for a long time, the inhibition effect can be slowed down by stopping the addition of organic matter and reducing the NRR [less than 0.43 kg/(m^3 · d)]. Therefore, in practical application, when Anammox is applied to treat wastewater containing organic matter, it is recommended to control the concentration of organic matter in the influent water at 100–200 mg/L or below, to ensure the activity of AnAOB.

The actual wastewater contains multiple types of organic matter, which could influence the Anammox activity. Different types of some organic matter have different effect on Anammox activity. The effect of different concentrations of organic matter on anaerobic ammonia oxidation reaction are shown in Table 8–2. Besides, some organic substances such as alcohols, aldehydes, phenols, and antibiotics can inhibit the activity of AnAOB through microbial poisoning or inhibiting enzyme activity. For example, methanol is the most potent inhibitor, which leads to complete and irreversible loss of activity at concentrations as low as 0.5 mmol/L, and this inhibition may be due to the conversion of methanol into formaldehyde by hydroxylamine oxidoreductase in Anammox cells, which leads to irreversible inactivation of proteins and enzyme. It is noteworthy that up to 6 kinds of AnAOB can oxidize volatile fatty acids into CO_2, and then be fixed via the acetyl–CoA pathway. And there are three acetic acid metabolic pathways was found in *Candidatus Brocadia* sinica, namely acetyl–CoA, ALDH and AckA Xfp, it was observed that the presence of certain small molecule of organic matter can improve the Anammox activity. The mechanism of small molecule organics on Anammox activity may be similar to propionate that Anammox bacteria can use it as substrate and be oxidized to CO_2 using nitrate or nitrite as the electron acceptor. Conversely, easily biodegradable organic compounds are required by the heterotrophic denitrifying bacteria as an electron donor for reducing nitrate to nitrogen gas. The effect of organic matter on denitrification, observed in the current work, was in accordance with the previous study. Inhibition effect of organic matter on Anammox could be explained by two different proposed mechanisms. First, organic compounds facilitate heterotrophic denitrifying microorganisms to grow faster and then have stronger competitiveness for common substrate than the autotrophic Anammox bacterias, leading to the decrease in Anammox activity. Second, Anammox bacteria have diverse metabolism that could also use organic compounds as a substrate. These findings present new inspiration for nitrogen removal by simultaneous Anammox and denitrification processes.

Table 8–2 The effect of different concentrations of organic matter on anaerobic ammonia oxidation reaction

Organic	Critical concentration/ (mg/L)	Impact on Anammox reaction
Sucrose	60	When it is maintained above 60 mg/L for a long time, the reaction is inhibited
	117	Increasing to 117 mg/L in the short term had no significant effect on the removal rate of ammonia nitrogen, nitrite and total nitrogen
	300	High concentrations of organic matter have an inhibitory effect on the reaction
Glucose	70	When the COD > 70 mg/L, the maximum specific reaction rate of Anammox sludge decreased to less than 0.05 kg NH_4^+–N/(kg VSS/d).
	120	When COD ≤ 120 mg/L, the organic matter has a promoting effect on the reaction, and vice versa
	300	The reaction is completely inhibited
Starch	200	The activity of AnAOB was promoted under the impact of 0–200 mg/L organic matter concentration
Methanol	0.5 mmol/L	The reaction stops
Phenol	—	The increase of phenol concentration in the short term leads to a rapid decrease in nitrogen removal rate
Acetate	1.16 mmol/L	Low concentrations of acetate have a certain promoting effect on the reaction
	200	Low concentrations facilitate the reaction
Domestic sewage treated by A/O process	100	When COD < 100 mg/L, it has little effect on the Anammox process
Degradable organics in landfill leachate	50, 150	When the concentration is less than 50 mg/L, the reaction deactivation effect is stable; when the concentration is more than 150 mg/L, the reaction collapses
	300, 500	A concentration of 300 mg/L inhibits the reaction, and a concentration of 500 mg/L stops the reaction

2. Nitrite

Anammox and denitrification work together, but there is competition. Under anoxic conditions, NO_3^- can be reduced by denitrifiers or other species promoting dissimilatory reduction of NO_3^- to NH_4^+ (well known as dissimilatory nitrate reduction to ammonia, DNRA) and release NO_2^- as a free intermediate. Subsequently, NO_2^- can be utilized by Anammox bacteria for the oxidation of NH_4^+ as per the stoichiometric. NO_2^-–N is the common substrate of Anammox and denitrification, and the denitrification reaction has lower free energy. In the presence of organic matter, denitrifiers will use organic matter and NO_2^-–N to preferentially grow, and the growth rate is much higher than that of AnAOB, which reduces the growth substrate and space of AnAOB, inhibiting the activity of AnAOB.

It is found that the activity of Anammox sludge is inhibited in the presence of a large amount of organic matter, so that the sludge shows high denitrification activity, but when the nitrite in the reaction system is sufficient, the effect of organic matter on the anaerobic ammonia activity of sludge is not obvious. It can be seen that when there is an organic carbon source, a moderate increase in the initial NO_2^-–N concentration can obtain a better effect of nitrogen and carbon removal. However, too high concentration of NO_2^-–N will inhibit the progress of Anammox and denitrification reaction. In order to ensure a better effect of nitrogen and carbon removal, COD should be controlled at 300 mg/L, NO_2^-–N should not exceed 500 mg/L.

3. C/N

Strict control of reaction conditions and improving the synergistic effect of AnAOB and denitrifiers is the primary condition for achieving efficient carbon and nitrogen removal, among which the main controlling factor is the C/N ratio. In order to make AnAOB and denitrifiers better synergistically carbon and remove nitrogen, it is recommended that the C/N ratio in the reaction environment should be maintained at (1 : 1)~(1 : 3). In the absence of organic matter in the reaction system, AnAOB was the main functional bacteria, and the denitrifiers were in an inhibited state. When the C/N ratio in the environment is maintained at (1 : 1)~(1 : 3), the activities of AnAOB and denitrifiers affect and restrict each other. It was found that when the C/N ratio was 1.7, the Anammox reaction effect in the reactor was the best. When the concentration of organic matter was higher, the activity of denitrifiers was much better than that of AnAOB, and the activity of AnAOB was limited by heterotrophic microorganisms.

To further decipher the role of C/N in the inhibition or recovery of Anammox, carbon and nitrogen metabolisms based on Kyoto Encyclopedia of Genes and Genomes (KEGG) are depicted as Figure 8–4 (a). Although Anammox bacteria is known to be sensitive to organics, the metabolic process of Anammox could be completed smoothly via key functional genes such as Hzs and Hdh at C/N of 0.5–1.5 (Normal stage). While the denitrification occurs independently through the complete chain of denitrifying genes including nar, nap, nirK/nirS, nor and nosZ which seems not directly affected Anammox. It could be inferred that harmonious living between Anammox bacteria (AnAOB) and denitrifying bacteria (DB) was developed under the low level of organic matter in Figure 8–4 (b). However, as the C/N was added to above 2.0, it is observed that there is another important metabolic pathway besides Anammox and denitrification. To be exact, high relative abundance of functional genes with glycolysis (eno, PK, glk, pgi, and pfk) indicated stronger expression of hydrolytic acidification. Therefore. It is reasonably speculated that such a good balance between AnAOB and DB was broken by hydrolytic acidification bacteria (HAB) proliferation. Notably, more low–molecular volatile fatty acids (VFAs) could be accumulated by

HAB and further utilized as carbon source for the proliferation of DB. Moreover, from the view of EPS property, the significant change in both the composition and the structure of EPS was also caused by the excess increase of C/N. On the one hand, excessive secretion of polysaccharides on the biofilm surface, reducing the adhesion of the carrier surface and then in favor of heterotrophic microorganisms. Importantly, more loose EPS induced by such high C/N ratios might redistribute microbial niche of function bacteria within the biofilm. The intensified inter-species competition and shrinking growth micro-environment triggered the disappearance of low-growing Anammox species. As a result, the Anammox function was deteriorated and even ceased, shown as Figure 8-4 (b). When the cutoff of influent COD was adopted, however, fermentation heterotrophs were unable to be propagated due to the lack of growth substrate. Instead, the abundance of Anammox-associated genes increased again, signifying the renewal of Anammox ability due to the inefficiency of the fermentation without organic substrates. Moreover, the realignment of properly tight EPS structure was also an essential signal for Anammox recovery. Therefore, it was vital that the C/N optimization for strictly suppressing the overgrowth of hydrolytic acidification species to ensure the stability of Anammox restart.

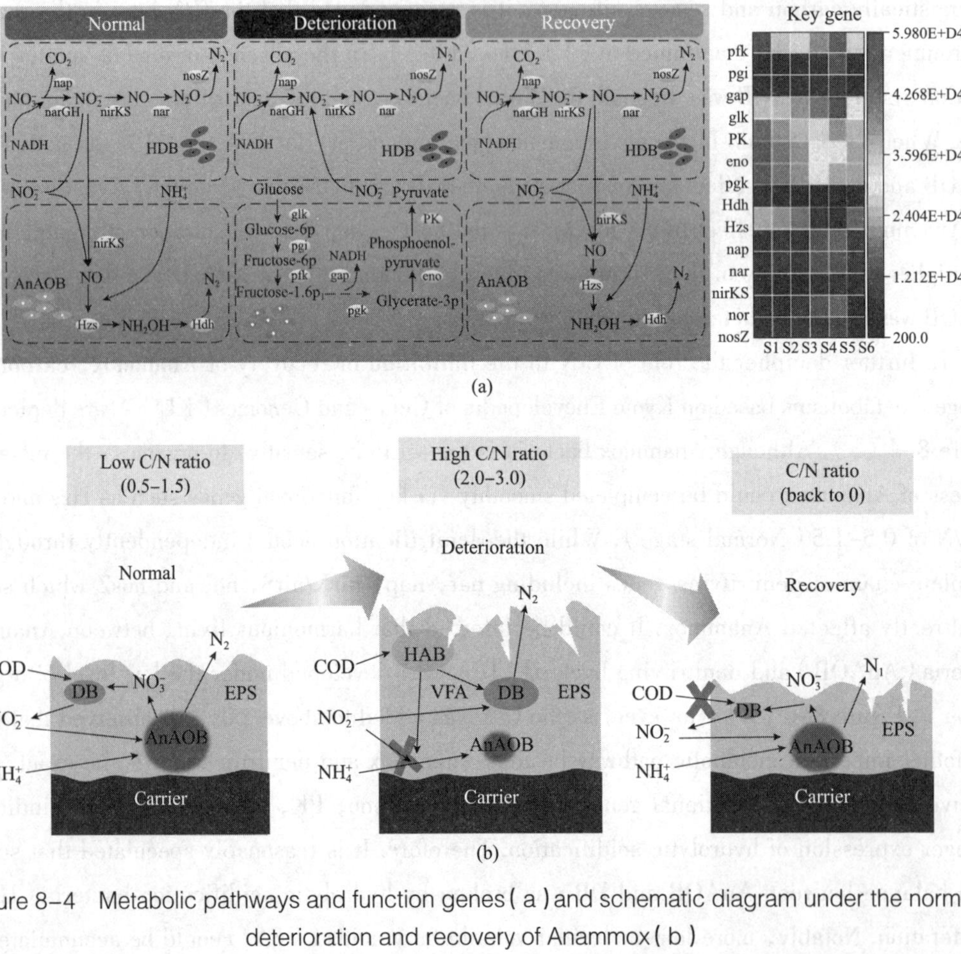

Figure 8-4 Metabolic pathways and function genes (a) and schematic diagram under the normal, deterioration and recovery of Anammox (b)

8.3 Microbial community of the simultaneous Anammox and denitrification process

The autotrophic Anammox bacteria grows in anaerobic conditions and is demanding to environmental conditions. Most denitrifiers are heterotrophic bacteria that use organic matter as the carbon and energy source for anaerobic respiration. Under the condition of organic matter, AnAOB and denitrifiers could coexist, but there are great differences in the dominant populations.

The microbial community structure in the SAD process was analyzed through constructing 16S rDNA clone libraries. The researchers established the phylogenetic tree construction of universal bacteria presented in Figure 8-5 (a) and the phylogenetic tree construction of Anammox bacteria presented in Figure 8-5 (b).

In terms of universal bacteria, *β-Proteobacteria* accounted for 52% and were predominant. Besides, *γ-Proteobacteria* accounted for 9%. Neither *α-Proteobacteria* nor *δ-Proteobacteria* clone were identified. The second most abundant phylogenetic group was *Acidobacteria* (18%), and others included *Chlorobi* (8%), *Chloroflexi* (6%), *Gemmatimonadetes* (4%), *Nitrospirae* (2%) and *Planctomycetes* (1%). In the Anammox bacteria clone library, *Candidatus Brocadia* sinica accounted for 96% (predominant), and uncultured *Planctomycetales* bacteria accounted for 4%. Compared to the universal bacteria, the microbial diversity of Anammox bacteria was less abundant. It was reported that there were large phylogenetic distances between different Anammox bacteria and different Anammox species were rarely found in the same ecosystem. This indicates that each Anammox species occupied a distinct ecological niche.

Among the microorganisms identified, denitrifying bacteria exhibited a good diversity and abundance, which mainly belonged to *β-Proteobacteria* (*Rhodocyclaceae* sp., *Thauera* sp., *Azoarcus* sp.). A previous study reported that *β*-subclass *Proteobacteria* played a key role in the activated sludge for nitrogen removal, which could contribute to both organic matter degradation and nitrogen removal. The *β*-subclass *Proteobacteria* (denitrifiers) played an important role in both Anammox-nitrification and Anammox-denitrification processes.

Granulation was achieved in the SAD process. Most of the denitrifiers identified in the present work were reported to facilitate the formation of aggregates thereby preventing biomass from getting washed out. *Thauera sp.* could produce lots of extracellular polymeric substances which contributed to the granulation process. Chloroflexi bacteria were facultative anaerobes and were important for granulation. Besides, Chloroflexi bacteria were present in a highly enriched Anammox biomass. Anammox bacteria grew slowly and were difficult to enrich in the reactor, which makes the Anammox process still a challenge in full-scale wastewater treatment. However, Anammox-based granulation could be achieved through synergy with denitrifying bacteria in the SAD process, which helped Anammox bacteria to enrich in the reactor (Figure 8-6).

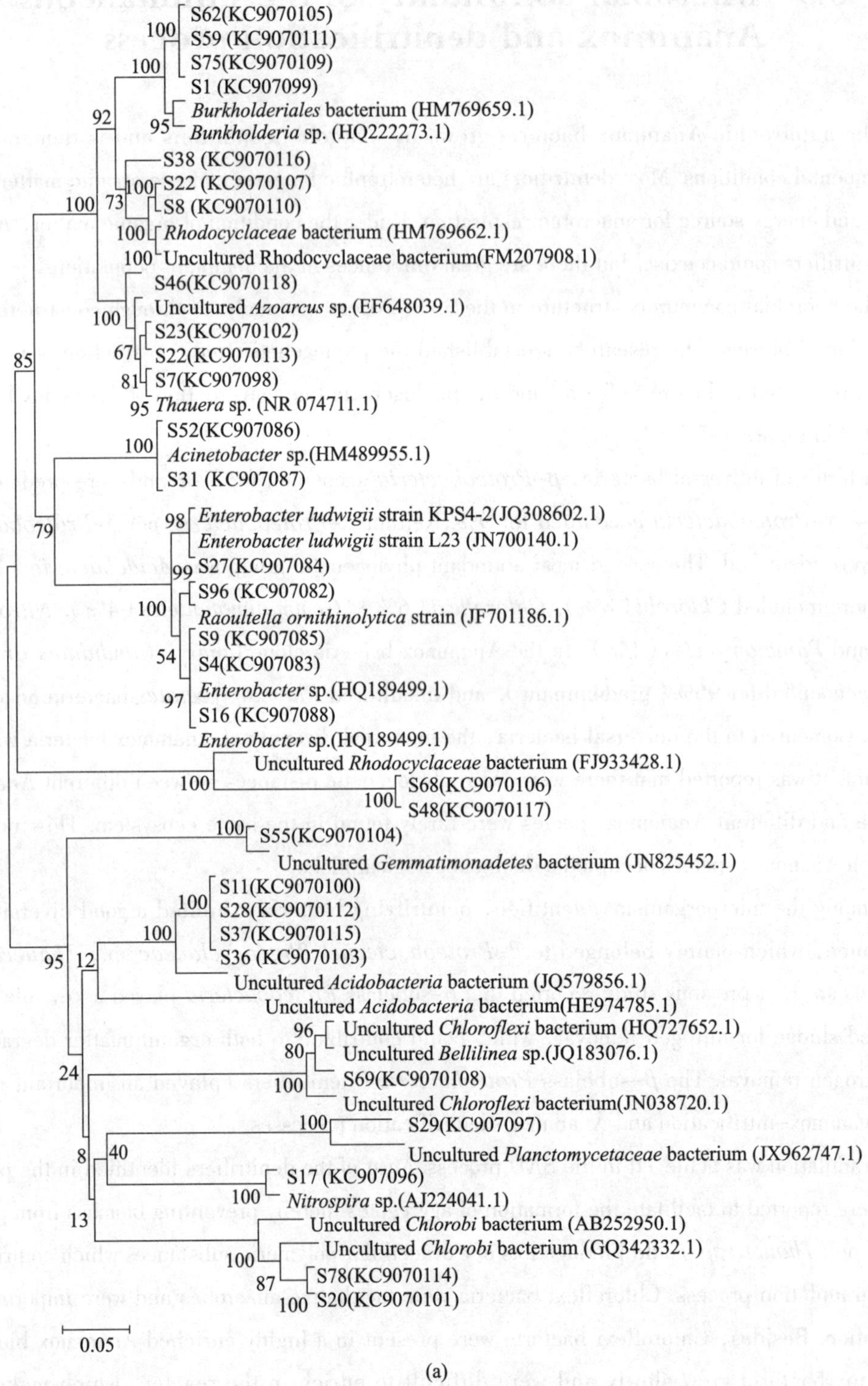

(a)

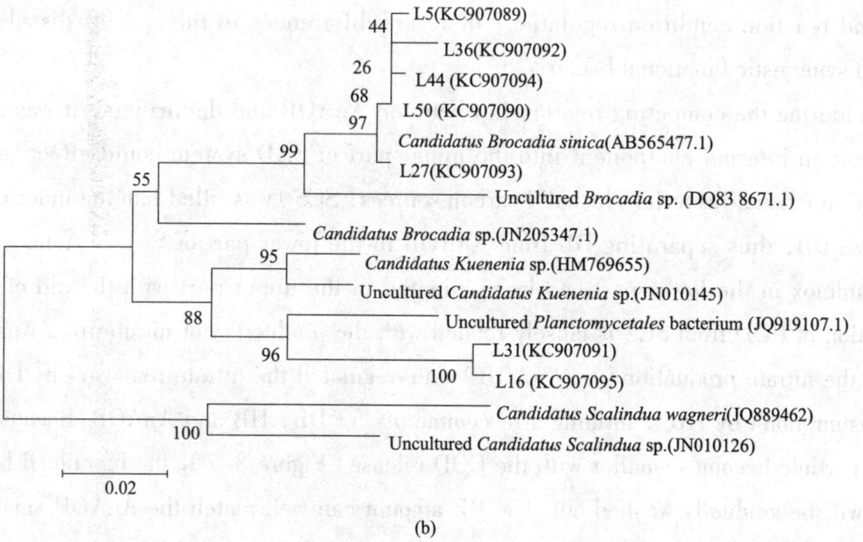

Figure 8-5 Phylogenetic trees constructed with the neighbor-joining method

(a) universal bacteria; and (b) Anammox bacteria. Reference sequences are obtained from GenBank with parenthesized accession numbers.

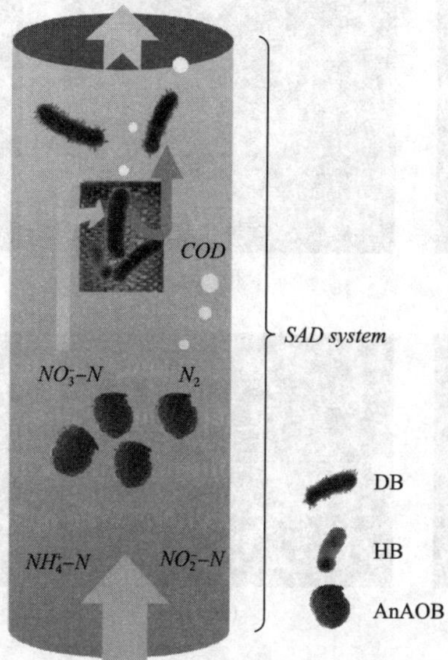

Figure 8-6 Spatial distribution of functional bacteria in the SAD system

AnAOB: Anammox bacteria, DB: denitrifying bacteria, HB: hydrolyzing bacteria.

The results show that in the UASB/SBR reactor that successfully starts Anammox, AnAOB and denitrifiers are mainly *Planctomycetes* and *Proteobacteria*, respectively. In the synergistic effect of Anammox and denitrification, molecular biology methods such as polymerase chain reaction and molecular cloning are used to analyze the microorganisms in the reactor, and the results showed that AnAOB are dominated by *Candidatus Brocadia* sinica of *Planctomycetes* and denitrifiers are mainly *β-Proteobacteria* of *Proteobacteria*. It can be seen that due to the differences in substrate supply and reaction condition regulation, there are differences in the species distribution and number of synergistic functional bacteria.

Considering the competing relationship between AnAOB and denitrifiers, it was found that introducing an internal component into the upper part of SAD system could effectively fix the location of functional bacteria. The solid carbon sources (SCS) was filled into the inner component to colonize DB, thus separating DB from AnAOB in the lower part of SAD system. The nitrate from Anammox in the lower part can be eliminated in the upper part with the aid of SCS. The consumption of COD from SCS is closely related with the production of nitrate from Anammox. In a sense, the nitrate production (by AnAOB) auto-regulated the nitrate removal (by DB) via the COD consumption (by HB), forming a fit community for DB, HB and AnAOB. Because the SCS like rice particle becomes smaller with the COD release (Figure 8-7), the functional bacteria on the SCS will be gradually washed out. The DB amount can well match the AnAOB amount with a proper allocation of reactor volume and a good cooperation between two bacterial groups although DB grow far more fast than AnAOB. In such a case, the microbial community was stable and robust, giving a strong support to the SAD system.

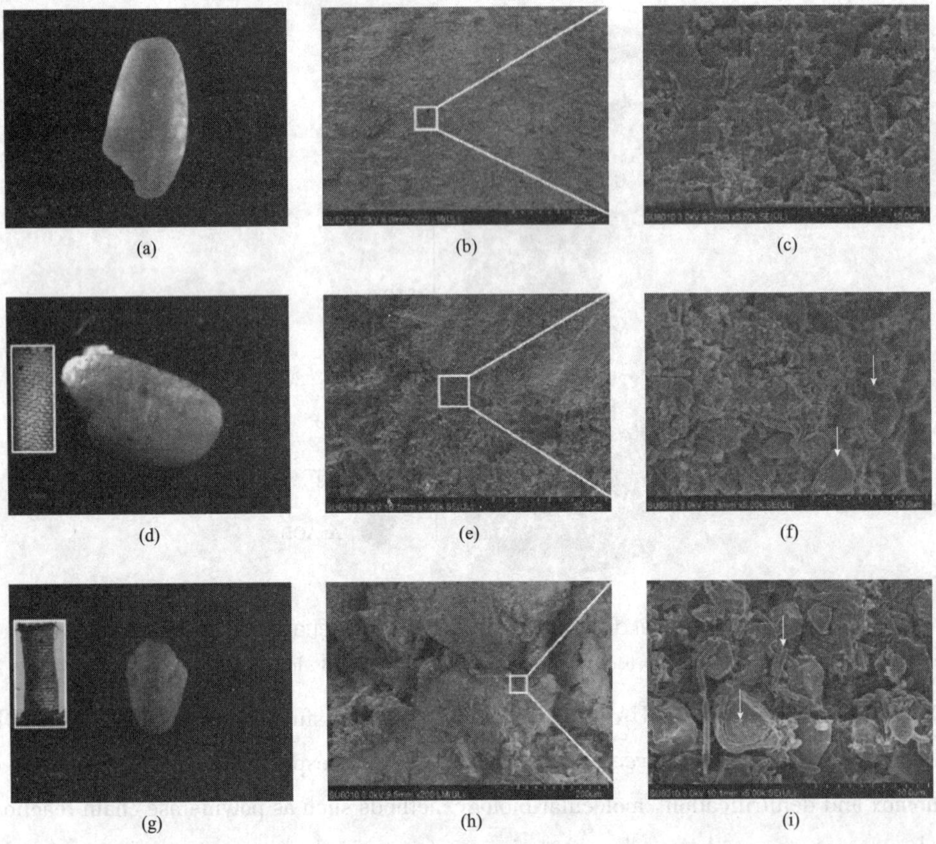

Figure 8-7 Surface properties of SCS (a, b, c: samples collected on day 20; d, e, f: samples collected on day 24; g, h, i: samples collected on day 28; The arrow pointed to starch grain and bacillus, respectively)

8.4 Emerging extensions of simultaneous Anammox-denitrification process

8.4.1 SAND

The simultaneous partial nitrification, Anammox and denitrification (SNAD) process couples ANAMMOX with denitrification in a single system to achieve a greater level of nitrogen removal by using the COD in the original feed.

$$NH_4^+ + 0.83O_2 \xrightarrow{AOR} 0.45NH_4^+ + 0.55NO_2^- + 1.98H^+ + 0.92H_2 \quad (8\text{-}5)$$

$$\&NH_4^+ + 1.31NO_2^- + 0.066HCO_3^- + 0.13H^+ \xrightarrow{Anammox} 1.02N_2 + 0.26NO_3^- + 0.066CH_2O_{0.5}N_{0.15} + 2.03H_2O \quad (8\text{-}6)$$

$$1.25CH_3COOH + 2NO_3^- \xrightarrow{HD_n} N_2 + 2.5CO_2 + 1.5H_2O + 2OH \quad (8\text{-}7)$$

SNAD process involves a cooperative and synergistic interaction among the functionally distinctive class of microbes i.e., AOB, Anammox bacteria, and denitrifiers, who perform their activities within the same reactor, as shown in Figure 8-8. Mechanism involves, first conversion of influent ammonium to nitrite via AOB [Eq. (8-5)], ANAMOX bacteria then convert the remaining ammonium and nitrite to N_2 gas and nitrate [Eq. (8-6)], and finally, the nitrate is converted to N_2 gas by the denitrifiers [Eq. (8-7)] to achieve close to 100% nitrogen removal efficiency. The AOB belongs to the genera of *Nitrosomonas*, *Nitrosovibrio*, *Nitrosococcus*, *Nitrosolobus*, and *Nitrosopira*, whereas NOB belongs to the genera *Nitrobacter*, *Nitrococcus*, *Nitrospira*, and *Nitrospina*. The Anammox bacteria belong to the *Candidatus* genera of phylum *Planctomycetes*. Whereas denitrifiers belong to *Denitratisoma*, *Haliangium* and *Thauera*.

Researchers developed the SNAD process under the limited DO concentration of 0.4–0.6 mg/L and C/N ratio of 1.2. The presence of low concentration of COD provided two major advantages: ① it enhanced the total nitrogen removal efficiency via denitrification, and ② it stimulated the growth of aerobic heterotrophs, which consumed the excess DO in bulk liquid and limited the growth of NOB. In another study, Wang et al.

It demonstrated that SNAD process has all the benefits of shortcut nitrogen removal process (CANON, OLAND) but exclusively has higher total nitrogen removal efficiency as given in Table 8-3 by investigating the efficacy of SNAD process under limited COD concentration (C/N ratio of 0.42–0.55). Due to the high-performance efficiency, SNAD process is effective for treatment of both side steams and domestic wastewater feeds. Existence of three independent processes within the same controlled niche theoretically saves the capital cost, whereas anoxic operation and partial nitrification reduces the aeration cost by 50%. Furthermore, the high efficiency it reduces the footprints of the targeted pollutants such as carbon, nitrogen, and phosphorus (Figure 8-8).

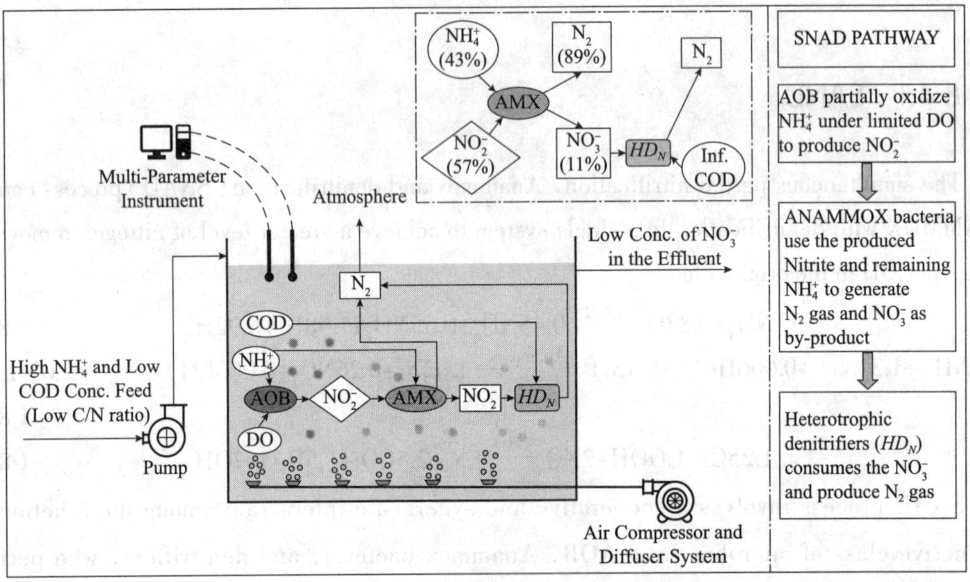

Figure 8-8 General schematic of single stage SNAD process and its stochiometric pathway

Table 8-3 Comparative assessment of different shortcut nitrogen removal processes

No.	Process	Stages	Application	Footprint Area	Process	Nitrogen removal Efficiency/%
1	SNAD	Single stage	Low C/N influent	Low	Yes	>90
2	CANON	Single stage	High NH_4^+ Side streams	Moderate	No	75 to 80
3	OLAND	Single stage	High NH_4^+ Side streams	Moderate	No	80 to 89
4	DEMON	Single stage	High NH_4^+ Side streams	Moderate	No	80 to 90
5	SHARON-ANAMMOX	Two stages	High NH_4^+ Side streams	High	No	90

The co-existence of nitritation, Anammox, and denitrification is complicated, however, several researchers have reported their successful linkage in laboratory- and pilot-scale systems. Studies have reported detailed operating conditions, significant experimental results, and important comments about the SNAD processes, as summarized in Table 8-4. In general, the SNAD process is performed in one reactor, i.e. a single-stage SNAD process. Table 8-4 shows that the forms of suspended sludge and biofilm reactors are adopted in most typical examples of the SNAD process. Because suspended sludge is widely used in full-scale wastewater treatment plants, it is of importance to study the SNAD processes operated with suspended sludge.

Table 8–4 SNAD process in different configurations/operating mode.

Configuration/operating mode	Operating conditions	Key experimental results	Remarks Reference
Lab–scale continuous–flow biofilm reactor (2 L) with 40% volume of suspended carriers	Anaerobic digester liquor of swine wastewater (ADLSW) with NH_4^+–N 140–460 mg/L, COD 120–380 mg/L; T 30 ± 1 ℃ and DO (0.3 ± 0.1) mg/L.	The average removal efficiencies of NH_4^+–N, total nitrogen (TN) and COD were 61.59%, 55.76% and 75.21% respectively.	By gradually increasing the influent NH_4^+–N concentration under low DO concentration, the feasibility of SNAD process could be achieved.
Lab–scale SBR (89.5 L) with K3 carriers operated in intermittent aeration (aeration/non–aeration) mode	Synthetic wastewater with NH_4^+–N 70–80 mg/L, COD 200–300 mg/L; T 30 ℃ and DO (0.7–3.5) mg/L.	① Nitrogen removal rate was 0.17–0.30 kg N/(m^3 · d); ② The average TN removal efficiencies was 87.5%–93.4%.	① Under the intermittent aeration mode, the increase of aeration time improved nitrogen removal rate; ② the low DO facilitated the AOB, Anammox and denitrification activities.
A pilot–scale Carrousel oxidation ditch (5.6 m^3) with suspended sludge and ribbon packing for biofilm	Real domestic wastewater with NH_4^+–N 16–20 mg/L, COD 90–120 mg/L; aeration/non–aeration time: 1 h/1 h, 2 h/1 h and 3 h/1 h.	The average nitrogen removal efficiencies were 64.7%, 73.2%, 76.1% and 67.6%, 77.5%, 78.8% (HRT 12 h, 15 h and 18.7 h) under aeration/non–aeration time 2 h/1 h and 3 h/1 h, respectively.	Both HRT (15 h) and aeration/non–aeration time (3 h/1 h) were important parameters for the biological treatment effect and effluent quality of the SNAD process.
Lab–scale SBR (10 L) with suspended activated sludge operated in anaerobic/aerobic/anoxic (A/O/A) mode	Real domestic wastewater with NH_4^+–N 60–80 mg/L, COD 120–270 mg/L, C/N 3–3.5; T (30 ± 1) ℃ and DO 0.2–0.4 mg/L.	The average TN and COD removal efficiency were 86.1% and 77% respectively; 89% of TN was removal by Anammox.	① The A/O/A operational mode with the ratio of anaerobic/anoxic time to aerobic time being 1/3 was an effective method to inhibit NOB. ② Micro–aeration could facilitate nitrogen removal mainly occurring in aerobic phase via the strong coupling of SNAD.
Lab–scale SBBR (Sequencing batch biofilm reactor, 10 L) with biofilm operated with real–time DO intelligent control	Synthetic wastewater with NH_4^+–N (2200 ± 20) and (50 ± 3) mg/L, COD 450 and 50 mg/L, C/N 0.2 and 1, DO 2.3–2.4 and 0.8–1.0 mg/L, respectively, in phase 1 and 6, and T (30 ± 1) ℃ in both phases.	The average effluent TN concentration was 32.9 mg/L and COD removal efficiency achieved 35%–44% in phase 1. The NRE was always above 90% in phase 6.	Forming biofilm or granules was a feasible way to improve biomass retention, and real–time DO intelligent control was suitable to SNAD system treating various influents in a range of 2200–50 mg N/L.
Lab–scale IFAS reactor (integrated fixed film activated sludge reactor, 2 L)	Synthetic wastewater with NH_4^+–N (400 ± 5) mg/L, COD 150 mg/L, C/N 0.37, DO (0.4 ± 0.1) mg/L and T (39 ± 1) ℃.	① The NH_4^+–N, NO_2^-–N and NO_3^-–N in the effluent were 15–120, 5–40 and 40–150 mg/L, respectively, resulting to 50%–83% TN removal efficiency. ② COD removal efficiency achieved 50%.	① The lower COD/N ratio could protect the activity of the Anammox bacteria. ② The conversion of ammonium into nitrite by AOB could occur mostly in the suspended sludge. ③ The Anammox bacteria could utilize ammonium and nitrite while the denitrifying bacteria utilize a carbon source and nitrate in the biofilm.

Continued

Configuration/operating mode	Operating conditions	Key experimental results	Remarks Reference
Full-scale MBBR (moving bed biofilm reactor, 832 m^3) with the combination of suspended sludge and biofilm	Sludge digester liquor pretreated by chemical precipitation with average NH_4^+-N 1550 mg/L, COD 658 mg/L, C/N 0.42, DO 0.2–0.4 mg/L and T 27–30 ℃.	① The total nitrogen concentration in the effluent was under 350 mg/L. ② The nitrogen removal was 68%–71%. ③ The COD removal efficiency was over 40%. ④ The SNAD process provided a removal capacity of about 530 kg N/d.	① AOB preferred to grow in suspended sludge. ② The denitrifying bacteria protected Anammox bacteria from the organic carbon or nitrite inhibition.
Pilot-scale SBR (89.5 L) with biofilm operated in sequencing intermittent aeration mode	Domestic wastewater with COD 150–250 mg/L, TIN 70–120 mg/L, NH_4^+-N 60–90 mg/L, C/N 1.2–3.5 and T 30 ℃. The average DO concentrations during aerobic period was 5.6 mg/L.	① The TIN removal efficiency varied between 80 and 90%. ② The average COD removal efficiency was 71%. ③ The effluent NH_4^+-N, NO_3^--N and TIN concentrations were 2, 7 and 11 mg/L respectively. ④ The average ΔNO_x^--N/ΔNH_4^+-N value was 0.10.	① The inhibition of DO on the Anammox bacteria could be alleviated by the biofilm and intermittent aerobic condition. ② Excessive aeration and influent organic matter loading should be avoided.
Lab-scale SBBR (10 L) operated in intermittent aeration mode of 8 h cycle sequentially	Landfill leachate with NH_4^+-N (1931 ± 47) mg/L, COD (1039 ± 45) mg/L, C/N 0.54, DO 2–2.4 mg/L and T 15 ℃.	① COD in the effluent was (331.0 ± 7.3) mg/L. ② TN removal efficiency fluctuated in the range of 82.1%–82.9%. ③ The average concentrations of NH_4^+-N, NO_2^--N and NO_3^--N were 165.4 mg/L, 20.8 mg/L and 164.6 mg/L respectively.	① The potential electron donors for denitrification could be served from the degradation of bio-refractory organics. ② The hydrolysis of macromolecular substance might play an important role in stimulating denitrification process.
Lab-scale NRBC reactor (non-woven rotating biological contactor, 2.5 L) with a hydraulic retention time of 5 h	Municipal sewage with COD 100 mg/L, NH_4^+-N 50 mg/L, C/N 2, DO 1.0 mg/L and T 35 ℃.	① The removal efficiencies of TN and COD were 70% and 64% respectively. ② The NO_2^--N and NO_3^--N in the effluent can be ignored.	① Starting-up partial nitrification, CANON and SNAD in turn could be an appropriate method. ② C/N and DO in the range of 1.4–2.3 and 0.2–0.8 mg/L could be optimum region for SNAD.
Lab-scale structured-bed reactor (consisting of 13 cylindrical structures of polyurethane foam, 6.1 L) with a hydraulic retention time of 24 h	Animal feed production effluent with COD (89 ± 40) mg/L, NH_4^+-N (231 ± 40) mg/L, C/N 0.28, DO 1.0 mg/L and T 30 ℃.	① The removal efficiency of TN was 48% ± 24%. ② The NH_4^+-N, NO_2^--N and NO_3^--N in the effluent were 48 ± 31, 37 ± 37 and 84 ± 56, respectively. ③ The average COD removal efficiency was 63%.	① Low aeration and recirculation applied in the structured-bed reactor to perform SNAD process has potentiality. ② Heterotrophic denitrifiers, AOB and Anammox bacteria coexisting with hydrolytic, SRB and methanogenic archaea in the system.
Lab-scale membrane bioreactor (MBR with submerged flat-sheet microfiltration membrane, 4.5 L) with a hydraulic retention time of 16 h	Synthetic wastewater with COD 100, 150 and 200 mg/L, NH_4^+-N 290 mg/L, C/N 0.33, 0.5 and 0.67, DO 0.1–0.3 mg/L and T 35 ℃.	① At C/N of 0.33, 0.5 and 0.67, the relative removal efficiencies of TN and COD were 88.7% and 96.2%, 92.5% and 97.9%, 86.4% and 87.2%, repectively. ② The biomass in the outer part of the SNAD granule was mostly AOB, while Anammox and denitrifiers were in the inner anoxic part.	The optimization of operating parameters, such as C/N, air flow rate, in the intermittent aeration MBR facilitates the efficient operation of the SNAD process.

Because biofilms, or attached growth systems, are effective for Anammox development even at slightly higher DO levels. The niche differentiation required for these aerobic and anaerobic processes can be achieved in different ways, such as biofilms on carriers or granules. In single-reactor SNAD biofilms, an outer layer of aerobic organisms consumes the available oxygen. This leaves the anoxic interior of the biofilm for Anammox. The high aggregate density serves to limit oxygen diffusion into the interior, allowing the proliferation of AnAOB and other Anaerobic bacteria. The AnAOB in a kernel layer can use NH_4^+ and produce NO_2^-. The surface coverage and O_2 consumption by AOB or denitrifiers protect the AnAOB from inhibition by O_2 and other environmental factors in the SNAD systems.

Moreover, when municipal wastewater is treated, different problems may remain, however, efficient performance is generally achieved treating influent with C/N ratios < 1.0 g/g. In many experimental cases, the optimal temperatures for SNAD processes have been controlled at 30 ℃; however, the temperatures of municipal wastewater consistently vary from 12 ℃ in summer to 32 ℃ in winter in different areas. Thus, in practical applications, temperatures may largely impact associated activity and community structures. The other major challenge for the SNAD system is the difficulty in suppressing NOB. Although real-time precision DO intelligent control, low DO concentration, free ammonia (FA) and free nitrous acid (FNA) were used in previous research, these strategies might be consistently affected by integrated and multiple factors. Understanding and analyzing the mechanisms involved in the SNAD process requires further study.

8.4.2 ADSF

Waste activated sludge (WAS) is an inexpensive, widely available carbon source, and can be generated on-site as a by-product of wastewater treatment at WWTPs. Recent studies have shown that readily biological degradable organic matter can be produced from WAS, providing electrons for denitrification. This removes the requirement for expensive electron donors, such as methanol, ethanol, or glucose in the SAD systems. Simultaneous Anammox, denitrification and sludge fermentation (ADSF), which stands for simultaneous Anammox, Denitrification and Sludge Fermentation, is a new proposed strategy for nitrogen removal. It relies on the interaction of above three groups of microorganisms and results in a generally complete conversion of NH_4^+ and NO_2^- (NO_2^- can be produced from partial nitritation) to dinitrogen gas along with sludge reutilization.

The proposal of the ADSF process is based on two scenarios. First, to provide electron donors for denitritation, sludge fermentation is introduced into the SAD process for the in-situ production of more readily accessible source of organics. Generally, the readily biodegradable organic substrates are mostly consumed after the aeration for partial nitritation. Second, to consume NH_4^+

released from the simultaneous sludge fermentation/denitritation system, Anammox is brought into the system. Figure 8–9 shows that sequential reactions are performed under anoxic conditions. A portion of the NO_2^- is used with NH_4^+ by AnAOB. After this, the produced NO_3^- as well as the remaining NO_2^- are reduced by denitrifiers and converted to dinitrogen gas. During this process, carbon sources required by denitrification are obtained from sludge fermentation, and the NH_4^+ released from sludge fermentation is removed through the Anammox process.

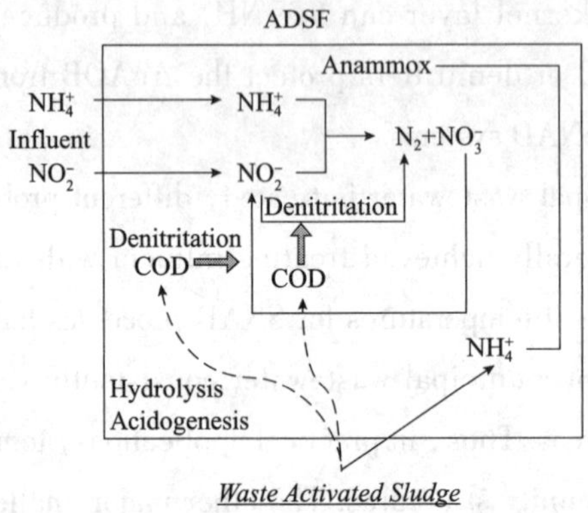

Figure 8–9 Flow diagram of ADSF processes

For municipal wastewater with low C/N ratios, the ADSF process has been shown to be feasible for treating the effluent pretreated by a partial nitritation process; the process can handle broad ratios of NO_2^- and NH_4^+ (2–10 g/g). This approach has achieved stable nitrogen removal, with a TIN of 9.7 mg/L in the effluent. ADSF showed comparable performance at ambient temperatures and at mesophilic conditions (e.g. 30 ℃), revealing its potential for practical application. Moreover, sludge reduction efficiencies of 38%–50% were realized due to sludge decay and the consumption of organics through denitrification. Syntrophic associations among Anammox, denitrifying, and fermentative microorganisms may have collectively facilitated their capabilities and growth. As previously shown in ADSF reactors, the dominance of functional genera *Thauera*, *Ca. Anammoximicrobium*, and *Pseudomonas* were, respectively, associated with denitrification, Anammox, and hydrolysis. This supports the theoretical relationship that the nitrogen was removed through denitrification and Anammox processes by using organic matter released from sludge degradation. Although the benefits of energy recovery may be lowered due to the adoption of acidogenesis instead of methanogenesis to reuse sludge in the ADSF process, the COD requirements and biomass production could be decreased by 30%–100% and 40%–91% respectively. Recently, similar TN removal and sludge reduction efficiencies of pilot-scale ADSF processes, compared to laboratory-scale processes, reflects the scaled-up effects (Figure 8–10). To date, however, the ADSF process is still considered a "black-box", and more investigations are needed to assess the microbial abundance, activity, and community structures.

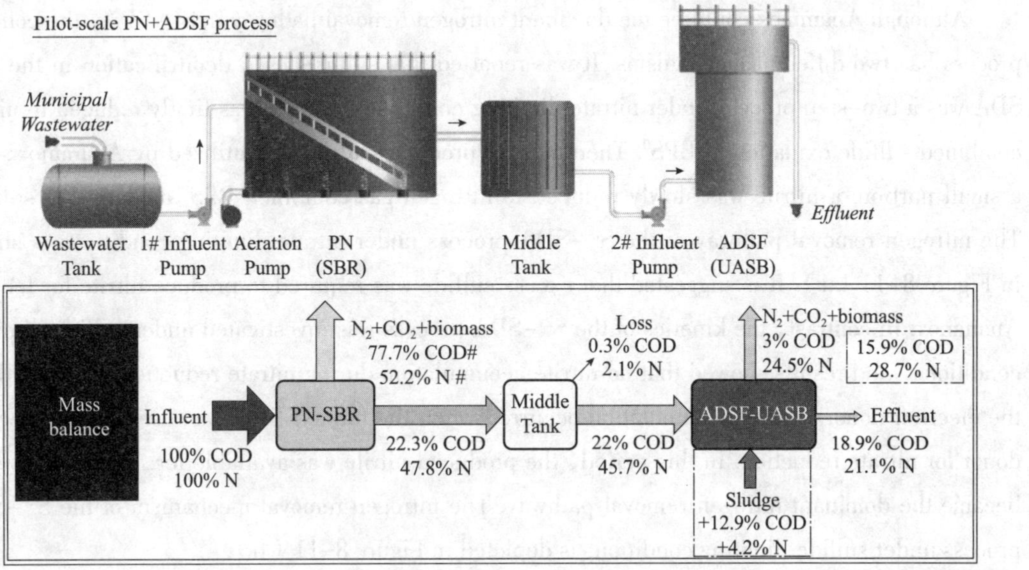

Figure 8–10 Schematic diagram of pilot-scale PN + ADSF process and the corresponding mass flows of COD and nitrogen

Note, # means the mass of COD and nitrogen taken away via wasting sludge from the PN-SBR is included. PN, partial nitritation; ADSF, simultaneous Anammox, denitrification and sludge fermentation.

8.4.3 SDA

It is researched that reduced sulfur compounds such as S^{2-}, S^0, and $S_2O_3^{2-}$ can serve as alternative electron donors for denitrification via sulfur-oxidising bacteria (SOB). This provides a new idea for mitigating the competition of SAD systems for nitrites.

The following analyzes the coupling sulfur-based denitrification with Anammox (SDA) process using sulfide-based denitrification as an example. Sulfide is the main product of sulfate respiration by sulfate-reducing organisms under anaerobic conditions, in which sulfate or other oxidised sulfur compounds are terminal electron acceptors, and hydrogen, ethanol, propionate, and lactate are electron donors. The produced sulfide is an efficient electron donor for denitrification. The sulfide-based denitrification comprises four bioreactions, as shown in Eqs. (8-8)-(8-11). The sulfide is initially oxidised to biologically produced elemental sulfur (BPS^0), which is further oxidised to sulfate. Stable nitrite accumulation in sulfide-based denitrification is a prerequisite for the implementation of the sulfide-based denitrification and Anammox (S^{2-}-SDA) process.

$$NO_3^- + S^{2-} + 2H^+ \longrightarrow S^0 + NO_2^- + H_2O \tag{8-8}$$

$$NO_2^- + 1.5S^{2-} + 4H^+ \longrightarrow 1.5S^0 + 0.5N_2 + 2H_2O \tag{8-9}$$

$$S^0 + 3NO_3^- + H_2O \longrightarrow SO_4^{2-} + 3NO_2^- + 2H^+ \tag{8-10}$$

$$S^0 + 2NO_2^- \longrightarrow SO_4^{2-} + N_2 \tag{8-11}$$

Although Anammox could be the dominant nitrogen removal pathway in S^{2-}–SDA, the coupled process has two different mechanisms. It was reported that sulfur–based denitrification in the S^{2-}–SDA was a two–step process under nitrate–limiting conditions. Nitrate was firstly reduced to nitrite combined sulfide oxidation to BPS^0. Then most of produced nitrite was utilized by Anammox, and a small portion of nitrite was slowly reduced to nitrogen gas combined BPS^0 oxidation to sulfate. The nitrogen removal pathway of the S^{2-}–SDA process under nitrate–limiting conditions is shown in Figure 8–11 (a). It is suggested that excess sulfide was required to produce nitrite for feeding Anammox. In contrast, the kinetics of the S^{2-}–SDA process was investigated under sulfide–limiting conditions, and results showed that no nitrite accumulation during nitrate reduction with sulfide as the electron donor. Nitrite accumulation occurred when the formed BPS^0 was used as the electron donor for nitrate reduction. In this period, the produced nitrite was available for Anammox, which became the dominant nitrogen removal pathway. The nitrogen removal mechanism of the S^{2-}–SDA process under sulfide–limiting conditions is depicted in Figure 8–11 (b).

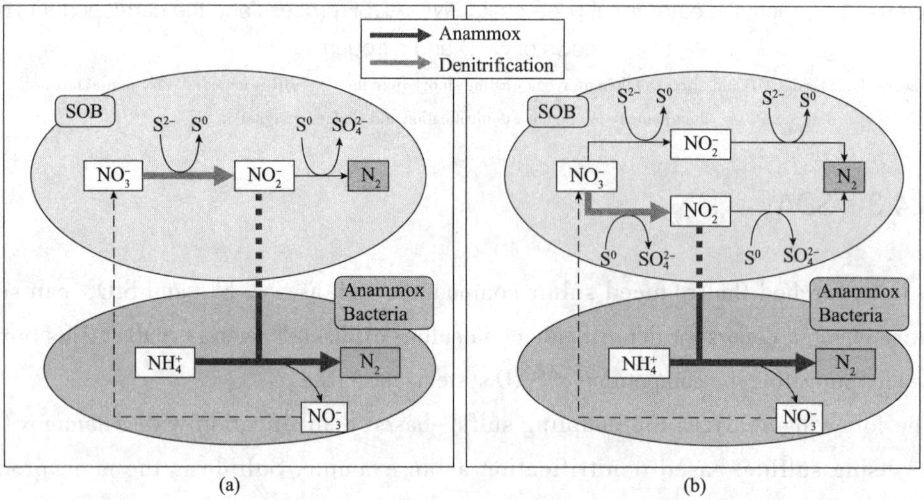

Figure 8–11 Nitrogen removal pathways in the S^{2-}–SDA process under (a) nitrate–limiting conditions and (b) sulfide–limiting conditions

The application of SDA for nitrogen removal from wastewater has several advantages. First, the combined system can alleviate the inhibitory effects of reduced sulfur compounds (i.e. S^{2-} and SCN^-) on Anammox. Though sulfide is a well–known inhibitor of various microorganisms including Anammox bacteria, Anammox bacteria have been found to survive in reactors with a continuous influx of sulfide, and they even dominated the nitrogen removal pathway in the SDA system. It is indicated that sulfide toxicity to Anammox bacteria can be mitigated in the S^{2-}–SDA process. Sulfide can be detoxified by SOB, which creates a suitable environment for Anammox bacteria. S^{2-}–SDA could convert sulfide into nontoxic substances (i.e. BPS^0 and sulfate), which can be applied for nitrogen removal from industrial wastewater without an external sulfur source. In addition, BPS^0 is an intermediate of sulfide oxidation, which is nontoxic, noncorrosive, and contains a high sulfur concentration, could be recovered as a mineral source in S^{2-}–SDA. The recovered BPS^0 can be used as industrial raw material, and it is an efficient donor for denitrification. BPS^0 can also be used to produce sulfide for heavy metal removal.

Despite the development of the SDA process, several challenges still limit its further implementation. The fluctuation in substrate concentration may cause sulfide or thiocyanate accumulation in the reactor, thereby inhibiting Anammox activity. Elemental sulfur accumulation may reduce biomass retention efficiency and the sulfide removal rate. Accumulated BPS^0 can also cause pipe blockage and secondary pollution. Meanwhile, accumulated BPS^0 in SDA reactors can hamper the mass transfer of substrates, thereby limiting the denitrification and Anammox activities. Moreover, the high sulfate production in the SDA process is undesirable. Sulfate is generally considered nontoxic, but discharged sulfate into freshwater is likely biologically reduced to toxic sulfide, which could harm the ecosystem. When the effluent of the SDA process is discharged into freshwater, the sulfate concentration should be controlled, and the sulfate discharge limits in industrial wastewater treatment typically vary from 250 to 1000 mg/L. Additionally, the low solubility of elemental sulfur limits the application of the SDA process. Elemental sulfur is particularly hydrophobic, with a solubility as low as 5 μg/L at 20 ℃, and the dissolution rate of solid sulfur is the limiting step in sulfur–based autotrophic denitrification. Further research is needed to explore the relationship between the low solubility of elemental sulfur and the efficient nitrogen removal in S^0–SDA.

Questions

[1] How does SAD achieve nitrogen removal? Write out the corresponding equations and illustrate it.

[2] What are the start–up methods of the reactor? What factors limit the start–up rate of a reactor? What are the commonalities and differences between the two methods?

[3] What are the factors that affect the efficiency of SAD system? How they work?

[4] Please identify the main microorganisms involved in the SAD process and analyze their community structure and mechanism of action.

[5] What are the new denitrification processes based on SAD? What are the advantages and disadvantages of each?

References

[1] Huang X X, Chen C J, Zhang R, et al. Research progress in Anammox–denitrification coupling progress [J]. Chinese Journal of Applied Ecology, 2012, 23 (3): 849–56.

[2] Kumar M, Lin J G. Co–existence of Anammox and denitrification for simultaneous nitrogen and carbon

removal—strategies and issues [J] . J Hazard Mater, 2010, 178 (1–3) : 1–9.

[3] Wang B, Gong X F, Peng Y Z. Simultaneous Anammox–denitrification process and its emerging extensions [J] . Chemical Engineering Journal, 2021, 415 : 128380.

[4] Zuo F M, Yue W H, Gui S L, et al. Resilience of Anammox application from sidestream to mainstream : A combined system coupling denitrification, partial nitritation and partial denitrification with Anammox [J] . Bioresour Technol, 2023, 374 : 128783.

[5] Jian J X, Liao X J, Li S P, et al. Nitrogen removal performance and sludge characteristics of wastewater from industrial recirculating aquaculture systems via Anammox coupled with denitrification [J] . Journal of Water Process Engineering, 2022, 49 : 103092.

[6] Wang L M, Hong Y G, Gu J D, et al. Influence of critical factors on nitrogen removal contribution by Anammox and denitrification in an Anammox–inoculated wastewater treatment system [J] . Journal of Water Process Engineering, 2021, 40 : 101868.

[7] Wang Y Q, Yu X L, Chen C J, et al. Research progress in influence of organic matter on anaerobic ammonia oxidation reaction [J] . Chemistry Bulletin, 2017, 80 (2) : 6.

[8] Chen J B, Hai Y, Zhang W, et al. Insights into deterioration and reactivation of a mainstream Anammox biofilm reactor response to C/N ratio [J] . Journal of Environmental Management, 2022, 320 : 115780.

[9] Li J, Qiang Z M, Yu D S, et al. Performance and microbial community of simultaneous Anammox and denitrification (SAD) process in a sequencing batch reactor [J] . Bioresour Technol, 2016, 218 : 1064–72.

[10] Singh V, Ormeci B, Mishra S, et al. Simultaneous partial Nitrification, ANAMMOX and denitrification (SNAD) – A review of critical operating parameters and reactor configurations [J] . Chemical Engineering Journal, 2022, 433 : 133677.

[11] Xu D D, Ying S Y, Wang Y H, et al. A novel SAD process : match of Anammox and denitrification [J] . Water Res, 2021, 193 : 116874.

[12] Deng Y F, Zan F X, Huang H, et al. Coupling sulfur–based denitrification with Anammox for effective and stable nitrogen removal : a review [J] . Water Res, 2022, 224 : 119051.

Chapter 9 Anammox Granular sludge

Emphasis: Characteristics, Influencing Factors, Limitations, Cultivation, and Applications of AnGS; Composition, characterization and role of EPS in the formation and operation of Anammox granular sludge; Practical engineering application of Anammox granular sludge.

Nut: Description of AnGS.

The long start-up time, susceptibility to excess substrate concentrations and other substances (sulphide) leading to inhibition and some operational problems of the Anammox process limits the extensive application of the technology thereby necessitating the need for more robust and efficient approaches. Granulation of Anammox biomass and reactor modifications geared towards complete biomass retention are some of the means employed by scientists to resolve the aforementioned problem.

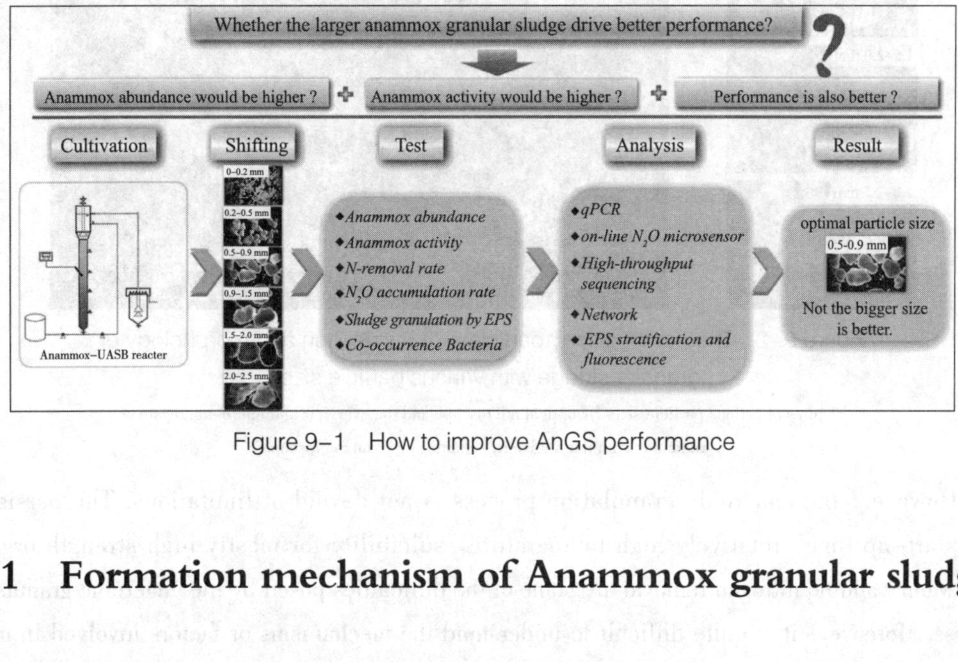

Figure 9-1 How to improve AnGS performance

9.1 Formation mechanism of Anammox granular sludge

Generally, granular sludges observed in bioreactors are reported to have high densities and regular outer shape. Comparatively, granules are denser and have a stronger microbial structure than flocs. Also, due to their high settling velocities, high hydraulic loads can be employed without consequential washout of the biomass. The granular sludge consists of microbial cells, extracellular polymeric substances (EPS), and perforated inorganic substances among others. The process of Anammox granular sludge formation is as illustrated in Figure 9–1.

Depending on the nature of the electron acceptors involved in the microbial metabolism, granular sludge could be classified as either aerobic or anaerobic. Anaerobic granules are mostly black and consist of methanogenic, syntrophic, acetogenic, and various hydrolytically fermentative bacteria (Figure 9–2).

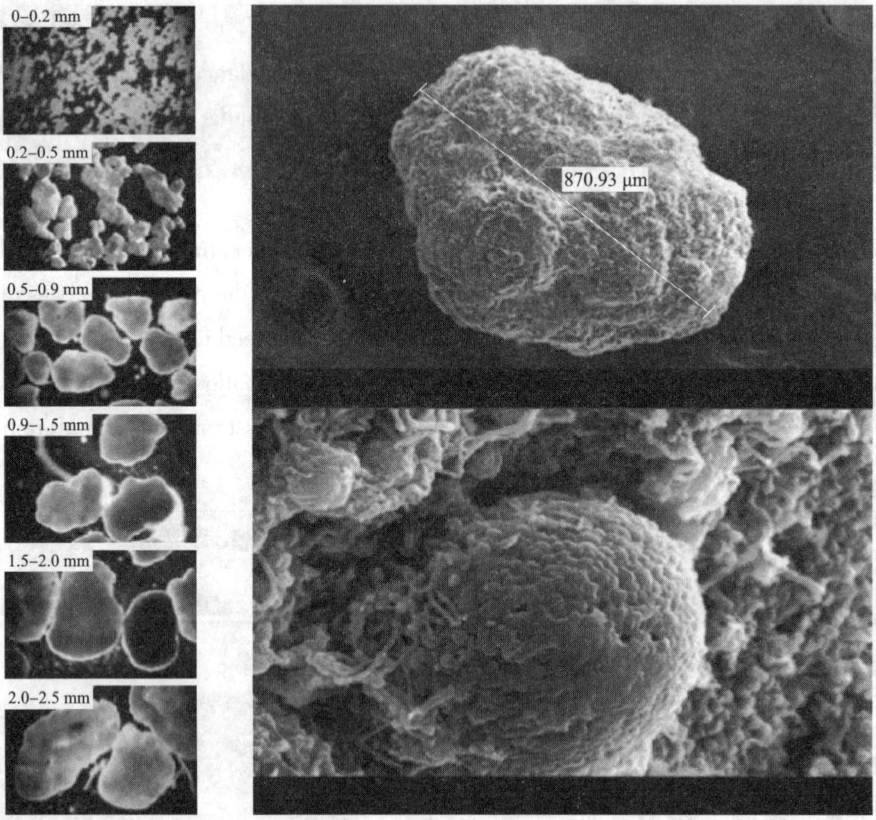

Figure 9–2 The composition, particle size distribution and morphology of granular sludge with various particle sizes

Pictures of various particle sizes by optical microscope and the overview and micro–aggregate of Anammox granular sludge by scanning electron microscope.

However, the anaerobic granulation process is not devoid of limitations. The persistent long start–up time, relatively high temperature, suitability for mostly high strength organic wastewater, and no nitrogen removal are some of the difficulties posed by the anaerobic granulation process. Moreover, it is quite difficult to understand the mechanisms or factors involved in niche differentiation in granular systems due to the aggregation of the biomass.

Anammox granular sludge (AnGS) – an anaerobic granular sludge, has the advantages of complete biomass retention and very low biomass yield. As mentioned earlier, Anammox bacteria have the capability of existing in the reactor as either biofilm, microbial aggregate, granules, or in entrapped forms such as gel carriers. However, contrary to the need for carriers in biofilm or gel entrapment systems, Anammox granular systems ensure complete biomass retention without the need for carriers.

The formation of Anammox granules involves different physicochemical and biological interactions. To help explain the granulation process, several models have been developed over the years- selection pressure model, extracellular polymers (ECP) bonding model, inert nuclei model, synthetic and natural polymer-bonding model, attrition model, secondary minimum adhesion model, multivalence positive ion-bonding model and surface tension model. Regardless of the state of Anammox activity (high or low), the selection pressure, multivalence ion, extracellular polymers (ECP), etc still aid in the granulation of Anammox biomass in the seed sludge.

The concentration of heme c is one factor that plays a vital role in the characterization and activity of Anammox granular sludge. Due to the presence of heme c, the mature active Anammox granules are mostly red. Moreover, heme c is an important constituent of the enzymes hydroxylamine oxidoreductase and hydrazine oxidoreductase-both of which play critical roles in the Anammox process (Figure 9-3).

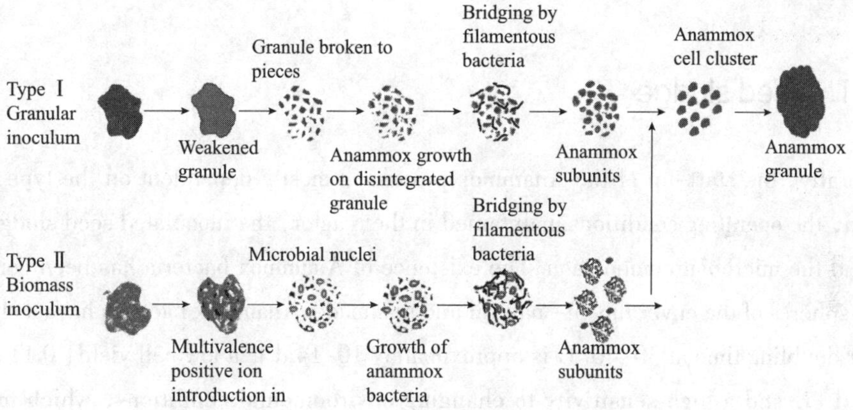

Figure 9-3 Mechanism of Anammox granulation

9.2 Factors influencing the development of Anammox granules

The granulation of the Anammox biomass just like the conventional Anammox process is greatly influenced by factors such as the substrate concentration, temperature, and pH, seed sludges. hydraulic retention time (HRT) and sludge retention time (SRT). There are however some factors whose influence are more pronounced in the aggregation of the Anammox biomass-extracellular polymeric substances (EPS), hydrodynamic shear force and stirring speed, presence of inorganic ions, and influence of nanoparticles present in wastewaters (Figure 9-4).

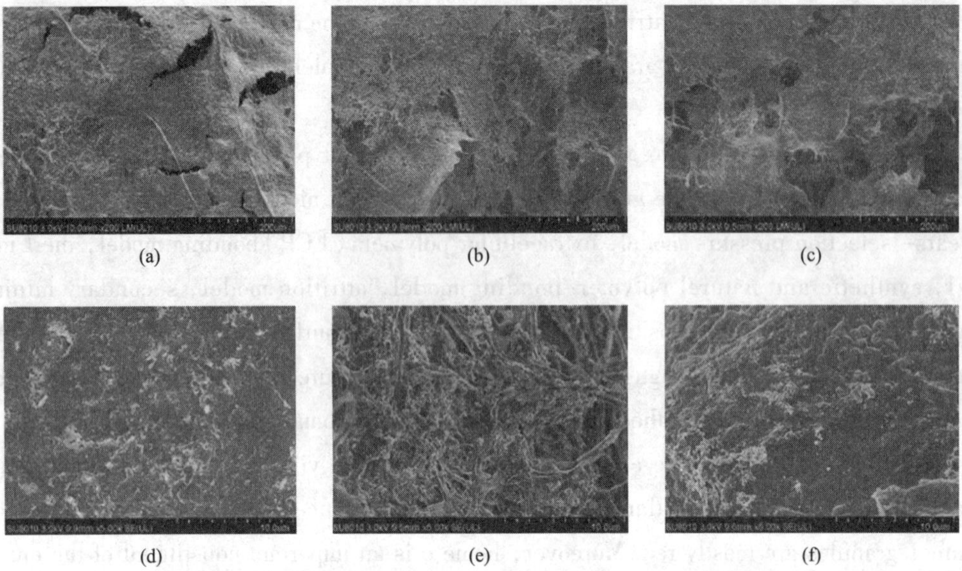

Figure 9-4 Surface microsurface of AnGS observed by SEM

a, b, c represented black, brown and red AnGS, respectively ×200;

d, e, f were the magnifying view of a, b, c ×5000

9.2.1 Seed sludge

Generally, the start-up of the Anammox process is mostly dependent on the type of reactor being used, the operating conditions maintained in the reactor, the inoculated seed sludge reaction ability, and the microbial composition. The existence of Anammox bacteria has been confirmed in almost all spheres of the environment- natural and manmade. Anammox bacteria have a slow growth rate (their doubling time at 30-40 ℃ is approximately 10-14 d), a low cell yield [0.11 g NH_4^+-N/(g VSS · d)], and a high sensitivity to changing environmental conditions, which make them extremely difficult to cultivate.

As such, extensive research has been carried out with some scientists varying the type of seed sludge (nitrifying sludge, aerobic granular sludge, leachate sludge, activated sludge, denitrifying sludge, anaerobic granular sludge and the sediments from freshwater and marine environments) used in the quest to optimize the performance of the Anammox process.

For instance, after experimenting with aerobic and anaerobic seed sludges in batch culture, it was concluded that the anaerobic seed sludge was capable of producing stable Anammox population and usually takes about 3-4 months. Due to the low level of activity of the Anammox bacteria, it often takes a long time for the Anammox biomass to aggregate. This is evidenced by the competition between the coexisting bacteria when using aerobic, anaerobic, or mixed activated seed sludge thereby suppressing the Anammox start-up. A number of granular Anammox reactors fed with either aerobic or anaerobic granular seed resulted in a faster Anammox start-up, as illustrated in Table 9-1.

Table 9–1 Anammox granular sludge performance

Inoculum	Type of Reactor	HRT	Working volume/L	Start-up period/d	NLR or NRR/ [kg N/(m^3·d)]
Nitrifying & Anammox sludge	UASB	19.4h	70000	800	9.5(NLR)
Anammox granules	SBR	1 d	1	180	0.3(NLR)
Methanogenic granules	UASB	—	200	153	6(NRR)
Inactive methanogenic granules	UASB	1.5–1.7 d	2.1	140	0.006(NLR)
Anammox sludge	SBR	0.25 d	1	140	1.0(NRR)
Anammox sludge	AnMBR	1 d	4	102	0.70(NRR)
Activated Sludge	UASB	0.28–0.12h	8	730	2.28–5.72(NRR)
Methanogenic granules	UASB	7.92h	1.65	65	52.6(NRR)
Activated sludge	SBR	3.6–1.9 d	20	60	1.06(NRR)
Methanogenic granules	UASB	0.16h	1.1	52	76(NRR)
Mature Anammox granules	UASB	<1 d	50	14	1.03(NLR)
Anaerobic granules	EGSB	1.5h	1	—	18.65(NRR)
Anaerobic granules	UASB	0.16h	1.1	214	74.3–76.7(NRR)
Anammox granules	EGSB	6–0.3h	1	30	25.68(NRR)

UASB–Upflow Anaerobic Sludge Blanket, MBR–membrane bioreactor, AnMBR–Anaerobic Membrane Bioreactor, SBR–Sequencing Batch Reactor, EGSB–Expanded Granular Sludge Bed Reactor, HRT–Hydraulic Retention Time.

9.2.2 Substrate concentration

Though they serve as substrates in the Anammox process, ammonium (NH_4^+), nitrite (NO_2^-) and even nitrates (NO_3^-) and free ammonia can sometimes also act as inhibitors. According to Sanchez Guillén's research, both the granular integrity and the efficiency of low nitrogen removal were negatively affected by the substrate in terms of nitrogen sludge loading rate (NSLR). Thus, ammonium and nitrite mostly act antagonistically to the Anammox process when the threshold limits are exceeded. The threshold limits are however not constant and mostly depend on the strategies adopted during the process. Table 9–2 summarizes the influence of these substrates on the Anammox process.

The prevalent microbial structure and community in the reactor are also greatly influenced by the substrate concentration. The substrate supply, thus, nitrogen loading rate (NLR) greatly influences the structure of the microbial community. Thus, a high substrate concentration positively influences the dominance of fast-growing anaerobic ammonium oxidizing organisms (AnAOB) thereby resulting in the high rate removal of nitrogen in Anammox setups.

Table 9-2 Substrate influence on the Anammox process

Compound	Concentration	Effects
Ammonium	1 g N/L	No effect
	90 mg NH_4^+-N/L	No effect
	55 mmol	50% inhibition
Nitrite	60 mg NO_2^--N/L(spiked)	No effect
	25 mmol	50% inhibition
	384–400 mg/L	50% inhibition
	100 mg NO_2^--N/L	Complete inhibition
	500 mg/L	Complete inhibition
	<1 μmol	Irreversible inhibition
	>10 mmol	Reversible inhibition
Nitrate	45 mmol	50% inhibition
Free Ammonia(FA)	38 mg/L	50% inhibition

9.2.3 Hydraulic Retention Time(HRT)and Sludge Retention Time(SRT)

The hydraulic retention time(HRT)and sludge retention time(SRT)are another set of factors that influence the aggregation of Anammox biomass. SRT plays an important role in the aggregation of Anammox biomass. It is favorable to maintain a longer SRT in anaerobic membrane bioreactors (AnMBR) systems.

According to Nicolella, granular sludge-based reactors are compact reactors that combine short HRT and long stable solids retention time. This is not surprising since obtaining the highest removal rate with minimum HRT is one of the main objectives of the Anammox process. In contrast, some researchers have reported that a longer SRT leads to less aggregation.

The optimum HRT for the enrichment of Anammox in an anaerobic membrane bioreactor (AnMBR) according to Suneethi & Joseph is 1.5 days. According to another study by Zhang et al, it took 1.8 h to achieve Anammox granulation. Tang et al. also achieved a nitrogen removal rate (NRR) of 76.7 kg N/(m^3 · d) at an HRT of just 0.1 h. Thus, maintaining a longer SRT and minimum HRT leads to Anammox granulation resulting in high removal rates.

9.2.4 Temperature and pH

According to scientists, the optimum temperature for the Anammox process ranges from 20–40 ℃. As such most granular Anammox reactors also operate within the same range. However, irreversible destruction of the Anammox activity ensues when the temperature is kept above 45 ℃.

At 37 ℃, a nitrogen removal rate (NRR) of 11.5 kg N/(m^3 · d)was achieved while 8.1 kg

N/(m^3 · d) was achieved at 20–22 ℃. According to An et al., decreasing temperature negatively influences specific Anammox activities. For example, in an SBR operating at 35–46 ℃, a removal rate of 0.25 g N/(L · d) was achieved whereas slow growth represented by 0.05 g N/(L · d) removal rate was recorded when another SBR was operated at 15 ℃. Some other scientists have however reported Anammox activity at lower temperatures.

Though there is evidence of stable Anammox reactor operation at a pH of 9.3, the optimum pH range for Anammox biomass in granular form is reported to be between 6.7–8.3. This is similar to values (7–8) reported by Tomaszewski et al. For instance, it was observed that a high pH corresponded with an increase in the concentration of free ammonia whereas decreasing the pH resulted in the accumulation of free nitrous acid in the setup.

To a large extent, the dominant Anammox bacteria in the Anammox process is determined by the prevailing pH. In Anammox systems operating between pH range of 7.8–8, *Candidatus Brocadia Anammoxidans* and *Kuenenia stuttgartiensis* were identified as the predominant species whereas at pH 6.8–7 *Brocadia Anammoxidans* and *Candidatus Anammoxoglobus propionicus* were the dominant species in the enrichment from aerobic granules and leachate sludge respectively. Thus, for efficient nitrogen removal, operating parameters such as temperature and pH need to be controlled at appropriate levels.

9.2.5 Hydraulic shear force and stirring speed

This is the physical movement resulting from mechanical stirring to aid in the initiation of bacterium–to–bacterium contact. Shear force is very key in producing a compact and stable granular structure. According to Reino and Carrera, liquid–induced shear forces greatly influence the granulation process of biomass. While a 0.8 m/h superficial upflow liquid velocity resulted in the development of Anammox granules, a higher superficial value of 1.3 m/h results in the flotation of the Anammox granules. Thus, the formation of granular sludge is greatly influenced by hydrodynamic shear force together with an interactive flow pattern. A sharp increase in the shear force as a result of the overproduction of proteins may result in the destruction of the granules.

Owing to the irregular movement of microbial aggregates in completely mixed reactors, irregularly shaped flocs are formed as observed in activated sludge process whereas, in column–type reactors, microbial aggregates move homogeneously in circular patterns along with the height of the reactor thereby producing regularly–shaped granules. A higher height to diameter (H/D) ratio results in a long flowing trajectory which is a requirement for effective granulation. This is evidenced by the experiment carried out by Tang et al. where a higher height to diameter (H/D) ratio of 13.6 used in an Anammox granular sludge bed reactor (AGSB) resulted in enhanced aggregation and accumulation of nitrogen gas. With a granule size of 0.60–0.78 mm and a stirring speed of 40–80 rpm, an SAA of 0.32 kg N/(kg VSS · d) which is similar to data reported by other studies was achieved.

According to a study, a stirring speed of 150–200 rpm had no significant effect on the size of Anammox granules whereas a 50 rpm stirring speed resulted in a stable granule formation of size 63–896 μm. Higher shear force, in terms of superficial upflow air velocity greater than 1.2 cm/s seems to stimulate the production of cell polysaccharides which in turn results in a balanced stable aerobic/anaerobic granular structure. Thus, suitable shearing stress created by upflow liquid flow or stirring mechanism creates a favorable environment for the granulation of Anammox biomass.

9.2.6 Presence of inorganic ions

Researchers have suggested that bacterial adhesion and adsorption to inert matters, inorganic precipitates, and to each other through physicochemical interactions and syntrophic associations might be responsible for initiating the granulation process, thereby serving as microbial nuclei (precursors) for new microbial growth. The microbial nuclei thus formed is as a result of the binding of bacterial cells with positive divalent ions such as Ca^{2+}, Mg^{2+}, and Fe^{2+}.

According to Rudd et al., the presence of calcium enhances granule formation with precipitates of calcium and magnesium reported to act as nucleation seeds in Anammox granulation. However, according to Yu et al, a calcium concentration of 60 mg/L resulted in relatively small-sized granules. Also, the precipitation of calcium phosphate resulted in the loss of Anammox biomass activity. The high salinity shock of the Anammox granules was alleviated with the addition of a calcium concentration of 152 mg/L resulting in the formation of high strength but small-sized granules. Also, the addition of 5 g/L NaCl, Anammox granules were successfully developed from flocculent Anammox biomass.

Hence, though the positive divalent ions aids in binding with the negatively charged bacterial surface thus promoting aggregation, however, it has a negative effect on the granular size.

9.2.7 Extracellular polymeric substances (EPS)

Also termed as "Extracellular polysaccharides", "Exopolymers", "Exopolysaccharides" or "Extra Cellular Polymers" (ECP), extracellular polymeric substances (EPS) are sticky materials consisting of polysaccharides, proteins, nucleic acids, lipids secreted by microbial cells and other macromolecules. According to Laspidou and Rittmann, sludge liquor consists of inert biomasses, EPS, and soluble microbial products (SMP). In bacterial cells, the SMP and EPS are closely related. Whereas EPS which are reported to be essential for the structural and functional integrity of cells in microbial aggregates are made up of insoluble materials such as sheaths, capsular polymers, condensed gel, attached organic material and loosely bound polymers, SMP, on the other hand, are considered to be soluble EPS with soluble macromolecules, colloids and slimes as some examples. The production of EPS facilitates the formation of biofilm. This is because EPS has a tendency to mediate the formation of dense Anammox biofilms. Many functions are assigned to

EPS, including functions related to structure and stability like promoting aggregation, maintaining the physical structure of granules, retaining water, and serving as a protective barrier for the cells. By acting as adhesive reagents and protecting the granules from toxic substances, EPS is able to maintain the stability of the Anammox granular sludge (AnGS). However, the production of EPS can interfere with the characterization of Anammox cell biology. In addition, unstabilized EPS are capable of causing impairment of the stability of the Anammox process.

Polysaccharides which are a major component of extracted EPS are reported to play important roles in the development of Anammox granules. A change in the microstructure of EPS greatly influences the macroscopic rheological property which in turn determines the mechanical properties of AnGS. During the excretion of intracellular polymers, molecules such as proteins, lipids, and nucleic acids are produced with polysaccharides produced only for some specific function.

In a one reactor autotrophic system, about 50% of the autotrophic space is occupied by EPS within the granules. The ratio of polysaccharides (PS) to protein in a produced EPS greatly influences biogranulation. The protein to polysaccharide ratio (PN/PS) is reported to also greatly influence the strength and settling ability of granules. The PN/PS ratio of aerobic or anaerobic or nitrifying granules is usually greater than 2. After reporting a PN/PS ratio of $\leqslant 1$ for denitrifying ad Anammox granules Ni & Yang concluded that polysaccharides play significant roles in the growth of Anammox granules.

On the contrary, there have been reports of PN/PS ratios >2 which were equally successful in the production of Anammox granules. For example, the reported PN/PS ratios of floating granules and settling granules were 4.15 and 2.57 respectively. Accordingly, a higher ratio of 15.3 resulted in a weaker settleability (57.8–63.5 m/h). Higher PN/PS ratio leads to low strength and weaker settleability of granules.

In summary, the PN and PS present in the produced EPS decide the fate of settling velocity of Anammox granules, hence the determinant factors in flotation.

9.2.8 Influence of nanoparticles present in wastewater

Nanotechnology has gained lots of attention over the years leading to its application in diverse fields of which wastewater treatment is not an exception. According to Gottschalk and Nowack, nanoparticles gain entry into wastewaters through products such as cosmetics, soaps, and paints. However, the impact of these nanoparticles on wastewater treatment processes and the Anammox process to be specific is very poorly understood. Generally, the application of nanoparticles is known to have both stimulating and inhibiting effects on the microbial activity of different microbes.

During the long-term addition of up to 200 mg/L of maghemite nanoparticles (MHNPs), an increase was observed in the SAA and EPS content of the Anammox bacteria. On the other hand, exposure of Anammox granules to copper nanoparticles (CuNPs) at 5 mg/g SS resulted in significant inhibition of the activity of Anammox bacteria and the release of lactate dehydrogenase (LDH) as well as the accumulation of N_2H_4 outside of the Anammox cells. Thus, CuNPs are toxic to Anammox bacteria. However, this toxicity can be remedied by the addition of ethylene diamine–tetra acetic acid (EDTA).

In an experiment to determine the toxicity effect of various nanoparticles (CuNPs, CuONPs, ZnONPs, and AgNPs) on Anammox bacteria both in the floc and granule form, it was concluded that the Anammox granules, unlike the floc, were protected against the toxic effect of the nanoparticles due to their multilayered structure.

9.3 Structure and microecology of Anammox granular sludge

As with any other process, different operating conditions lead to the formation of diverse granular sludge with regards to form. Some of the factors that greatly influence the properties of Anammox granular sludge include nitrogen loading rate (NLR), the shape of the reactor, hydrodynamic shear force, etc. Though enough has been reported on the characteristics of heterotrophic aerobic granules, anaerobic granules, denitrifying granules and hydrogen–producing granules, relatively little has been reported on Anammox granules with settling property, density, diameter, color, morphology and specific Anammox activity (SAA) being the chief characteristics investigated.

9.3.1 Properties of the Anammox granular sludge

9.3.1.1 Diameter

The diameter of the Anammox granules is indicative of the performance of the Anammox reactor and greatly influences the settleability of the Anammox granules. Generally, Anammox granules could be classified as either settling or floating with the average diameter of the settling granules [(2.96 ± 0.99) mm] reported to be smaller than that of floating granules [(4.58 ± 1.22) mm]. There is a positive correlation between settleability and the overall performance of the reactor. Thus, the higher the settleability of the granules, the better the performance of the reactor. Large Anammox granules are relatively less affected by nitrogen concentration as compared with smaller Anammox biofilms. A diameter of 1.75–4.00 mm was reported to be the optimum diameter for Anammox granules for the efficient operation of the Anammox process based on their settling models. Also, to prevent flotation of the granules, it is advised to always keep the diameter of the Anammox granules under 2.20 mm. However, the diameter of Anammox granules is inversely proportional to its density. Hence, increasing the diameter leads to a decrease in the density of the Anammox granules which may cause flotation of the biomass (Figure 9–5, Figure 9–6). This is illustrated in Table 9–3 where a larger granule diameter results in overall poor reactor performance. In addition, the shape of the reactor, hydrodynamic shear force, and NLR greatly influence the diameter of AnGS.

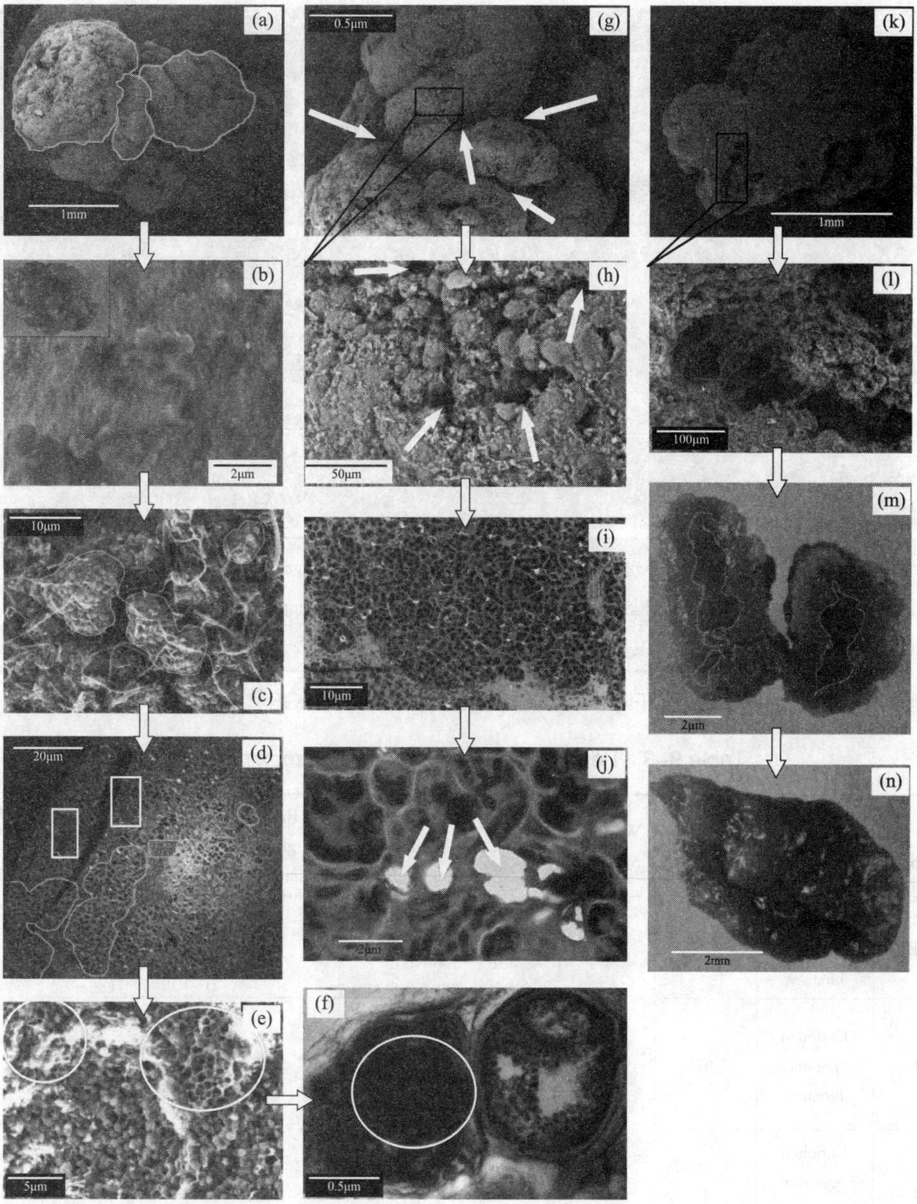

Figure 9-5 Structures of Anammox granules DMP (b, m-n), SEM (a, c, e-h, k and l), TEM (d, f, i-j)

(a) One Anammox granule composed of subunits (marked with closed curve) and ditches (indicated with arrows) acted as the boundaries between subunits ; (b) An Anammox granule (on the top left corner) broken into several subunits between two glass slides at a pressure of 40.5 Pa ; (c) Microbial cell clusters (marked with closed curve) connected by the filamentous bacteria ; (d) Microbial cell clusters (marked with closed curve) separated by filamentous bacteria-EPS bands (indicated with rectangles) ; (e) The honeycomb-like structures (indicated with circles) without microbial cells ; (f) Anammox bacteria-like cells with an anammoxosome-like structure (marked with a circle) ; (g) Ditches (indicated with yellow arrows) between subunits in the exterior of granules ; (h) Interstitial voids (indicated with yellow arrows) between microbial cell clusters in the exterior of granules ; (i) Interstitial voids (indicated with rectangles) between microbial cell clusters in the interior of granules ; (j) Gas tunnels between microbial cells (indicated with arrows) ; (k) Small gas pocket (marked with rectangle) ; (l) Gas pocket of a settling granule formed by inflation of dinitrogen gas ; (m) Gas pocket of a floating granule (indicated with closed curve) ; (n) A broken floating Anammox granule formed by the burst of dinitrogen gas

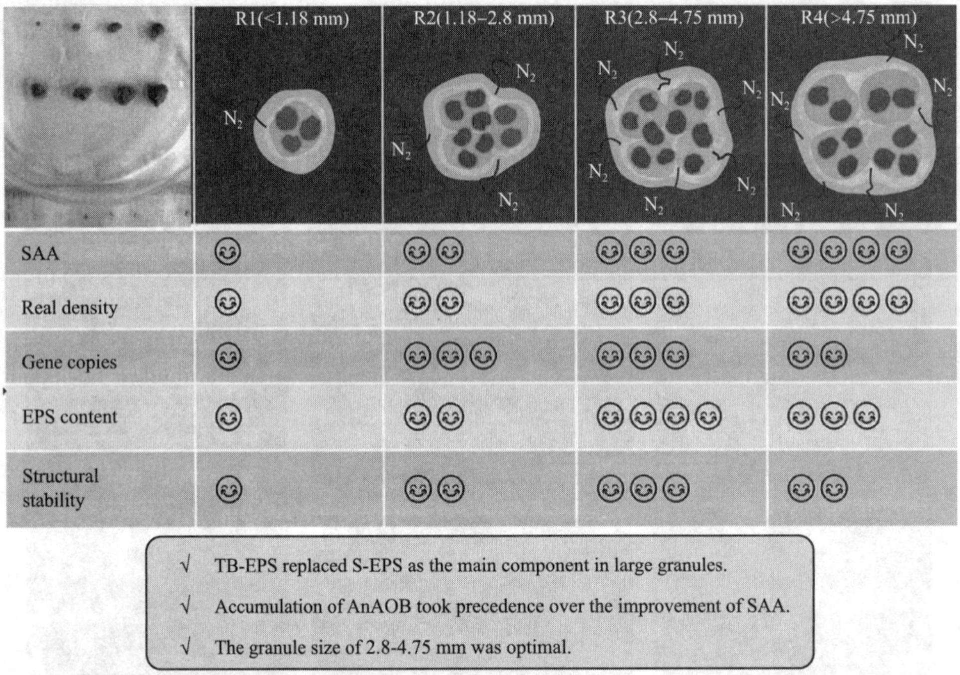

Figure 9–6 Effects of size on AnGS

Table 9–3 Performance of different Anammox reactors

Granulation reactor	Inoculum	HRT/d	Working volume/L	Operating conditions	Reported granules diameter/mm	Total nitrogen efficiency/%	NLR/[kg N/(m^3·d)]
SAnMBR	Enriched Anammox biomass	2	4	Stirring speed—50 rpm SRT—16–10 d	0.28	88	4.1
SBR	Enriched Anammox biomass	0.625	1	6 h cycle Feed—330 min Settling—20 min Decant—10 min	0.82	78	0.7
SBR	Enriched granular Anammox biomass	1	1	6 h cycle Feed—330 min Settling—20 min Decant—10 min	0.64	—	0.3
SBR	Mixed activated sludge	1.9	20	8 h cycle N_2 gas addition—4 min Feed—356 min Reaction—80 min Settling—20 min Decant—20 min	2.7 ± 0.7	99.5	1.6
MBR	Enriched granular Anammox biomass	2	15	Stirring speed—160 rpm SRT—16–12 d	—	—	—

Continued

Granulation reactor	Inoculum	HRT/d	Working volume/L	Operating conditions	Reported granules diameter/mm	Total nitrogen efficiency/%	NLR/ [kg N/ (m^3 · d)]
MBR	Aerobic activated sludge	2	4.8	Stirring speed—100 rpm SRT—16–12 d	—	90	—
UASB	Enriched Anammox biomass	0.25	11	—	1.39–1.65	83	1.2
UASB	Enriched granular Anammox biomass	—	6.4	Upflow velocity—2.3–4.7 m/h	0.9–1.3	92.3	—
UASB	Aerobic granular biomass	1	2.1	—	1.0–4.5	94	0.5

SAnMBR–Submerged Anaerobic Membrane Bioreactor, UASB–Upflow Anaerobic Sludge Blanket, MBR–membrane bioreactor, SBR–Sequencing Batch Reactor, NLR–Nitrogen Loading Rate, HRT–Hydraulic Retention Time.

In an experiment to determine the nitrogen removal efficiency of different–sized granules in a one–stage PN–A reactor, it was observed that an increase in the total autotrophic nitrogen removal rate (TANRR) corresponded with an increase in the size of the granule.

9.3.1.2 Settling Velocity

As mentioned earlier, Anammox granules are effective at retaining biomass in the Anammox reactor mostly due to their good settleability. Stoke's equation establishes the relationship between the settling velocity, diameter, and the density of the Anammox granules as explained above. For example, during the cultivation of AnGS in an SBR, after 72 days of operation, it was observed that the average granular diameter had increased from 1.2 to 3.69 mm while the settling velocity also increased from 107.68 to 118.49 m/h. However, since the floating granules do not settle, only the settling velocities of settled Anammox granules could be measured. An increase in the density and diameter leads to a corresponding increase in the settling velocity of the Anammox granules. The settling velocities of anaerobic granular sludge and Anammox granules are 60 m/h and 73–88 m/h (in a UASB reactor) respectively. It is reported that AnGS with poor settling ability could lead to unstable operation in the Anammox systems. Other factors such as sphericity, roundness, and mass of the Anammox granules are also of great influence on the settling velocity of Anammox granules. For example, it is widely accepted that smaller particles have a larger surface area to mass ratio, resulting in a slower flow velocity due to frictional drag than larger particles. According to Komar and Reimers, the more spherical the grain, the higher its settling velocity within the fluid compared to non–spherical grains of the same weight.

9.3.1.3 Morphology

Generally, Anammox granules are reported to have a sphere-like three-dimensional shape. In examining the morphology of Anammox granules, Scanning Electron Micrograph (SEM) and Transmission Electron Micrograph (TEM) have been used extensively. From experiments conducted by Lu et al., it was discovered that Anammox granules were irregular in shape and uneven on the surface with structures including granule, subunit, microbial cell cluster, and single cells. This is however in contrast to observations made by Jianlong and Jing where Anammox granules from an EGSB reactor operated for about three months were observed to have regular shapes (sphere-like) and smooth surfaces. They were also reported to have a number of cavities which could be the probable site for nitrogen production. Anammox granules have a cauliflower-like shape with surfaces covered with spherical and elliptical bacteria though no bacilli or filamentous bacteria were found in the reactor. The presence of the bacteria could probably be indicative of the dominance of Anammox bacteria.

9.3.1.4 Density

Similar to the diameter of the Anammox granular sludge, the density is also indicative of the performance of the Anammox reactor. Anammox granules have been reported to have a higher specific density (91-120 g VSS/L granules) compared with aerobic granules (40-70 g VSS/L granules). The density and morphology of Anammox granules are likely related to its diameter. This is demonstrated in the inversely proportional relationship between density and diameter of the Anammox granules, where an increase in the diameter results in a decrease in the density of the Anammox granules. Though the influence of diameter on the settleability of Anammox granules far exceeds that of density during experiments, the overall settleability is however determined by the density of the Anammox granules (Figure 9-7).

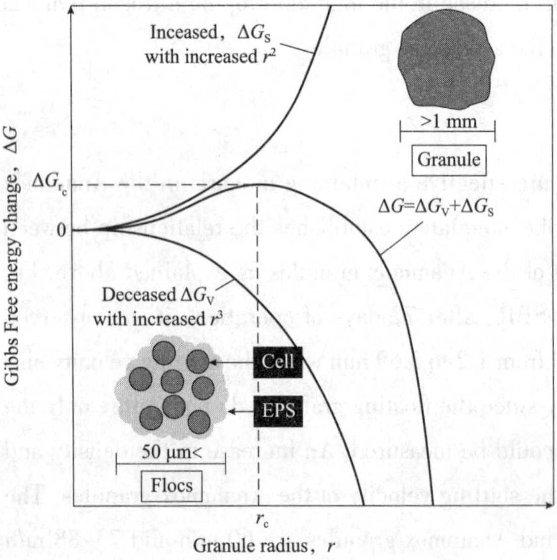

Figure 9-7 Thermodynamic theory of granulation process from floc to granule

ΔG : the sum of free energy change ; ΔG_V (volume free energy) : change on bulk free energy to form a new biofilm layer ;

ΔG_S (surface free energy) : interfacial free energy between new and old biofilm layers during the formation of a new biofilm layer ;

r_c : a critical size in the maximum of the total free energy ;

ΔG_{r_c} : maximum of the total free energy

9.3.1.5 Specific Anammox Activity (SAA)

The granular sludge characterized by high activity plays an important role in the performance of high-rate bioreactors. A nearly constant SAA [0.4 g N/(g VSS · d)] was recorded in an SBR with an average Feret diameter of 0.64 mm for the formed granules. The average maximum specific activities reported in batch experiments in a gas-lift reactor and an SBR were 0.90 and 0.44 g N/(g VSS · d) respectively. However, the system was destabilized and the biomass began to float after the maximum specific activity was exceeded. The positive correlation between temperature and SAA was reported by Jin et al. In an experiment to investigate SAA at different temperatures, the maximum SAA of 6.12 mg N/(g VSS · h) was recorded at 35 ℃.

9.3.1.6 Color

The unique characteristic carmine color of Anammox granules is also indicative of the high activity of the Anammox bacteria. Though high load Anammox granules are reported to be usually carmine in appearance, their color could still vary from carmine to brownish or black. This is largely attributed to the heme c concentration in the Anammox setup. Thus, Anammox granules with higher heme c concentrations are grey whereas black Anammox granules lack heme c. In the cultivation and granulation of AnGS in an SBR, after 72 days of operation, it was observed that the color had changed from khaki to red. Figure 9-8 below illustrates the surface microstructure of three AnGS observed by SEM during the color characterization of AnGS. The three AnGS depicted were operated under typical biological loading rates. It was discovered that the color values correlated positively with the specific Anammox activity and the biological loading rates of the

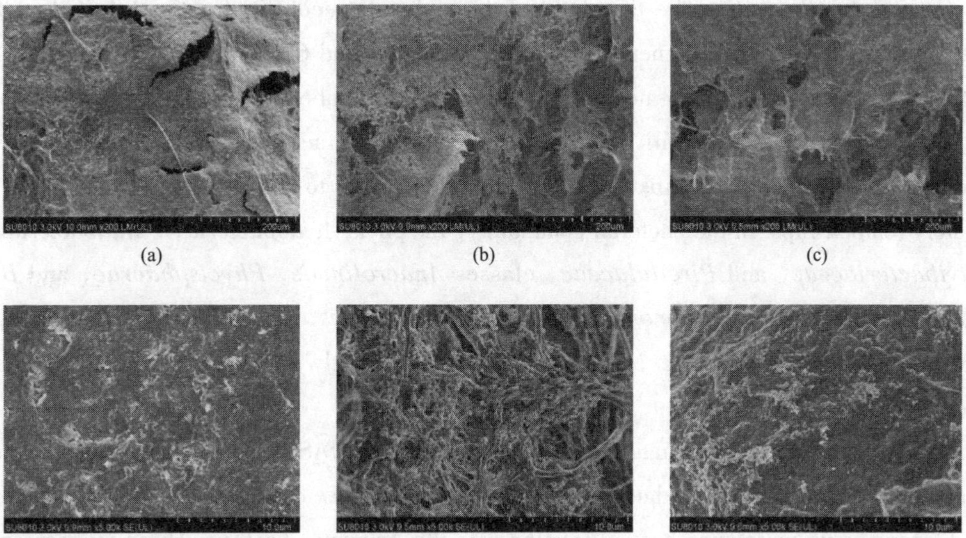

Figure 9-8 Surface microsurface of AnGS observed by SEM

A, B, C represented black, brown and red AnGS, respectively x200; D, E, F were the magnifying view of A, B, C x5000

AnGS. In addition, it was found that the difference in the surface microstructure greatly affected the subjective color of the granular sludge as seen by the naked eye.

9.3.2 Microbial ecology of granular sludge

The co-occurrence of Anammox bacteria with other microorganisms in the granular sludge was investigated using the high-throughput sequencing approach [Figure 9-9 (a), 9-9 (b)]. A total of 16 phyla were detected, including the most abundant *Chloroflexi* (36.7%), *Planctomycetes* (27.2%), *Proteobacteria* (14.7%), and *Chlorobi* (9.1%), and the other less abundant phyla, *Acidobacteria*, *Actinobacteria*, *Armatimonadetes*, *Bacteroidetes*, *Caldithrix*, *Cyanobacteria*, *Elusimicrobia*, *Firmicutes*, *Gemmatimonadetes*, *Nitrospirae*, *Spirochaetes*, *Verrucomicrobia*. The dominant *Chloroflexi* was mainly composed of *Anaerolineae* (35.3%), *Chloroflexi* (0.2%), and *Thermomicrobia* (0.1%). *Planctomycetes* was the second largest bacterial phylum. Specifically, *Brocadia* (17.1%), *Phycisphaerae* (7.9%), and *OM190* (2.0%) were the most abundant classes. For *Proteobacteria*, at the class level, *Betaproteobacteria* (10.1%) dominated in the granular sludge, followed by *Gamma-Proteobacteria* (2.0%) and *Beta-Proteobacteria* (1.7%). We also observed some bacteria with very low relative abundance (< 0.1%), such as *Gemmatimonadetes*, *Verrucomicrobia*, *Cyanobacteria*, and *Nitrospirae*, consistent with the qPCR results on AOB. Network analysis based on significant and strong correlations was used to explore the co-occurrence patterns of Anammox bacteria [Figure 9-9 (c)]. Four bacterial classes (*Gemmatimonadetes*, *Acidobacteria*, and double *Anaerolineae*) and four families (Caldilineaceae, *Comamonadaceae*, *Rhodospirillaceae*, and *Patulibacteraceae*) were correlated with *Candidatus Brocadia*. The threshold of the molecular ecological network based on RMT calculation was 0.990, and the network included 66 nodes and 627 links, in which 26.2% were positive interactions and 73.8% negative. Moreover, the number of links to Anammox bacteria was 120, including 22.5% positive and 77.5% negative interactions. Positive interactions may represent mutualism relationships, while negative ones refer to competitive, antagonistic, and predatory relationships. In the bacterial community, the phyla *Acidobacteria*, *Patulibacteraceae*, *Ignavibacteriaceae*, and *Pirellulaceae*, classes *Anaerolineae*, *Phycisphaerae*, and *Beta-Proteobacteria*, orders *Fimbriimonadales* and *Phycisphaerales*, family *Phyllobacteriaceae*, and genera *Tepidibacter*, *Fusibacter*, and *Rubrivivax* significantly co-occurred with the Anammox bacteria.

High-throughput sequencing showed that although the UASB was inoculated with enriched Anammox bacteria granular sludge, Anammox *Planctomycetes* only accounted for 27.2% of the total bacteria, with more microbes co-existing with the Anammox bacteria. These microorganisms had different metabolic processes and activities, resulting in several potential competitive and symbiotic interactions with Anammox bacteria, as suggested by the results of the network analysis.

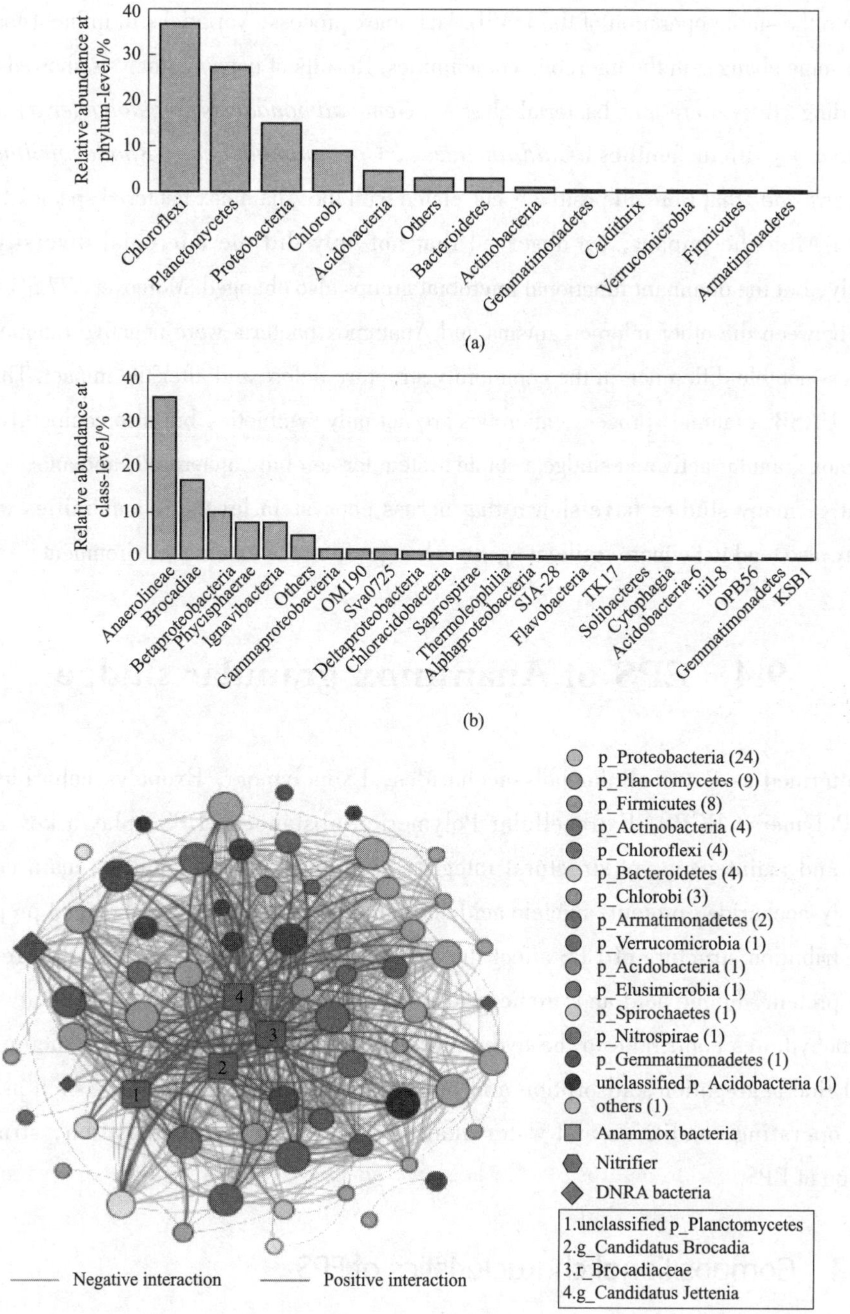

Figure 9-9 The community composition, diversity and microbial interaction among bacteria associated with Anammox granular sludge

The abundance of bacteria co-occurring with Anammox bacteria at the (a) phylum- and (b) class levels; (c) The ecological networks of co-occurring Anammox bacterial OTUs at different particle size. Each node in the network signifies a microbial taxon. Colors of the nodes in the right indicate different major Classes. A blue edge refers to a positive interaction between two individual nodes, while red edges indicate negative interaction. The nodes in the square frame represent microbial populations at the level of class correlated with the Anammox bacteria, and the numbers represent each correlating Class. The labels surrounding the network are the specific name of the microbial groups correlated with the Anammox bacteria. "c": Class, "f": Family, and "g": Genus.

Despite the stable operation of the UASB Anammox process, variations in influent loadings still resulted in some changes in the microbial communities. Results of network analysis showed that before impact loading, there were four bacterial classes (*Gemmatimonadetes*, *Acidobacteria*, and double *Anaerolineae*) and four families (*Caldilineaceae*, *Comamonadaceae*, *Rhodospirillaceae*, and *Patulibacteraceae*) that were all positively correlated with the Anammox bacterial species *Candidatus Brocadia* . After the impact, we observed that not only did the microbial diversity increase significantly, but the dominant functional microbial groups also changed. Moreover, 77.5% of the total 120 links between the other microorganisms and Anammox bacteria were negative interactions. This was the most notable difference in the community structure before and after the impact. This suggests that in the UASB Anammox process, microbes are not only symbiotic, but also competitive, making the Anammox granular activated sludge a stable system for aerobic, anaerobic, and anoxic organisms. Consistently, many studies have shown that across ecosystem levels, communities with higher biodiversity also tend to be more resilient against changes from the external environment.

9.4 EPS of Anammox granular sludge

Also termed as Extracellular polysaccharides, Exopolymers, Exopolysaccharides or Extra Cellular Polymers (ECP), Extracellular Polymeric Substances (EPS) play a key role in the formation and maintenance of structural integrity of granular sludge, and its main components include polysaccharide, protein, nucleic acid and humic acid. The physicochemical properties and spatial distribution structure of EPS affect the structure and function of microbial aggregates. For example, protein, humic acid and uronic acid in EPS contribute to the hydrophobicity of sludge, while carbohydrates contribute to the hydrophilicity. Some protein secondary structures promote biological floc aggregation, adsorption and biofilm formation. Many factors, such as treatment process, operating conditions and water quality, will affect the composition, structure and composition of EPS.

9.4.1 Compositional characteristics of EPS

The content of the EPS components varied considerably between the different types of Anammox granular sludge (Table 9-4). This may be related to reactor type, inoculated sludge and incubation strategy, although strong evidence is lacking for the underlying mechanism of the differences. EPS is mainly composed of proteins, nucleic acids, polysaccharides, lipids and other polymeric compounds found within various microbial aggregates. Of these, polysaccharide (PS) is the main component of EPS, followed by humic substances and proteins (PN)(Figure 9-10).

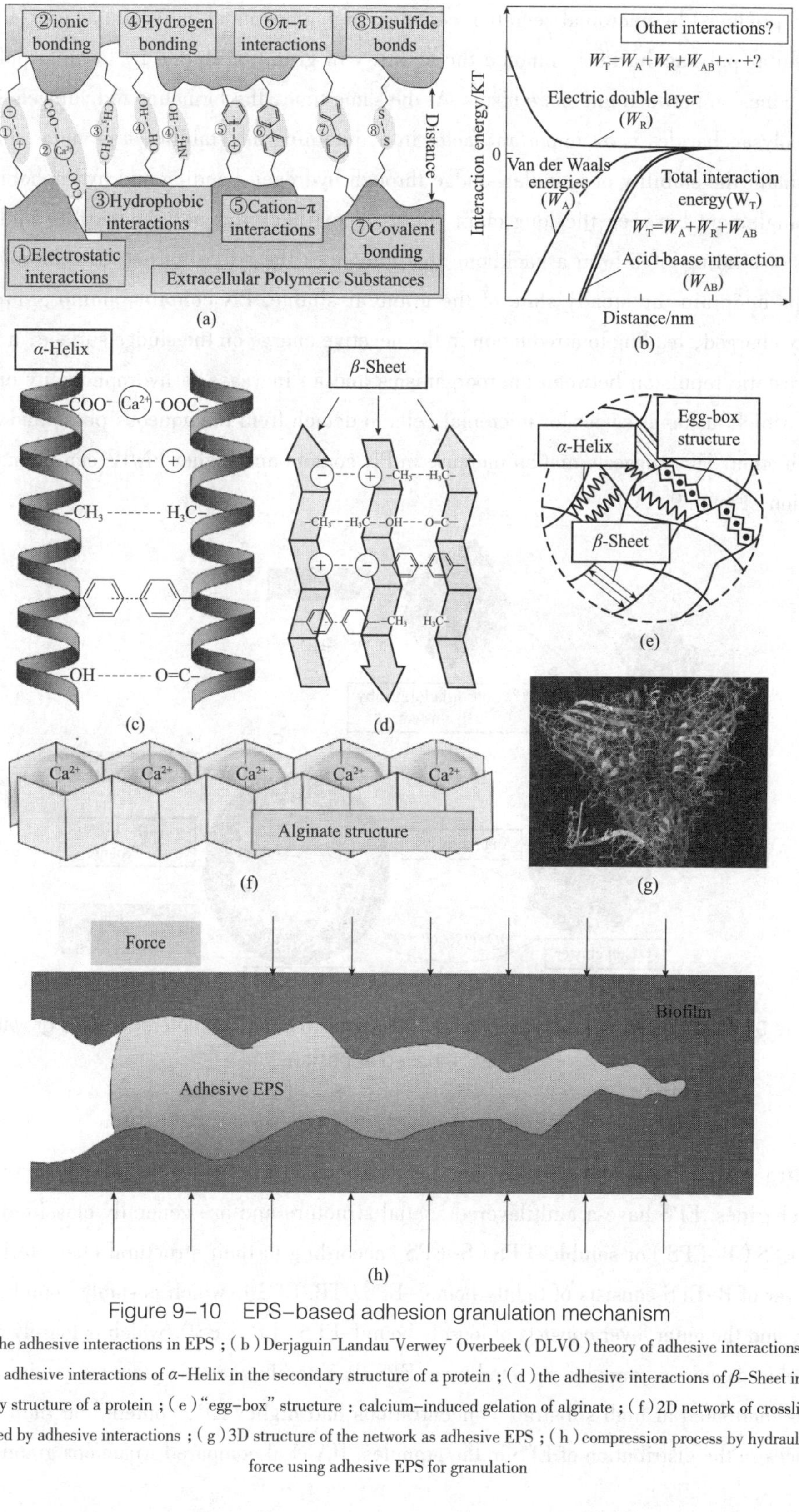

Figure 9-10 EPS-based adhesion granulation mechanism

(a) the adhesive interactions in EPS; (b) Derjaguin-Landau-Verwey-Overbeek (DLVO) theory of adhesive interactions; (c) the adhesive interactions of α-Helix in the secondary structure of a protein; (d) the adhesive interactions of β-Sheet in the secondary structure of a protein; (e) "egg-box" structure: calcium-induced gelation of alginate; (f) 2D network of crosslinking formed by adhesive interactions; (g) 3D structure of the network as adhesive EPS; (h) compression process by hydraulic force using adhesive EPS for granulation

EPS produced by microbial secretion contains large amounts of extracellular polysaccharides. Extracellular polysaccharides enhance the stability of granular sludge by forming polymers to promote adhesion between microorganisms. At the same time, the formation of hydrogels by cross-linking polysaccharides is an important factor in maintaining the stability of granular sludge. The gels maintain the stability of granular sludge through hydrogen bonding and hydrophobic effects. The entanglement between the long main chains of extracellular polysaccharides and the rich bridging of binding sites form a backbone that enhances the adhesion between microorganisms and helps maintain the steady state of the granular sludge. PN contains amino groups and is positively charged, leading to a reduction in the negative charge on the sludge surface, a reduction in electrostatic repulsion between microorganisms and an increase in hydrophobicity on the cell surface, which makes it easier for microbial cells to detach from the aqueous phase and aggregate with each other. This suggests that an increase in PN content and higher PN/PS can promote sludge granulation (Figure 9-11).

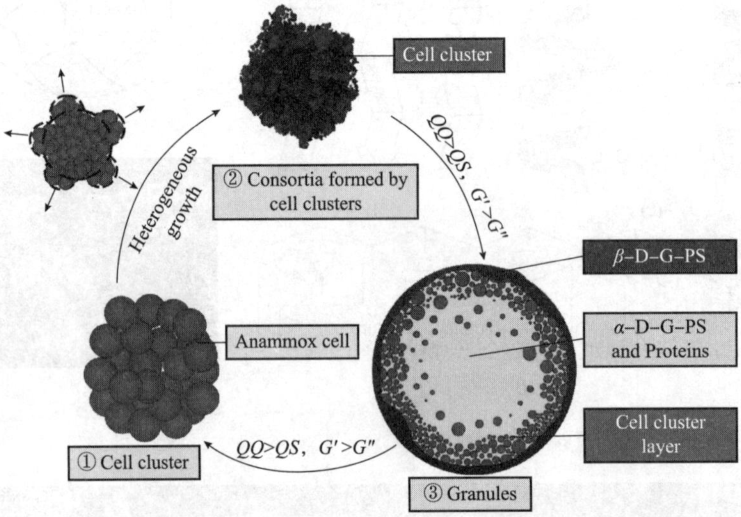

Figure 9-11 Granulation cycle mechanism. Mechanism based on heterogeneous growth and EPS-based adhesion

β-G-PS: β-glucopyranose polysaccharides; α-G-PS: α-d-glucopyranose polysaccharides; *QS*: quorum sensing; *QQ*: quorum quenching; G': storage modulus; G'': loss modulus

Extracellular polysaccharides can be divided into bound and soluble extracellular polysaccharides. EPS have a multilayered spatial structure and are generally classified as either bound-EPS (B-EPS) or soluble-EPS (S-EPS) according to their structural characteristics. The inner layer of B-EPS consists of tightly bound-EPS (TB-EPS), which is stably bound to the cell surface, and the outer layer consists of loosely bound-EPS (LB-EPS), which is loosely dispersed. Autotrophic Anammox granules secreted more EPS than aerobic/aerobic granules, while Anammox granules incubated at high substrate concentrations had higher EPS content and there are large differences in the distribution of EPS in the granules. JIA et al compared Anammox granular sludge produced under 18 different conditions and found that TB-EPS was closely related to granular sludge morphology and could be used as an indicator of the viability of anaerobic Anammox microorganisms and the morphology of microbial aggregates. Also, TB-EPS contains a large number of protein hydrophobic groups that promote microbial aggregation. One researcher used amylase, a polysaccharide-degrading enzyme, to enzymatically digest Anammox granular sludge and found that the α-amylase-treated group showed swelling on the outer edges of the granular sludge, while the β-amylase-treated group showed no significant change on the surface of the granular sludge, but showed fragmentation and a significant decrease in stability. β-D-furanose was concentrated in the outermost layer of the granular sludge, while proteins, lipids, α-furanose and α-mannose were distributed throughout the granular sludge, mainly in the outer part of the granular sludge. Proteins and lipids form the skeleton of anaerobic ammonia granular sludge, and Anammox bacteria are distributed in the middle of proteins and lipids. HOU et al found that EPS content and chemical composition are mainly influenced by the metabolic effects of bacteria, and the metabolic nature of bacteria also determines the aggregation capacity of EPS.

The diameter of Anammox granular sludge directly affects the EPS content and composition. The 3-layer structure of EPS (S-EPS, TB-EPS and LB-EPS) varies with particle size. When the particle size was <0.5 mm, the content of S-EPS (in terms of the mass of EPS per gram of VSS) accounted for more than 48% of the total EPS, and decreased with the increase of particle size. The content of S-EPS (in terms of EPS mass per gram of VSS) accounted for more than 48% of the total EPS, ranging from 121.0 to 302.3 mg/g, and decreased with the increase of particle size, accounting for 35% of the total EPS at a particle size of 2.5 mm. The content of TB-EPS increased with increasing particle size, from 95.6 mg/g at particle size < 0.2 mm to 334.1 mg/g at particle size 2.5 mm, and TB-EPS gradually replaced S-EPS as the main component of EPS with increasing particle size of Anammox granular sludge. With the increase of particle size of Anammox granular sludge, TB-EPS gradually replaced S-EPS as the main component of EPS. In contrast, the LB-EPS content fluctuated less with particle size and remained below 24% of the total EPS content. The difference in diameter also caused differences in the content of each component of EPS, with particle size increasing from 0.5–1.4 mm to >2.8 mm, PN content increasing from (56.88 ± 0.86) mg/g to (98.59 ± 2.10) mg/g, and total EPS amount increasing from (68.05 ± 0.97) mg/g to (94.26 ± 2.20) mg/g. Therefore, the operational stability of granular sludge can be enhanced by controlling the reactor operating conditions, regulating the appropriate particle size, and enhancing the secretion of EPS from Anammox granular sludge.

9.4.2 Influencing factors of EPS

9.4.2.1 Effect of PN/PS in EPS on its stability of EPS

The two major components of EPS, PN and PS, play an important role in the formation and operation of Anammox granular sludge. The EPS content of Anammox granular sludge was significantly different from PN/PS under various reactor types and denitrification loading conditions. Higher denitrification loads tend to PN/PS is larger, but there is no linear correlation between the two, which may be directly related to the operating environmental conditions. When the nitrogen removal load is 0–1 kg/(m^3 · d), The total amount of EPS (in terms of mass of EPS per gram of VSS) of Anammox granular sludge under different reactor operating conditions ranged from the total EPS (in terms of EPS mass per gram of VSS) ranged from 13.35 to 850 mg/g and PN/PS ranged from 0.54 to 7.66 under different reactor operating conditions. The PN/PS of Anammox granular sludge formed by homogeneous inoculated sludge is relatively similar. However, different inoculated sludge and culture strategies can have a significant effect on the PN/PS of granular sludge, but studies to date have not provided a reasonable explanation for it. However, the differences in the components of EPS have not yet been explained and need to be further investigated.

The PN/PS of granular sludge EPS can be used to characterize its stability properties and settling performance. The PN/PS size is directly related to the hydrophobicity of sludge surface, chargeability and formation of granular sludge. The relative hydrophobicity of the sludge gradually increases with the increase of PN/PS, thus promoting the aggregation of Anammox granular sludge, and the relative hydrophobicity of the sludge gradually increases with PN/PS, thus promoting the aggregation of Anammox granular sludge and a large increase in PN/PS with the formation of granular sludge. The proteins in the EPS of Anammox granular sludge are the main components that determine hydrophobicity. Therefore, with the increase of PN/PS, the higher the protein percentage, the more electronegativity of the cells and the more hydrophobic and granular microbial surface. Although there was no significant correlation between EPS content and particle size of granular sludge, PN/PS was greatly influenced by particle size. CHEN et al. studied the EPS content of four different particle sizes of Anammox granular sludge and found that the small particle size (1.0–1.5 mm) had a PN/PS of 3.81, a higher anaerobic ammonia oxidation activity and higher EPS secretion. The PN/PS of small particle size (1.0–1.5 mm) granular sludge was found to be 3.81, with higher anaerobic ammonia oxidation activity and higher EPS secretion. PN/PS for diameter of 300–500 μm tends to increase with larger particle diameter, but when the particle diameter increases to a certain extent (> 500 μm), the granular sludge secretes less EPS to the extracellular, and PN/PS decreases, the aggregation between small particles weakens, the affinity decreases, and the sedimentation capacity decreases.

9.4.2.2 Effect of other components in EPS on the properties of granular sludge

Analysis of the elemental composition of Anammox granular sludge showed that the composition of C, H, N, O, and S elements accounted for 44.4%, 6.6%, 9.0%, 35.7%, and 1.4% of the total, respectively. In addition to the main components, the EPS extract of granular sludge also contains Na, K, Ca, Mg Al, Fe, Mn, P, Si, S and other mineral particles. Among them, Na, K, Ca, and Mg account for the major components, and their content size is K > Na > Ca > Mg. K, Ca and Mg in Anammox granular sludge EPS exist in two forms, ionic and nonionic, but most of them are in ionic form. The percentages of metal ionic forms of K and Ca were 68.6% and 56.2%, respectively, and the highest percentage of ionic forms of Mg was up to 94.7%. Among the non-ionic forms, K and Mg accounted for 31.4% and 5.3%, respectively. Among the metallic elements, Ca has the highest percentage of non-ionic forms with 43.8%. As the particle size increases, the PN content of Anammox granular sludge EPS increases, which can bind more metal elements, resulting in an increase in the metal element content and Ca^{2+} content, which in turn neutralizes the negative charge on the bacterial surface and polysaccharide molecules, elevates the polysaccharide level and significantly shortens the formation time of particles, accelerating the formation and maintaining the stability of granular sludge. Metal ions play different roles in granular sludge aggregation, for example, Na^{+} in EPS will promote the aggregation of granular sludge by compressing the bilayer, while Na^{+} and K^{+} will compete with Ca^{2+} for binding sites in EPS by ion exchange. The Na^{+} and K^{+} will compete with Ca^{2+} for the binding sites in EPS through ion exchange. Therefore, the Na^{+} has an opposite trend to Ca^{2+} and Mg^{2+} content. It has also been suggested that metals such as Ca, K, Mg, Na, Al, and Fe in EPS metal elements affect the surface properties of Anammox granular sludge. For example, Ca can induce the formation of $CaCO_3$ and $Ca_5OH(PO_4)_3$ precipitation. As the initial nucleation of Anammox granular sludge and contribute to the formation of granular sludge. The role of other metallic elements in the formation of Anammox granular sludge and the structure of EPS composition remains to be further investigated.

Table 9-4 The main components of Anammox granular sludge EPS under different operating conditions

Reactor Type	Nitrogen removal load / [kg/(m^3 · d)]	Total EPS/(mg/g)	EPS component content /(mg/g)		PN/PS
			PN	PS	
UASB	0.24	387.23	226.9	140	1.62
UASB	0.14	105.15 ± 6.63	93.01 ± 6.23	12.15 ± 0.40	7.66
UASB	0.17	265.2 ± 4.6	164.4 ± 9.3	71.8 ± 2.3	2.29
UASB	0.30	850	500	350	1.43
UASB	0.06	174.2	49.2	91.1	0.54
UASB	0.40	13.35	11.61	1.74	6.67
UASB	0.19	220.20	114.66	99.71	1.15
UASB	0.36	334.1	162.37	158.36	1.03
UASB	—	133.7	55.6 ± 3.2	70.8 ± 6.5	0.79
UASB	5.64 ± 0.2	275.4	171.5	103.9	1.65
EGSB	1.55	143	97.5	45.5	2.14
EGSB	0.17	333.03	234.25	90.78	2.58

Continued

Reactor Type	Nitrogen removal load / [$kg/(m^3 \cdot d)$]	Total EPS/(mg/g)	EPS component content /(mg/g)		PN/PS
			PN	PS	
SBR	1.12	131.82	94.01	37.81	2.49
SBR	0.24 ± 0.02	165	140	25	5.60
SBR	0.125	200	140	60	2.33
SBBR	0.19	290.92	132.05	106.32	1.24
A^2O	1.01	146.69	90	23	3.91

Note: The content of each component of EPS is based on the mass of that component per gram of VSS.

9.5 Application of Anammox granular sludge

For over two decades, the Anammox process as a biological nitrogen removal (BNR) process has been researched extensively. The continuity in researches in this direction has mostly been due to the pertinent issue of long start up time. Not surprisingly, a wide range of Partial Nitritation-Anammox (PN/AMX) system configurations have been transformed into full-scale plants over the past two decades with China, Europe and North America among the current leading states in its application. The very first full-scale facility in Asia is reported to have been constructed in Japan and took 60 days before start-up.

Though, earlier PN/AMX systems were designed on the two-step system model, recent focus has moved to the single-stage reactor systems mainly due to the low investment costs involved with SBR being the most commonly used system type (>50% of all PN/AMX systems), followed by granular sludge processes and MBBRs in that order. It has also been reported that a relatively small number of full-scale RBCs and activated sludge reactors are also in operation.

To resolve the issue of long start-up as mentioned earlier, various techniques as discussed earlier have been employed: biomass retention being at the core of each of these techniques.

Granulation of the Anammox biomass is one such means of retaining biomass within the Anammox reactor. So far, the application of Anammox granules to the treatment of ammonium-rich wastewater has been successful at all levels: laboratory-scale, pilot scale, and full scale.

The first full-scale AnGS was successfully started in Rotterdam, the Netherlands in 2002. Unlike the conventional Anammox process which took two years, this took 3.5 years with a reported final NRR of 9.5 kg N/($m^3 \cdot d$) after the UASB reactor had been modified. Ever since the worldwide application of the first full-scale AnGS system, the number of reactors has risen to about 20% of the partial nitritation/Anammox (PN/A)(about 40), according to the application survey with an average N-load of upto 3000 kg N/d per plant. This happened concurrently with the successful full-scale implementation of other granular sludge reactors over the past decade. Up until now the highest reported NRR is 74.3–76.7 kg N/(m^3 · d). In order to improve the effluent quality of industrial wastewater, the granular sludge nitrification/Anammox reactor employed was stably operated on full-scale for 3 years. Consequently, there was a 17 times reduction in the required reactor volume and a 1.5 GWh per year savings in electric power.

In the treatment of low-strength nitrogen wastewater in sequence batch advanced nitrogen removal (ANR) systems, three types of granular sludge separately collected from expanded granular sludge bed (EGSB) reactors treating nitrogen wastewater at high, moderate and low-nitrogen loading rates (NLR) respectively were employed. It was concluded that the moderate granular sludge from moderate ANR system was the most efficient, recording an effluent TN of < 1 mg/L, 92.8% TN removal efficiency, 0.039 kg N/(m^3 · d) nitrogen removal rate (NRR), with an influent TN of 15 mg/L, cycle length of 8h, and initial biomass concentration of 2.43 g VSS/L.

Using a mixture of different activated sludge biomasses as inocula for an Anammox SBR, Anammox granular sludge was formed under the operating conditions of 6 h anaerobic feed, low mixing reaction phases (to minimize the shear stress) with final settling phases of 20 min. The Anammox biomass granulated naturally which supports reports that Anammox biomass usually grow over fixed biofilms or in a granular suspended form. The granular sludge formed had a sludge volume index (SVI_5) of 34 mL/g with a low thickening process verified by an SVI_{10} to SVI_{30} ratio of 1.0 as reported. In another study to determine the effect of salinity on the stability and aggregation of Anammox granular sludge in saline wastewater treatment, it was discovered that a salinity of 30 g/L NaCl produced the highest Anammox activity registering average removal efficiencies of 91.9% ± 1.4%, 97.3% ± 0.4% and 86.3 ± 0.9% for NH_4^+–N, NO_2^-–N and TN respectively.

The removal of flocky biomass and the careful increment of the upflow velocity in the reactor are two of the most important factors to consider during Anammox granulation. The formation of Anammox granules with high density and settling velocity of 100 m/h and diameter 1.4 mm resulted in a stable operation even at loading rates exceeding 10 kg N/(m^3 · d).

9.6 Main factors affecting EPS secretion and sludge granulation

9.6.1 Organic concentration

Anammox granular sludge is an autotrophic denitrifying granular sludge, the main functional bacteria are Anammox bacteria, which have slow growth and long multiplication time. When organic matter is present in the wastewater, heterotrophic bacteria such as denitrifying bacteria will proliferate rapidly and compete with Anammox bacteria for the reaction substrate NO_2–N, resulting in the inhibition of Anammox bacteria growth, thus affecting the performance of Anammox granular sludge or even disintegrating it. Study found that lower organic matter mass concentrations (<100 mg/L) would make the functional microorganisms in the simultaneous partial nitrification, Anammox and denitrification (SNAD) reactor more likely to accelerate the rapid granulation of the sludge and increase the particle size, reaching a maximum size of 1,103 μm; after about 140 d of incubation, the maximum EPS content reached 140 mg/g (in terms of the mass of EPS per gram of VSS contained). Similarly, the EPS content reached a maximum value of (482.69 ± 9.83) mg/g at a COD of 100 mg/L; while it decreased to (451.80 ± 9.79) mg/g as the COD increased to 150 mg/L; and began to decrease sharply as the COD continued to increase to 200 mg/L. LI et al found that conditions with COD of 200 mg/L promoted the secretion of EPS from Anammox granular sludge, which increased from 33 mg/g to 181 mg/g, but decreased to 155 mg/g when COD increased to 300 mg/L. From the confocal laser scan images, it can be found that PN increases significantly in the presence of organic matter, but PS decreases. A moderate amount of organic carbon source (<130 mg/L) increased the EPS content of Anammox granular sludge, with PN and PS content increasing from the initial 63 and 45.1 mg/g to 140 and 60 mg/g, respectively (in terms of PN or PS mass per gram of MLSS). At this time, the settling performance of the sludge was significantly improved. However, the EPS content decreased at COD higher than 130 mg/L. This may be due to the fact that excess organic matter can cause partial disintegration of Anammox granular sludge, which in turn causes the release of EPS.

9.6.2 Nitrogen concentration and loading in the reaction system

Nitrogen as the main substrate for the Anammox process, its concentration and loading directly affect the formation of Anammox granular sludge and EPS secretion. ZHANG et al found that the EPS content of Anammox granular sludge increased by about 60% when the nitrogen load increased from 4.8 g/(L · d) to 10.0 g/(L · d); and when the nitrogen load reached a maximum value of 20.0 g/(L · d), the total EPS increased to 77.2 mg/g (in terms of the mass of EPS per gram of VSS contained), but Anammox granular sludge disintegrated considerably due to density reduction. WANG et al also found that with increasing nitrogen load [from 0.09–0.26 kg/(m^3 · d) to 0.21–0.97 kg/(m^3 · d)], the EPS content of Anammox granular sludge increased from 11.09 mg/g to 83.38 mg/g and PN/PS increased from 5.09 to 8.84. NI et al regulated the nitrogen loading of the reactor by setting different residence times (HRT) and found that: there were differences in the characteristics of EPS under different nitrogen loading operating conditions, with PN/PS of 0.8 for low HRT (22 h) high nitrogen loading Anammox granular sludge and 2.1 for high HRT (36 h) low nitrogen. Under low nitrogen conditions, PN was the main component of extracellular polysaccharides; under high nitrogen conditions, EPS content was higher and PS was the main component. Huilei Jin found that the EPS content of Anammox granular sludge was 529.13 mg/g at high nitrogen load [4.97 kg/ (m^3 · d)]; as the nitrogen load decreased to 1.01 kg/ (m^3 · d), the EPS content decreased to 256.52 mg/g.

9.6.3 External mediator

The addition of external mediators can directly promote the activity of Anammox bacteria or its functions such as enhancing electron transfer, and also directly affect the secretion of EPS in Anammox granular sludge. The results of FTIR analysis showed that the addition of graphene nanosheets (GNs) could increase the content of PN and PS in Anammox granular sludge EPS, enhance its hydrophobicity, and promote the formation of granular sludge. The addition of (γ-Fe_2O_3 NPs) resulted in the complete disappearance of the PN peak. It was found that when the dosage of MnO_2 nanoparticles (MnO_2 NPs) was gradually increased to 200 mg/L, in order to resist the intrusion of NPs, the EPS content in Anammox granular sludge increased to (481.5 ± 13.4) mg/g (based on the mass of EPS contained in each gram of VSS). However, adding excess mediators also inhibited EPS secretion. Under the stress of low concentration (5 mg/L) La (Ⅲ) in Anammox granular sludge, EPS secretion increased from (292.9 ± 17.5) mg/g to (398.3 ± 26.8) mg/g. However, as the concentration of La (Ⅲ) increased to 50 mg/L, a large amount of La (Ⅲ) inhibited the activity of related metabolic enzymes after EPS adsorption and transmembrane transport, resulting in a decrease in EPS secretion. CHENG et al. investigated the changes of EPS of Anammox granular sludge under the dual influence of copper nanoparticles (CuNPs) and oxytetracycline (OTC) and found that the low mass concentration (the mass concentration of CuNPs and OTC was 0.5 mg/L), the EPS content increased by 18.7% compared with the initial level; when the mass concentrations of CuNPs and OTC were increased to 1 mg/L, the EPS content increased by 21.4%; while at high concentrations (the mass concentrations of CuNPs and OTC were 2.5 mg/L and 2.0 mg/L, respectively), the EPS content decreased rapidly again, only 77% of the initial level. When the concentration of the external mediator was low, in order to resist the intrusion of the external mediator, a large amount of PN secreted in the Anammox granular sludge would increase the EPS content. However, when the concentration of the external mediator is too high, the EPS secretion is saturated. Under the long-term influence of the external mediator, the mediator is transported into the cell through the transmembrane, resulting in the inhibition of the metabolic function of the related enzymes, and may cause the shedding of the outer EPS (PS separation).

9.6.4 Other influencing factors

In addition to particle size, concentration of organic matter and nitrogen, and mediators, the effects of salinity, pulsed electric field, and signal molecules also affected EPS secretion from Anammox granular sludge. Studies have shown that adding a certain salinity can stimulate the secretion of EPS from granular sludge to avoid cell dehydration and inactivation. The EPS characteristics of Anammox granular sludge were analyzed under the conditions of different salinity (the mass concentration of 0, 15, 30 g/L NaCl), and it was found that the EPS content changed under different salinity stress. With the increase of salinity (NaCl mass concentration from 0 to 30 g/L), the PN content (calculated by the mass of EPS per gram of MLSS) decreased from (30.58 ± 2.50) mg/g to (18.11 ± 2.1) mg/g, while PS increased from (1.48 ± 0.09) mg/g to (10.52 ± 0.50) mg/g. Under the stress of salinity, the polysaccharide C–O groups of granular sludge EPS increased, which affected synthetase activity and metabolic flux reorganization. In addition, it was found that under the condition of intermediate frequency pulse voltage (1000 Hz, 1.5 V), the PN/PS of Anammox granular sludge increased by 28.46% and 54.20%, respectively, compared with the two conditions of low frequency and high frequency. ZHANG et al. inhibited the excessive secretion of EPS from Anammox granular sludge under high nitrogen loading by adding exogenous signal molecules (C8-HSL, *N*-octanoyl-DL-homoserinelactone). The PN/PS decreased from 4.09 to 2.14, and the addition of exogenous signal molecules directly affected the sedimentation performance of granular sludge.

9.7 Limitations of Anammox granulation

Notwithstanding the overwhelming benefits of the AnGS system as earlier discussed, it still has some limitations. These limitations discussed below to a large extent warrant the fervent research efforts it has seen in the past decade.

9.7.1 Flotation

This phenomenon is common in Anammox granular sludge systems due to high loading rates that cause the granules to lose their stability. Thus, higher loading rates leading to increased production of nitrogen gas as well as increased hydrodynamic conditions are the probable causes of flotation of Anammox granules. However, flotation could be the result of the formation of gas pockets inside the granules with mechanical breakage the effective means of counteracting it.

In the formation of Anammox granules, EPS combines Anammox clusters leaving a void space between them. Consequently, some of the nitrogen gas produced thereof gets accumulated in the voids thus formed between the Anammox clusters forming a nitrogen gas tunnel in the Anammox granules during high loading rates. The increasing size of the Anammox granules means a longer distance to be traveled by nitrogen gas in order to leave the granule. The production of EPS, however, obstructs the path of nitrogen gas leaving it trapped inside the granule. Hence, the gas pockets formed in the Anammox granule. The gas pockets expand causing the granule to float. This may easily result in biomass washout thereby minimizing the performance of the system. Also, a larger granule size allows for the formation of a dead zone in the granule which could result in sludge flotation. Thus, reducing the efficiency of the mass transfer between AnGS and pollutants (Figure 9-12).

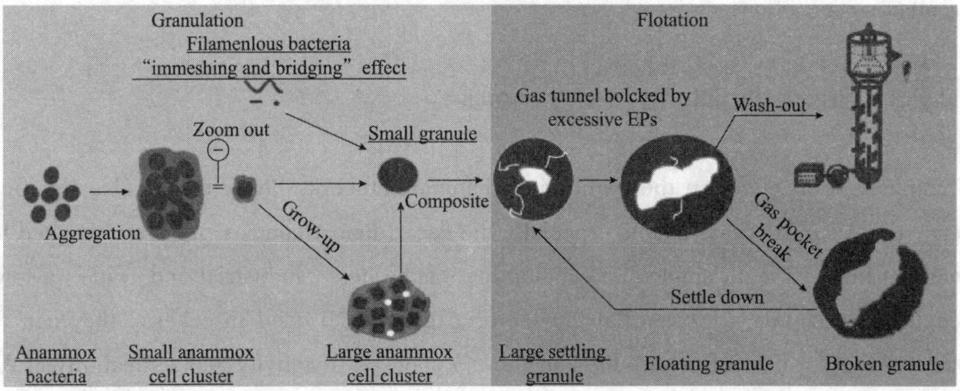

Figure 9-12 The hypothesized mechanism for granulation and flotation of Anammox biomass

In order to control the flotation and also optimize the performance of the system, a myriad of techniques have been studied. The technique of "collecting-breaking-returning of floating granules" was an effective strategy and resulted in a high nitrogen removal rate of 16.5 kg N/($m^3 \cdot d$). The common result of removal of only the gas bubbles attached to the surfaces of the granules but not the bubbles inside the granules was reported for studies based on both shear force enhancement and stirring speed. According to study, the highest activity was recorded by Anammox granules with a diameter of 1.0-1.5 mm. Though it could be practically difficult, it is advised to keep the diameter of the granule below 2.20 mm for optimum system performance. Hence, the Anammox granule density could be increased by precipitating the ions from the nutrient supplement without altering the activity and nitrogen removal rate.

9.7.2 Storage stability of Anammox granules

The first Anammox biomass storage demonstrated by using Anammox biomass from an oxygen-limited autotrophic nitrification and denitrification (OLAND) process reported that the operating temperature condition of -20 ℃ temperature was unsuitable for their survival. Over the years, researchers have employed various preservation techniques: lyophilization (freeze-drying), refrigeration, freezing (-20 to -200 ℃) and gel encasement with or without a cryoprotectant.

Pre-freezing at -200 ℃ with liquid nitrogen followed by lyophilization resulted in the successful reactivation of Anammox biomass. However, Anammox biomass was reported to have lost its activity after being immobilized in polyvinyl alcohol (PVA) gel at -8 ℃ for 17 h. In contrast, after 45 days, Anammox biomass entrapped in polyvinyl alcohol-sodium alginate beads at room temperature with 3 mmol molybdate was successfully reactivated registering a 95% specific Anammox activity. It is also possible to store Anammox sludge at room temperature with the regular addition of nutrients. Though there are quite a few Anammox sludge storage methods, new alternatives need to be devised seeing that the application of cryoprotective agents is very costive among other complications.

9.7.3 Susceptibility to heavy metals

Similar to their effect on the traditional Anammox process, heavy metals also significantly influence the AnGS process. This is especially the case when Anammox process is applied to the treatment of high-strength ammonium-containing wastewaters- industrial and swine wastewater. Although heavy metals are essential for bacterial growth, if used in excess, they can cause biological toxicity which influences the specific morphology, reactivity and settleability of AnGS. Discussed below are some of the heavy metals that have been reported to have greatly influenced the AnGS process.

Ferrous iron (Fe^{2+}) plays an essential role in heme-c proteins biosynthesis which in turn is reported to be a major part of AnAOB. An Fe^{2+} concentration of 0.09 mmol is reported to have greatly enhanced the specific Anammox growth rate by 45.8%. Similar report was corroborated., where 0.09 mmol Fe^{2+} concentration resulted in a shorter start-up time and a 32.2% increase in nitrogen removal rate. However, an Fe^{2+} concentration of 0.18 mmol is reported to deteriorate the reactivity of AnAOB.

Zn^{2+}, another heavy metal that influences the activity of AnAOB, serves as an enzyme cofactor and is typical in many ammonium-rich wastewaters. AnAOB are reported to be capable of adapting when exposed to a Zn^{2+} concentration between 1-10 mg/L, but are inhibited once the concentration exceeds 20 mg/L. It was also observed that the half maximal inhibitory concentration (IC50) was 25 mg/L which resulted in a decrease in the diameter of the sludge while the settling property was enhanced. In addition, in an experiment to determine the start-up characteristics of a granule-based Anammox UASB reactor seeded with anaerobic granular sludge, the precipitation of Zn-Fe was observed on the surface of the granules during operation was reported to be responsible for the inhibition of microbial activity thereby elongating the start-up time of the reactor.

Cu is an essential component of nitrite reductase of AnAOB. The IC50 of Cu^{2+} on the Anammox activity is reported to be in the range of 1.9-12.9 mg/L. Similarly, in investigating the long-term effects of Cu^{2+} on the Anammox process, it was observed that a 5 mg/L concentration could strongly inhibit the Anammox activity reducing it by about 94%. Accordingly, due to the influence of Cu^{2+}, the Anammox sludge had a lower settling velocity and a larger diameter. Observations by electron microscope revealed that Cu^{2+} could induce EPS secretion and cause cell membrane damage. The joint long-term application of Cu and Zn has also been reported to have resulted in a similar outcome. This causes the AnAOB to shrink and large amounts of precipitation around aggregates (Figure 9–13).

Thus, different heavy metals suppress the AnGS formation to varying degrees.

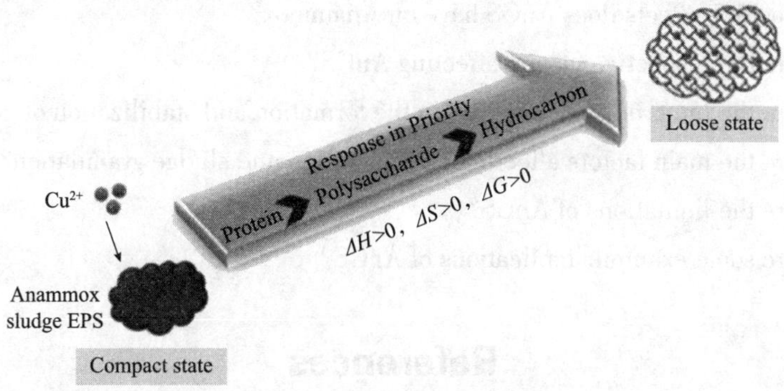

Figure 9–13 Schematic of Anammox sludge EPS evolution after the addition of Cu^{2+}

9.8 Conclusions

Foaming, scaling and flotation have been identified as persistent problems in the AnGS set up. The use of high NLR could lead to flotation of the granules which in turn could result in destabilization or even collapse of the system. Though the density of the granules could be improved significantly by precipitating the minerals from influent phosphate and calcium cations, this unfortunately leads to the decrease in the volatile suspended solids (VSS) to suspended solids (SS). Therefore, a lot more research still needs to be done in understanding the causation of flotation as well as the control strategies of granule flotation.

In light of what has been presented in this review, it stands to reason that more comprehensive research work be done in understanding the underlying principles of the biogranulation models. There is also the need to investigate the survival mechanism adopted by Anammox bacteria in the Anammox granular system owing to the initial competition between aerobic/anaerobic/inactive granules organisms. More research is also required to settle the issue of PN/PS ratio of Anammox granules (many contradictory reports). The potential environmental applications of nanoparticles should be further investigated and their effects on wastewater treatment processes (Anammox activity) ascertained.

Future Anammox processes could also focus on developing a visible indicator for monitoring the microbial community within the Anammox setup since they are essentially responsible for the process. The information to be gathered would greatly help in the optimization of Anammox process. Likewise, moving away from the use of lab-scale research reactors to the treatment and analysis of actual wastewaters such as pharmaceutical wastewater, landfill leachate will give a better understanding of the inner-workings of the Anammox process.

Questions

[1] What positive effects does AnGS have on Anammox?

[2] What are the respective factors affecting AnGS?

[3] What are the important roles of EPS in the formation and stabilization of granular sludge?

[4] What are the main factors affecting EPS secretion and sludge granulation?

[5] What are the limitations of AnGS?

[6] What are some example applications of AnGS?

References

[1] Baeten J E, Batstone D J, Schraa O J, et al. Modelling anaerobic, aerobic and partial nitritation-Anammox granular sludge reactors-A review [J] . Water Research, 2019, 149 : 322-341.

[2] Adams M, Xie J, Kabore A W J, et al. Research advances in Anammox granular sludge : a review [J] . Critical Reviews in Environmental Science and Technology, 2022, 52 (5) : 631-674.

[3] Wang P, Lu B, Liu X, et al. Accelerating the granulation of Anammox sludge in wastewater treatment with the drive of "micro-nuclei" : a review [J] . Science of the Total Environment, 2023, 860 : 160238.

[4] Jin R C, Yang G F, Yu J J, et al. The inhibition of the Anammox process : a review [J] . Chemical Engineering Journal, 2012, 197 : 67-79.

[5] Ran X, Zhou M, Wang T, et al. Multidisciplinary characterization of nitrogen-removal granular sludge : a review of advances and technologies [J] . Water Research, 2022, 214 : 118214.

[6] Izadi P, Eldyasti A. Holistic insights into extracellular polymeric substance (EPS) in anammosx bacterial matrix and the potential sustainable biopolymer recovery : a review [J] . Chemosphere, 2021, 274 : 129703.

[7] Chen C J, Cao Q F X, Zou X Y, et al. Advances in the study of the role, composition and influencing factors of EPS in anaerobic ammonia oxidized granular sludge [J] . Journal of Environmental Engineering, 2022, 16 (2) : 381-389. (In Chinese)

[8] Jia F X, Yang Q, Liu X, et al. Stratification of extracellular polymeric substances (EPS) for aggregated Anammox microorganisms [J] . Environmental Science & Technology, 2017, 51 (6) : 3260-3268.

[9] Zhu G, Wang S, Ma B, et al. Anammox granular sludge in low-ammonium sewage treatment : not bigger size driving better performance [J] . Water Research, 2018, 142 : 147-158.

[10] Chen J, Ji Q, Zheng P, et al. Floatation and control of granular sludge in a high-rate Anammox reactor [J] . Water Research, 2010, 44 (11) : 3321-3328.

Chapter 10 Application of Anammox

Emphasis: Application of Anammox in various wastewater treatments; Process-influencing factors; Enhancing Engineering Practices in Anammox Wastewater Treatment; Nitrogen removal performance; Research on nitrogen removal microorganisms.

Nut: Nitrogen removal performance.

Anaerobic ammonia oxidation (Anammox) means under the condition of the anaerobic or anoxia, anaerobic ammonia oxidation bacteria (AnAOB) to NO_2^--N as electron acceptor, NH_4^+-N is the biological process of nitrogen (N_2). Its advantages are as follows: ① oxygen demand is reduced by 62.5% compared with the traditional, which can effectively reduce oxygen charging power consumption; ② without organic carbon source, operation cost is reduced; ③ Without involving heterotrophic denitrifying bacteria, a significant reduction in residual sludge production can be achieved. With the popularity of the Anammox process, this new nitrogen removal technology has been successfully applied to treat a variety of wastewaters, including urban domestic sewage, landfill leachate with high NH_4^+-N concentrations and low C/N ratios, and swine wastewater.

10.1 Urban domestic sewage

10.1.1 Realization of a compatible Anammox process

Deep denitrification of urban domestic sewage has become one of the important ways to curb water eutrophication. Denitrification by traditional nitrification and denitrification needs to go through two processes: aeration nitrification and anoxic denitrification. In the process of nitrification, most organic matter is removed by aerobic oxygen, resulting in the lack of organic carbon source in denitrification process. At the same time, the traditional denitrification wastewater treatment process requires a large amount of aeration, high energy consumption, release a large amount of greenhouse gases in the operation process, and the amount of surplus sludge is large, which is contrary to the theme of today's green, energy saving and sustainable development economic model.

With the development of urbanization, the regeneration of urban sewage and energy recovery have increasingly become the focus of research. The application of the autotrophic Anammox process is expected to enable municipal wastewater plants to achieve energy self- sufficiency. In addition, the Anammox process has a high volume efficiency of up to 125 kg/ ($m^3 \cdot d$), which can greatly reduce the reactor volume and significantly reduce the investment cost. Therefore, if Anammox process can be applied to denitrification in urban wastewater treatment plants, it can significantly reduce the aeration rate during nitrification and the added amount of denitrification organic carbon source, thus reducing the treatment cost.

10.1.2 Influencing factor

Anammox has strict requirements on water quality, reaction conditions and other influencing factors. Whether the technology can be applied to urban domestic sewage is the prerequisite. The Anammox process includes two processes: short-cut nitrification and Anammox, and environmental factors can affect both stages. It is generally believed by researchers that the main difficulty in realizing Anammox process in urban domestic sewage is stable short-cut nitrification, that is, NH_4^+-N and NO_2^--N in appropriate proportions should be provided at the front of Anammox. Secondly, in the Anammox stage, AnAOB has very strict requirements for condition control, such as organic matter, sludge retention time (SRT), dissolved oxygen (DO), temperature and sludge interception will affect the stable operation of Anammox process.

10.1.2.1 Organic matter

In the process of nitrosation, when the concentration of organic matter in the influent was higher and the chemical oxygen demand (COD) was higher, it would promote the growth of heterotrophic bacteria and cause them to interact with ammonia-oxidizing bacteria (AOB) competes for living space and inhibits AOB activity. Therefore, many scholars have conducted researches based on the appropriate concentration range of COD. The nitrosation rate decreased with increasing COD concentration. When the COD concentration ranged between 60-120 mg/L, AOB remained dominant, forming a stable partial nitritation reaction. As the Anammox reaction utilizes inorganic carbon as the carbon source, the presence of organic matter negatively impacts the Anammox process, promoting heterotrophic denitrification and inhibiting the Anammox process. In experiments studying the effect of COD concentration on the initiation of Anammox, it was observed that higher COD concentrations led to increased denitrifying bacteria activity, decreased AnAOB growth activity, and prolonged the initiation time of Anammox. AnAOB was unaffected at a COD concentration of 120 mg/L. The COD concentration in the readily biodegradable section of Anammox should not exceed 150 mg/L.

10.1.2.2 SRT

SRT is an important factor in Anammox process. In the process of nitrosation, when the temperature is controlled at 20–35 ℃, the growth rate of AOB is higher than that of nitrite oxidizing bacteria (NOB), and the SRT of the control system can be larger than the generation time of AOB and smaller than that of NOB. NOB is discharged from the system while AOB is maintained at a high concentration, thus achieving a stable accumulation of NO_2^-–N. Many studies have shown that different researchers use different SRT to achieve stable nitrosation. For example, the step-feed activated sludge (SFAS) process of Singapore Changi Recycling Water Treatment Plant uses a shorter SRT (about 2.5 days) to achieve stable nitrosation. SRT has several effects on AnAOB and is a crucial parameter for inducing changes in population structure. Shortening the SRT has been found to be beneficial for the Anammox reaction and efficient nitrogen removal. The impact of gradually reducing the SRT on NO_2^-–N removal load and NO_2^-–N sludge load was investigated in a set of SBR reactors. The results revealed that the overall nitrogen removal performance of the system decreased by 16.4% when the SRT gradient was reduced from 21 days to 12 days. However, the nitrogen removal efficiency per unit mass of AnAOB increased by 66.8%. A short SRT (approximately 3 days) was implemented at the Changi Reuse Water Treatment Plant in Singapore to achieve the Anammox effect in the anoxic zone. This study suggests that a shortened SRT is a promising approach to realize mainstream Anammox process for urban domestic sewage treatment.

10.1.2.3 DO

AOB has a higher half-saturation coefficient for oxygen than NOB, and NOB has a weaker ability to compete for oxygen under low DO. Controlling the concentration of DO and adjusting the aeration mode is one of the control means to achieve short-cut nitrification. According to the DC Water study, transient hypoxia can effectively inhibit NOB. According to the study, the rate of NO_2^-–N accumulation gradually increased as DO concentration decreased from 1.4 mg/L to 0.7 mg/L. In the SBR reactor, the NO_2^-–N accumulation rate reached 92% with an aeration mode of 4 minutes aeration and 2 minutes de-aeration, coupled with a DO limitation strategy (1.3–1.7 mg/L), resulting in the successful achievement of nitrification. This success can be attributed to the faster recovery of AOB compared to NOB after anoxic disturbance, allowing for partial nitrification (PN) during intermittent aeration. The control of DO concentration is a prerequisite for the application of Anammox process in urban domestic sewage treatment. The effect of DO on Anammox was investigated using an SBR reactor, revealing that the activity of AnAOB was completely inhibited at DO levels of 0.5%, 1.0%, and 2.0% of air saturation. However, after transitioning to anaerobic conditions, the activity of the previously inhibited AnAOB was found to recover, indicating that the inhibition of AnAOB by DO is reversible. In conventional municipal wastewater treatment, the

DO level is typically maintained at 2 mg/L. In such cases, the inhibition of AnAOB by DO in urban domestic sewage can be mitigated through the regulation of DO and the utilization of bio-carriers. When AOB and AnAOB coexist in the system, AnAOB can maintain normal activity even if the DO concentration exceeds 0.2 mg/L. Moreover, in an integrated Anammox reactor with a thick biofilm, AnAOB can thrive even at high DO levels (5–6 mg/L), demonstrating stable and high nitrogen removal performance. Therefore, the use of intermittent aeration and a DO limitation strategy may serve as a promising approach to facilitate the mainstream implementation of Anammox process for urban domestic sewage treatment

10.1.2.4 Temperature

The effect of temperature on nitrosation is significant. Higher temperature can not only improve the activity of AOB, promote the growth rate of AOB, but also further expand the difference of activity and growth rate between AOB and NOB. The increase in nitrite accumulation at temperatures above 30 ℃ was attributed to the inhibition of NOB activity and the enhancement and dominance of AOB activity. Dutch Delft university of technology and development of SHARON (single reactor for high ammonium removal over nitrite) process is based on this principle in the medium temperature (30–40 ℃) under the condition of realizing the stable shortcut nitrification, and this process has been engineering application. The temperature of urban domestic sewage is generally kept at 10–16 ℃ in winter and 24–30 ℃ in summer. How to achieve nitrosation at low temperature is an important research content. When the temperature was lowered from 35 ℃ to 21 ℃, the activity of AOB decreased by only 8%, and when the temperature was lowered to 15 ℃, the activity of AOB was still able to maintain 75% of the optimal value, suggesting that nitrosation can be achieved at room temperature. In the Anammox segment, the optimal growth temperature of AnAOB is 30–37 ℃. The study revealed that in an Anammox treatment reactor for urban domestic sewage, the activity of AnAOB significantly decreased, dropping from 0.465 mg N/(L · d) at 29 ℃ to 0.046 mg N/(L · d) at 12.5 ℃. Achieving stable Anammox operation at room temperature has become a significant area of research focus. In a study employing UASB reactors and spanning natural periodic temperature variations from 9 to 28 ℃ across four seasons, the research demonstrated that the reactor maintained a high nitrogen removal capacity, particularly at medium and low temperatures (11–25 ℃). The enrichment rate of AnAOB in UASB granular sludge reached 53.8%, and a transition in dominant bacterial species occurred, shifting from *Candidatus Brocadia* to *Candidatus Kuenenia*. In another independent study, Anammox was effectively conducted using an EGSB reactor at 13 ℃. *Candidatus Kuenenia* was identified as the prevailing strain, comprising 57.0% of the relative abundance. Furthermore, it was noted that AnAOB retained its activity even under conditions of low temperature and reduced load. Researchers concluded that, within the temperature range characteristic of urban domestic sewage, temperature had only a marginal impact on AnAOB. They recommended that an effective approach to ensure operational efficiency involves reducing the inlet nitrogen load. All the studies showed that the inhibition of AnAOB activity by moderate and low temperature could be effectively alleviated by microbial strengthening and nitrogen loading regulation.

10.1.2.5 Sludge interception

AOB and AnAOB are autotrophic bacteria with long generation cycle, and sufficient microbial biomass must be guaranteed to ensure good nitrogen removal performance. There are two main methods for sludge interception of AOB. One is to use the characteristics of high NH_4^+–N and high temperature water quality in the side flow section of urban sewage treatment process to promote the growth of AOB, and the side flow supplement the mainstream process to supplement microorganisms. Another method is to retain it in the form of biofilm or granular sludge, which mainly depends on the attachment of AnAOB to the innermost layer of the filler and AOB to the outer layer of the filler. Through a comparison of suspended sludge and biofilm systems, it was observed that when the DO concentration was maintained at 3.0 mg/L, the abundance of AOB on the biofilm surpassed that of the suspended sludge, reaching 20.43%. Results from the cultivation of granular sludge showed that the nitrosation characteristics remained highly stable, with AOB accounting for 17.8% of the relative amount, while NOB made up less than 0.6%. Bio–enhancement of AnAOB was complemented by bio–enhancement of the sidestream. The hydrocyclone played a central role in separating AnAOB and AOB. This allowed AnAOB in the sidestream to be introduced into the mainstream process via the cyclone, while the overflow rich in AOB also entered the mainstream process. The operation experience of Strass sewage treatment plant in Austria shows that after the use of hydrocyclone technology to supplement AOB and AnAOB from side–flow process to mainstream process, the activities of AOB and AnAOB in side–flow process are not reduced, but the abundance and particle size of AnAOB in mainstream process are significantly increased. It was found that the circulation of hydrocyclone could also prevent microbial enrichment on the surface of AnAOB, which was conducive to reducing matrix diffusion resistance and maintaining high microbial activity. In addition, the side–flow supplementation of the mainstream can also solve the problem that the affinity of AnAOB to NO_2^-–N is lower than that of NOB, making it easier for AnAOB to obtain NO_2^-–N. In addition, the use of biofilm and granular sludge to improve the biomass of AnAOB has also become a research hotspot. An analysis of the microbial community showed that in a Moving Bed Biofilm Reactor (MBBR), the primary AnAOB accounted for 4.53% of the total bacteria in the biofilm and 0.75% of the total bacteria in the activated sludge. This indicates a higher abundance of AnAOB in the biofilm compared to the activated sludge. Additionally, it was found that the strength of sludge particles was positively correlated with AnAOB when operating the Anammox process at medium and low temperatures. This finding holds significant promise for the implementation of mainstream Anammox process in the interception of urban sewage sludge.

In conclusion, the application of Anammox technology in urban sewage still faces a series of difficulties. However, the Anammox reaction can be stably operated by maintaining COD concentration below 150 mg/L, providing suitable proportions of NH_4^+–N and NO_2^-–N by achieving stable short-cut nitrification, carrying out intermittent aeration while maintaining appropriate DO, adopting microbial enhancement, nitrogen load regulation and other measures. At the same time, the sludge can be retained by supplementing the mainstream or in the form of biofilm or granular sludge to ensure good denitrification performance.

10.1.3 Treatment process

At present, Anammox treatment technology is adopted for urban domestic sewage treatment plants, which mainly includes two kinds: ① Side flow process, that is, the treatment of sludge digestion solution, pressure filtrate and other sewage, the water volume only accounts for about 2% of the total influent of the sewage plant, and the water quality is usually characterized by high temperature and high NH_4^+–N, as shown in Figure 10–1; ② Mainstream process, that is, Anammox process is applied to the main line of urban sewage treatment process, as shown in Figure 10–2.

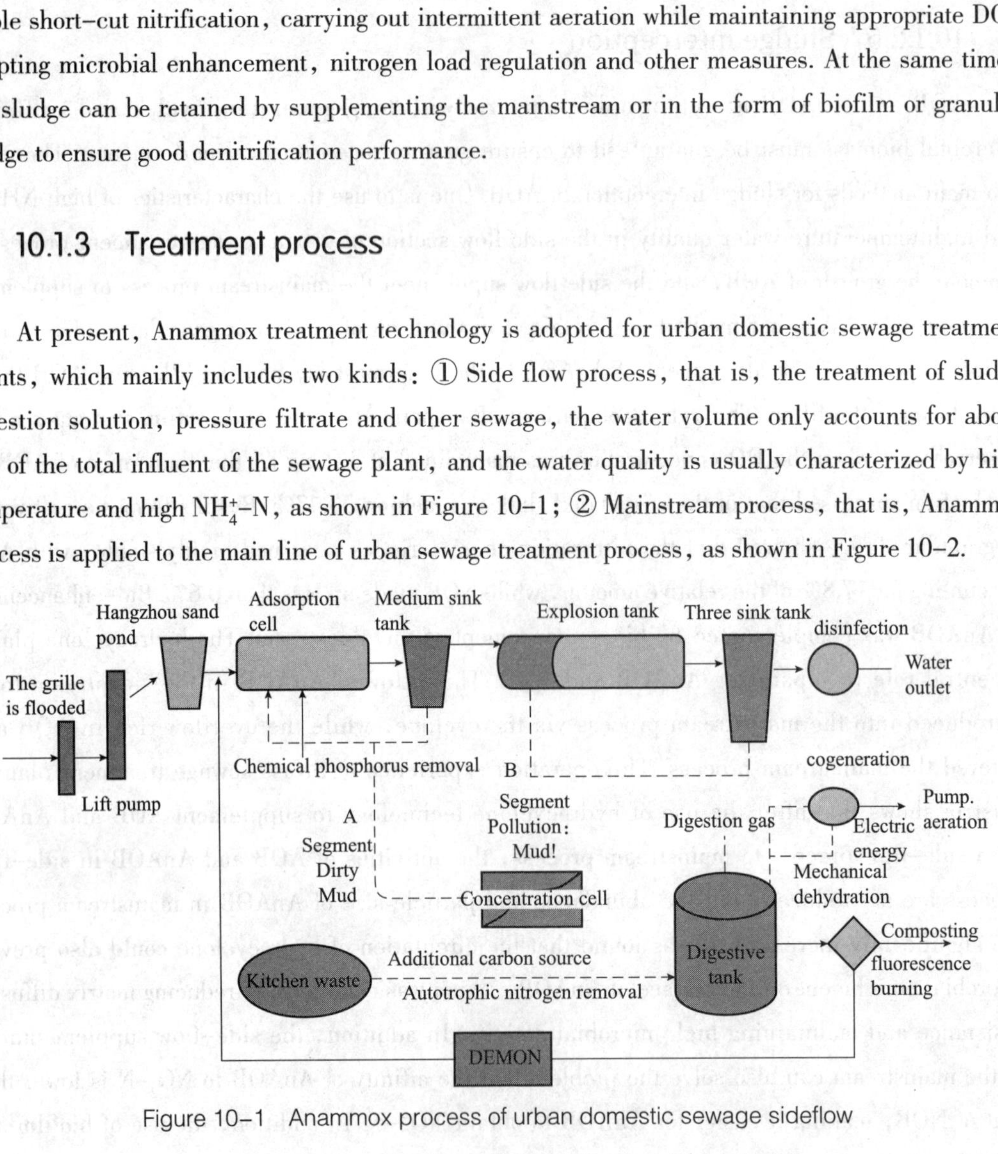

Figure 10–1 Anammox process of urban domestic sewage sideflow

10.1.3.1 Mainstream treatment process

The nitrogen content of urban domestic sewage is low, with TN in the range of 20–75 mg/L, and the operating temperature is low, which is kept at 10–16 ℃ in winter and 24–30 ℃ in summer. There are limitations to quickly start Anammox. Researchers have successfully started Anammox by using urban domestic sewage, using different process forms, inoculating different sludges and regulating operation conditions. See Figure 10–2 for details.

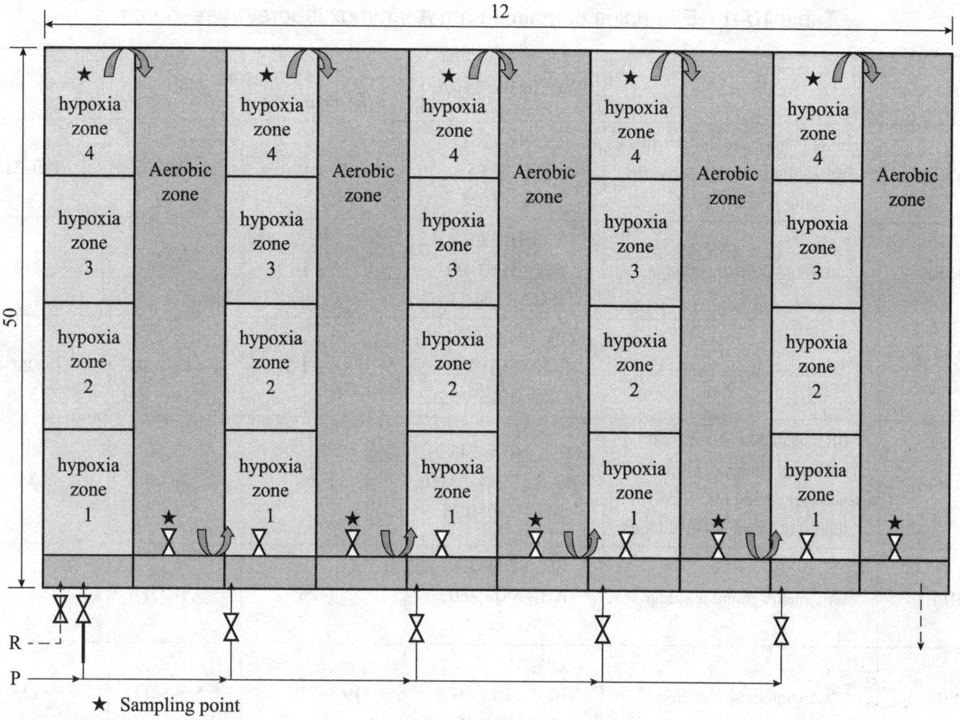

Figure 10–2 Singapore Changi Water Reuse Plant Stage Feed Water reactor (mainstream Anammox) process

Table 10–1 lists the successful start–up examples of seven mainstream Anammox laboratory studies of urban domestic sewage. Most researchers use SBR as the start–up process of Anammox. Before the treatment of urban domestic sewage, AnAOB is firstly cultured and enriched to make it stable. It has been determined that inhibition of Anammox by temperature and other factors can be mitigated, ensuring stable operation within the mainstream of urban domestic sewage. In laboratory settings, Anammox has consistently demonstrated effective operational stability. To address this, a process involving front–end carbon removal in an Anaerobic Biological Reactor (ABR), coupled with nitrosation and Anammox, was applied to treat urban domestic sewage. The objective was to maintain the COD concentration of ABR effluent at 120 mg/L, preventing adverse effects on the subsequent coupled system. Additionally, the ratio of effluent from the nitrosation system and ABR effluent was controlled to meet Anammox requirements. The results revealed a TN removal rate exceeding 86% and a COD removal rate exceeding 85%. This front–end carbon removal method presents a feasible concept for engineering applications. Furthermore, the implementation of processes such as denitrification and Partial Denitrification/Anammox (PD/A) within the context of urban sewage mainstream Anammox processes has shown promising prospects. Through experiments using granular sludge, successful implementation was achieved. Under conditions with an influent TN concentration of 42.56 mg/L and a C/N ratio of 1.48, the average effluent TN was 8.74 mg/L, resulting in a nitrogen removal rate of 80%.

Table 10-1 Examples of mainstream Anammox laboratory research

Process form	Sludge source	Water intake/(mg/L)	Process temperature/ ℃	pH	Startup time/ d
ASBR	Anaerobic granular sludge: black river sludge: aerobic activated sludge = 2 : 2 : 1	COD: 125; NO_2^--N: 25; NH_4^+-N: 25	34 ± 1	7.8	180–210
ASBR	Aerobic nitrifying sludge	COD: 280; NH_4^+-N: 63.45; NO_2^--N: 84.53	35 ± 1	7.58	130
A/O + Anammox	Anammox sludge containing	SCOD: 155.28 ± 26.70; NH_4^+-N: 57.31 ± 3.36; NO_2^--N: 0.15 ± 0.24	28.9 ± 1.3	7.44 ± 0.10	35
SBR+SBBR	SBR, secondary sedimentation tank reflux sludge of municipal sewage treatment plant; SBBR, short-cut nitrification sludge	COD: 160.4–243.9; NH_4^+-N: 32.58–58.00; NO_2^--N: 0–0.26	22–28	7.3–7.6	80
SDAD	Anammox granular sludge	COD: 200–300; NH_4^+-N: 60–80; NO_2^--N <1	24–26	7.5–8.0	74
SBR	Anammox sludge floc	COD: 61.3–144.7; NH_4^+-N: 18.1–31.7; NO_2^--N: 0–0.4	19.5–31.7	5.6–8.5	133
CANON	Nitrifying sludge: Anammox granular sludge =1 : 1	COD: 96.9–325.6; NH_4^+-N: 41.2–56.9	30	7.01–7.33	110

Analysis of the microbial community in the reactor showed that AnAOB accounted for 7.79% of the total number of bacteria in the particles, showing excellent biological retention ability. From the genus analysis, *Thauera* (17.46%) and *Candidatus Brocadia* (6.24%) were the most abundant. *Thauera* also has the function of partial denitrification (PD), co-promoting Anammox reaction, and 90% of denitrification contribution is attributed to AnAOB. Compared with the traditional nitrification and denitrification process, PDA process can save 45% aeration and 79% carbon source. In recent years, the research on the mainstream process, process combination, reactor design, and optimization of control means have all provided reference for the application of the mainstream Anammox process.

10.1.3.2 Sidestream treatment process

The water pH of sludge digester and sludge pressure filtrate is generally 7.0–8.5, the temperature is 30–37 ℃, the concentration of NH_4^+-N can reach 500–1300 mg/L, and the C/N ratio is about 1. It is a typical wastewater with high NH_4^+-N and low C/N ratio. The treatment of traditional sludge digester and pressure filtrate is generally returned to the grate and mixed with raw water for treatment, which will increase the nitrogen loading of the overall process and ultimately lead to higher TN of the effluent. However, due to the competition mechanism between AOB and NOB and the survival habit of AnAOB, the water quality conditions of medium and high temperature and high

NH_4^+–N of sludge digester and sludge pressure filtrate can guarantee the dominant population status of AOB and inhibit the activity of NOB, which provides conditions for the application of Anammox.

In 2002, Dokhaven Sewage Treatment plant in Rotterdam, the Netherlands was the first to implement the application of Anammox process for urban domestic sewage sideflow. Its production Anammox reactor uses the SHARON–ANAMMOX system to treat sludge digesters. The Anammox reaction was the dominant reaction in the reactor.

Up to now, about 75% of the world's Anammox engineering installations are mainly used to treat sludge digesters and pressure filtrates, and the processes often used include integrated DEMON (deammonification, Aerobic reverse ammonification), CANON (completely autotrophic nitrogen removal over nitrite) process, split SHARON–ANAMMOX process, etc. The specific technical form, reactor volume, sludge concentration, hydraulic retention time, sludge nitrogen load, TN removal rate and energy consumption are shown in Table 10–2.

Table 10–2 lists the operation efficiency of 11 Anammox engineering applications to treat sludge anaerobic digestion solution of urban domestic sewage treatment plants. It is found that SBR is the most commonly used reactor system in the side–flow Anammox process, and DEMON is the most commonly used configuration in the SBR system. It is found that DEMON adopts a hydrocyclone to regulate the SRT of AnAOB and AOB, which can effectively separate the slow–growing AnAOB and make the Anammox reaction run effectively.

The large–scale engineering application of Anammox in urban domestic sewage sideload has also exposed problems such as long start–up time and low AnAOB biomass. Laboratory research and engineering exploration on related optimization are also being carried out.

Table 10–2 Example of Anammox engineering application of sludge digestion solution

The name of project	Process form	Reactor volume/m^3	Sludge concentration/ (g/L)	Hydraulic retention time/ h	Sludge load/ [g/(kg · d)]	Ammonia nitrogen removal rate /%	The energy consumption/ (kW · h/kg)
Amersfort	DEMON	780	4.5	26	145	>64	—
Apeldoorn	DEMON	2400	3.5–4	58	155	>84.5	1.10
Balingen	DEMON	750	1.2	94	35–95	>69.8	0.92
Heidelberg	DEMON	2 × 570	1.0–2.5	114	119	>89.5	1.67
Ingolstadt	SBR	2 × 560	2.0–4.0	75	71	>74.9	1.92
Nieuwegein	DEMON	450	—	42	—	>68	—
Plettenberg	DEMON	134	3	40	149	>69	1.11
Zurich	SBR	2 × 1400	3.5–4.5	45	134	>89.9	1.11
Malmo	ANITA Mox	4 × 50	16	24	64	>79.5	1.45–1.75
Landshut	TERRANA	288+495	—	25,42	101,108	>89.9	—
Rotterdam	SHARON–ANAMMOX	1800+70	0.27,7–10	36,6	238	>84.5	4.17

In an experiment, tourmaline was utilized to enhance the Anammox process. In comparison to the Anammox system without tourmaline, this resulted in an 11% reduction in startup time, and the nitrogen removal rate reached 85.5%. Tourmaline proved to be highly effective. It is believed that tourmaline enhances the metabolic capacity and activity of AnAOB due to its electrical polarity and spontaneous polarization effects. Additionally, tourmaline acts as a filler, promoting the growth of AnAOB. A separate study employed a MBBR to treat sludge digestate, successfully achieving the coupling of short-cut nitrification and the Anammox process. The maximum NH_4^+-N removal rate reached 96%. Analysis of microbial communities revealed that AOB accounted for 10.46% and 21.46% of the total bacteria in the MBBR biofilm and activated sludge, while AnAOB accounted for 4.13% and 0.71%. This suggests that primary nitrosation and Anammox were respectively completed in the activated sludge and biofilm. The results demonstrate the feasibility of applying Anammox process in an integrated reactor for treating sludge digestion solutions, providing a critical foundation for engineering applications. In practical engineering exploration, a team at the Beijing Gaobeidian Wastewater Treatment Plant successfully implemented a full-size Partial Nitrification/Anammox (PN/A) process for sludge digestion liquid using an integrated fixed-biofilm activated sludge (IFAS) reactor. This reactor has effectively initiated PN/A and is operating steadily. It has a processing capacity of up to 2500 m^3/d, and the nitrogen removal efficiency exceeds 85%. This will save US $3.5 million in operation costs every year. Analysis of microbial communities showed that the abundance of AOB in flocculating sludge (1.2%) was higher than that in biofilm (0.2%). In contrast, the abundance of AnAOB in biofilm was significantly higher (17.9%) than that in flocculating sludge (0.02%). AOB and AnAOB co-existed in the reaction system to complete autotrophic nitrogen removal together.

10.1.4 Engineering practice

Urban sewage mainstream Anammox process also is less, the practical application of engineering is currently the most successful Singapore Changi recycle water, Austria Strass sewage treatment plant, and xi'an fourth sewage treatment plant in China, the three mainstream sewage plant Anammox process of practical application for the process of large-scale promotion provides a good support.

The Singapore Changi Reuse Water Plant was the first to implement the engineering application of the mainstream Anammox process with a five-point segmentalized inlet activated sludge process (SFAS) with a hydraulic retention time of 5.6 h and a total SRT of 5 days, as shown in Figure 10-3. Due to the geographical location and climate, the water temperature can be kept at about 30 ℃, which is conducive to the survival and reproduction of autotrophic bacteria. Through anoxic aerobic exchange, short SRT and on-line control, AOB was the dominant flora. The analysis showed that the accumulation rate of nitrite could reach 76%, and stable short-cut nitrification was achieved. The content of AnAOB in the reactor could reach 105 to 106 copies/mL, and the dominant bacteria were AOB and suspended free AnAOB. The contribution rate of the autotrophic denitrification process was 37.5%, and the effluent quality was good, TN <5 mg/L, NH_4^+-N < 3 mg/L. In addition, the SFAS process at Changi Reuse Water Treatment Plant reduces energy consumption by 10% to 30% compared to other reuse water treatment plants in Singapore.

Figure 10-3 Singapore Changi Reuse Water Plant

Strass wastewater treatment plant in Austria adopts AB process, which can achieve energy self-sufficiency rate of 108%, as shown in Figure 10-4. In 2004, the DEMON process was first adopted in the side flow. The designed inlet nitrogen load of DEMON unit is 300 mg N/(L · d), and the volume load is 0.6 mg N/(L · d). The DO is 0.3 mg/L. On the basis of side flow, the plant began to explore the application of mainstream Anammox process in 2011. The specific measure is to install a hydrocyclone on the mainstream process, and use the hydrocyclone to adjust the SRT of AnAOB and AOB. The lighter AOB overflow from the top and the heavier AnAOB gather at the bottom and return to the reactor. AnAOB continues to flow back into the mainstream process. In terms of aeration control, good nitrification and denitrification reactions can be achieved through flexible aeration control mode, that is, automatic adjustment of aeration volume and aeration frequency according to instantaneous load of influent water, so as to achieve stable operation of mainstream Anammox. The Strass wastewater treatment plant achieved an autotrophic nitrogen removal efficiency of more than 85%.

Figure 10-4 Strass wastewater treatment plant

Xi'an No. 4 Sewage Treatment Plant is the first large-scale sewage treatment plant (250 thousand tons/day) to realize Anammox under temperate climate conditions worldwide, which fills the gap in the research of production plant at room temperature in the world, as shown in Figure 10-5. Anammox is mainly implemented in the first phase of 250000 tons/day engineering transformation, and the main process is positive A_2/O^+MBBR (moving bed biofilm reactor). HRT is 1h in the anaerobic section, HRT is 3.6h in the anoxic section, and fillers are added in the anaerobic and anoxic section. The first phase of Xi'an No. 4 Sewage Treatment Plant has been transformed since November 2012. Since August 2013, the effluent quality has been stable and reached the Grade A standard, among which the effluent TN reaches 10 mg/L. By high-throughput analysis of microorganism in filler and suspended sludge, anaerobic zone and anoxic zone, the carrier has a high Anammox activity of 52 mg N/(L · d). The dominant AnAOB is *Ca.Kuenenia* (K-type AnAOB), which is highly concentrated on the carrier in the anoxic zone. In 2018, the Anammox reaction in anoxic environments was further confirmed by isotopic tracer method, which confirmed that Anammox is a very important pathway in anoxic regions, and the quantitative results showed that Anammox process accounted for about 30% of nitrogen removal. The abundance analysis of carrier AnAOB from 2015 to 2018 showed that the abundance of AnAOB on the filler in the anoxic zone gradually increased and reached nearly 10% in 2017. Mainstream Anammox is running stably.

Figure 10-5 Xi'an No. 4 Sewage Treatment Plant

The analysis of the mainstream Anammox process in the above three sewage treatment plants shows that the abundance of K-type AnAOB and B-type AnAOB (*Ca. Brocadia* sp.40) in the fillers is very different. The dominant AnAOB in Singapore Changi sewage treatment plant is B-type AnAOB. However, K-type AnAOB was found to be dominant in Xi'an Fourth Sewage Treatment Plant, which may be mainly caused by temperature. The water temperature in Singapore Changi Sewage Treatment Plant is 30 ℃, while the water temperature in Xi'an Fourth Sewage Treatment plant is 11-20 ℃. The above further shows that the mainstream Anammox process can be realized at room temperature.

10.2 Industrial wastewater

10.2.1 Landfill leachate

Landfill leachate contains high concentration of NH_4^+-N and lack of organic carbon source, which makes it difficult to reach the standard treatment, which is the biggest challenge facing the way of urban solid waste landfill. Currently, a combined approach involving physical, chemical, and biological processes is the predominant method for treating landfill leachate. In practice, the traditional nitrification denitrification process is currently the most commonly used biological nitrogen removal method, but the process suffers from high aeration energy consumption, production of large amounts of greenhouse gases and high costs as well as carbon sources in long-term operation, which limits its further application. In particular, the low COD/N and BOD_5/COD ratios of mature landfill leachate do not lend themselves to the application of the process. Anammox processes with granular sludge systems, biofilm systems, single-stage systems and multi-stage systems have all been applied to treat mature landfill leachate. The sidestream Anammox process has been successfully applied in engineering and more than 100 high ammonia wastewater treatment plants have been built.

10.2.1.1 Characteristics of landfill leachate

The composition of landfill leachate shifts depending on the age of the landfill; in particular, leachate that has been contained in a landfill for more than ten years is referred to as mature leachate. This section discusses the constituents of landfill leachate. As shown in Table 10.3, the literature covers a vast array of landfill leachate components that can be broadly classified into four groups: dissolved organic matter, inorganic macromolecules, heterogeneous organic matter, and suspended solids. The primary synopses are as follows:

- Dissolved organic matter, quantified as chemical oxygen demand (COD) or total organic carbon (TOC), volatile fatty acids and more difficult to degrade compounds [fulvic acid compounds and humic acid (HA) compounds], COD is a critical parameter for treatment.
- Inorganic ions: include heavy metals and dissolved salts, classified as Cadmium (Cd^{2+}), chromium (Cr^{3+}), copper (Cu^{2+}), lead (Pb^{2+}), nickel (Ni^{2+}), zinc (Zn^{2+}), calcium (Ca^{2+}), magnesium (Mg^{2+}), sodium (Na^+), potassium (K^+), ammonium (NH_4^+), iron (Fe^{2+}), manganese (Mn^{2+}), chloride (Cl^-), sulphate (SO_4^{2-}) and hydrogen carbonate (HCO_3^-), where the concentration of NH_4^+ is typically >1000 mg/L.
- Heterogeneous organic compounds: derived from household or industrial chemicals in

relatively low concentrations (<1 mg/L), including various aromatic hydrocarbons, phenols and plasticizers.

- Suspended solids: Suspended solid waste particles, such as soil particles and rubbish debris.

In summary, high concentrations of nitrogen and organic matter are the main pollutants in landfill leachate, and high concentrations of organic matter can reduce the removal of TN and cause significant difficulties in leachate treatment. Therefore, it is essential to dilute these wastewaters with significant concentrations of organic compounds.

Table 10-3 Analysis of landfill leachate composition and properties (unit: mg/L, except for pH)

Number	1	2	3	4	5	6	7	8	9
pH	6.5–7.5	7.2–7.9	7.2–8.0	7.45	7.29–8.61	7.61	8.85 ± 0.42	8.8	8.1
COD	13212–52334	7830–28120	14000–33000	7000	580–1821	482	920 ± 115	2880	7109.8
BOD_5	5765–19280	7720	6600–21500	4800	76–701	29.5	202 ± 56	65	2313
NH_4^+-N	2122–3213	1728–4998	2070–2730	17.51	66–364	367	0.33 ± 0.12	33.6	2331.7
TN	2312–3706	2240–6048	2100–3130	28	95–401	—	0.46 ± 0.07	3020	2494.5
TP	29–35	—	17–37	<0.1	1.4–15.7	3.84	—	58500	30.9
Ca	525–2933	120–721	250–900	256.51	192–430	1930	—	1850	35.49
Mg	311–670	195–340	420–600	—	126–419	640	—	651	207.45
Fe	15–369	2.7–12.6	30–100	9.75	—	—	0.154 ± 0.016	—	8.45
Co	0.04–0.2	—	0.5–1.3	—	—	—	—	—	—
Ni	0.2–0.9	0.3–0.4	1.2–1.7	1.01	—	—	0.140 ± 0.004	—	0.41

10.2.1.2 The Nitrogen removal performance of Anammox for leachate treatment

Combined with a large number of current studies, processes based on Anammox treatment of landfill leachate are shown in Figure 10-6, which mainly include the following: one-stage PN/A, two-stage PN/A, and other combined processes. The SHARON-ANAMMOX process can effectively treat landfill leachate. When the temperature was (30 ± 1) ℃, the NH_4^+-N load of landfill leachate ranged from 0.27 to 1.2 mg N/(L · d), and the DO was 0.8–2.3 mg/L. The SHARON section consistently produced a suitable ratio of NO_2^--N to NH_4^+-N between 1.0 and 1.3, which was ideal for subsequent Anammox treatment. Under controlled conditions of (30 ± 1) ℃ temperature and an inlet ammonia load of 0.06–0.11 mg N/(L · d), the Anammox stage removed 60% of NH_4^+-N and 64% of NO_2^--N. The removal rates for NH_4^+-N, TN, and COD reached 97%, 87%, and 89%, respectively.

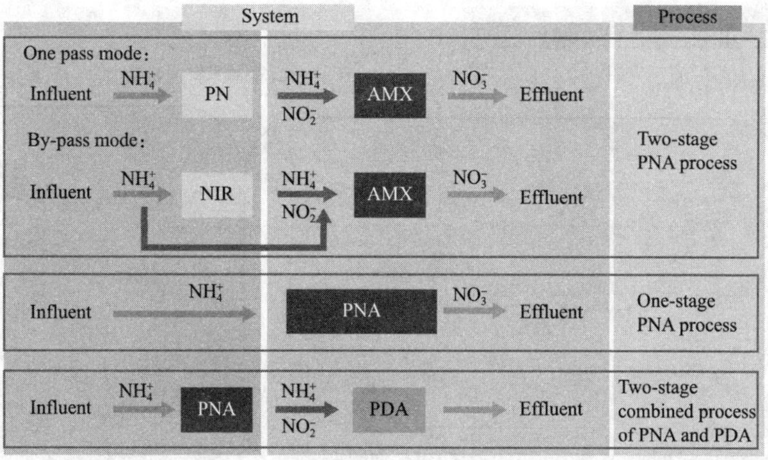

Figure 10–6 Different process routes for the treatment of landfill leachate by Anammox

In 2012, the SHARON–ANAMMOX process was applied to treat landfill leachate that had undergone prior treatment using UASB and MBR. The overall removal efficiency for both COD and Total Kjeldahl Nitrogen (TNK) exceeded 90%. In 2011, a SBR was utilized to achieve PN of landfill leachate, which served as a pretreatment solution for Anammox. The research highlighted the significance of pH as a critical factor influencing nitrification. The mass concentration of free ammonia was measured at (605.48 ± 87.18) mg/L (quantified using N–NH_3 mass), and the concentration of free nitrite was measured at (0.49 ± 0.09) mg/L (quantified using N–HNO_2 mass). These concentrations were indicative of conditions for the partial inhibition of ammonia oxidation. The mass concentration of bicarbonate was found to be (0.01 ± 0.16) mg/L (quantified using C mass). In the treatment of landfill leachate, a synergistic effect of Anammox and denitrification primarily occurs in the Anammox section. In 2008, the Anammox process achieved a nitrogen removal rate of (85.1 ± 5.6)%, while anaerobic denitrification achieved a nitrogen removal rate of (14.9 ± 5.6)% in the SHARON–ANAMMOX process.

A novel combined process based on PN/A and PD/A enables deep nitrogen removal from landfill leachate. Since NO_3^-–N is the primary product of PN/A and the primary substrate of PDA, the stable operation of PN/A is necessary for the PDA process to be realized. The combined NO_2^-–N supply from PN and PD enables AnAOB to attain profound nitrogen removal in both aerobic and anoxic phases. Figure 10–7 demonstrates an innovative two-stage PNA– partial denitrification Anammox (PNA–PDA) combined treatment system, which achieved a NRE of (97.6 ± 1.5)% after successful start-up of the PNA–SBR and sodium acetate-induced PD of the coupled process. The novel single-stage PNA–PDA process in a SBBR increased NRR and NRE to (2.83 ± 0.06) mg N/(L · d) and 98.6 ± 0.2%, respectively, by stepwise increase of dissolved oxygen DO (from 0.5 to 3.5 mg/L) and continuous carbon source feed.

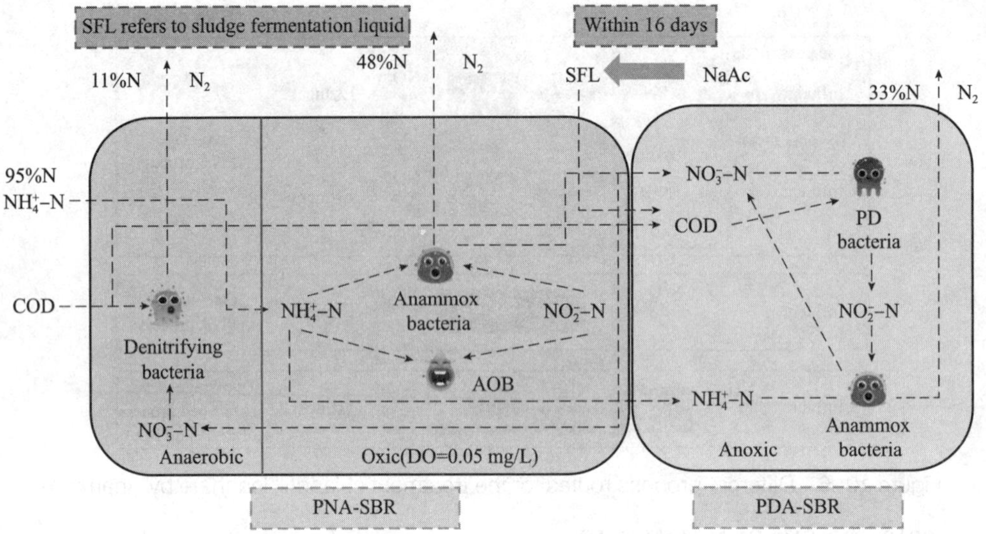

Figure 10–7 Nitrogen removal pathway for landfill leachate treatment in a two-stage PNA–PDA combined system

10.2.1.3 Microbiological analysis of landfill leachate treatment with Anammox

Leachate from waste includes a great deal of organic matter that is difficult to digest as well as hazardous compounds. In order to defend themselves, some bacteria will adapt to the toxic environment and create an increased amount of extracellular polymeric substances (EPS). It is possible for EPS to keep the activity of the AnAOB stable due to the fact that it will supply more nutrients to microorganisms in such a hostile environment. In addition, AnAOB have the ability to release huge volumes of EPS; hence, the composition of EPS changes in a manner that is analogous to the abundance of *Brocadiaceae*. When it comes to the process of SNAD nitrogen removal, the presence of HA in landfill leachate may have both positive and negative impacts (shown in Figure 10–8). During the process of nitrification, HA may enhance the cell permeability of the bacteria that are doing the nitrifying and also function as an electron shuttle to lower the number of membrane transporter protein genes. It also has the potential to boost denitrification performance. On the other hand, HA causes a reduction in the number of metabolic genes active during Anammox, which leads to a worse performance of nitrogen removal. *Anaerolineaceae*, linked to hydrophobic reactions, played a pivotal role as the principal degraders of rDOM. In landfill leachate, the heterotrophic bacteria known as *Eutropha* is able to breakdown aromatic compounds. In addition, the bacterium known as *Azoarcus tolulyticus* is able to digest some harmful chemicals. In SNAD systems, the bacterium known as *Acidobacter* may create huge quantities of EPS for the removal of HA, which can assist stabilize the system.

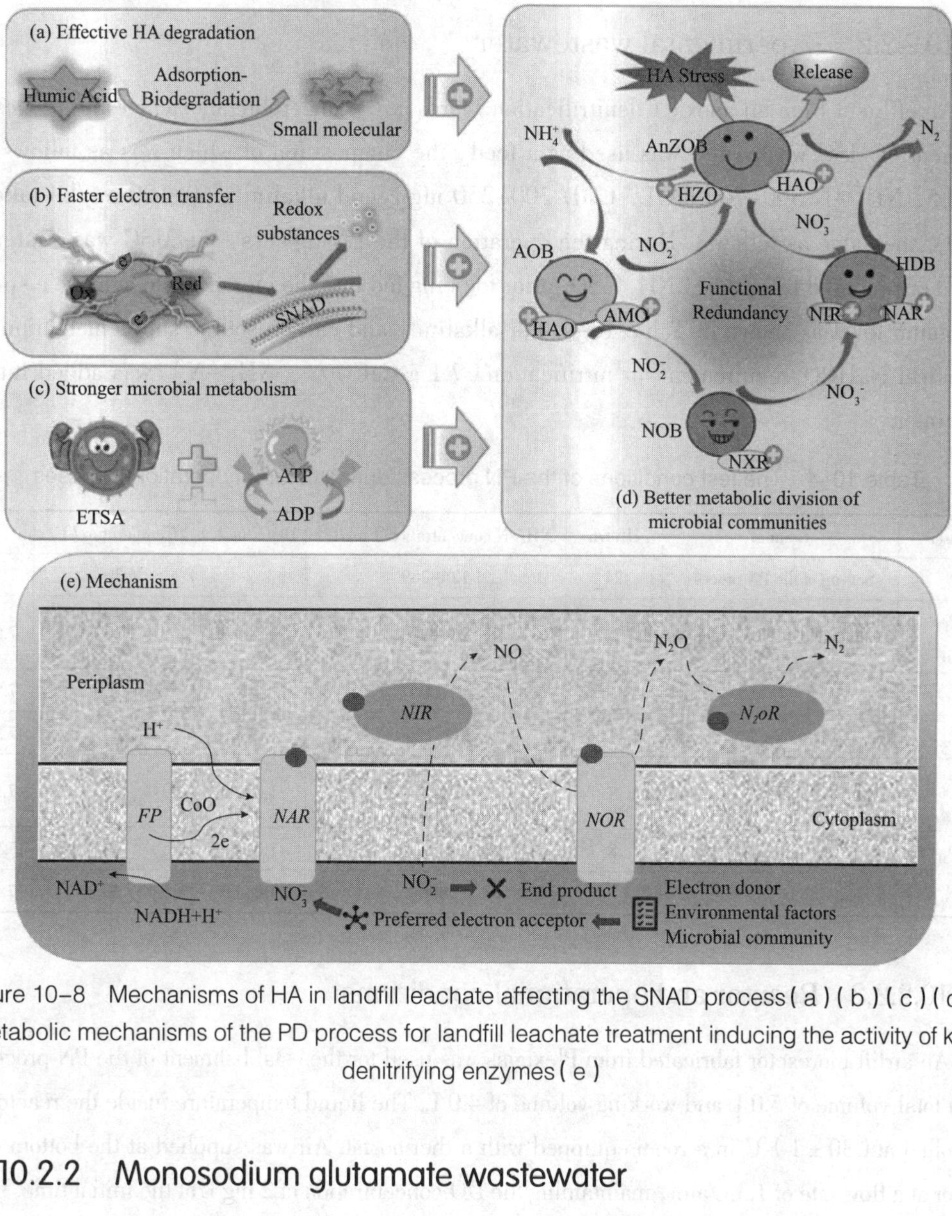

Figure 10–8 Mechanisms of HA in landfill leachate affecting the SNAD process(a)(b)(c)(d), Metabolic mechanisms of the PD process for landfill leachate treatment inducing the activity of key denitrifying enzymes(e)

10.2.2 Monosodium glutamate wastewater

10.2.2.1 Seed biomass

Aerobic activated sludge samples to serve as strains were collected from an MSG wastewater treatment plant, a municipal wastewater treatment plant and a landfill leachate treatment plant. The aim of using a mixture of three sludges was to widen the microbial diversity of the natural nitrifying bacteria. The collected sludge samples were mixed at a volumetric ratio of 1 : 1.2 with final suspended solid (SS) and volatile suspended solid (VSS) concentrations of 20.9 and 13.5 g/L, respectively.

10.2.2.2 Experimental wastewater

An effluent from an anoxic (denitrification) tank in the anaerobic/anoxic/aerobic process for treatment of MSG wastewater was used as a feed, the composition of which was as follows: pH 5.5–6.5, NH_4^+–N 500–700 mg/L, COD 200–250 mg/L and alkalinity 60–100 mg/L (taken as $CaCO_3$, the same as follow). During the operation of the PN process, the MSG wastewater was diluted or prepared by adding $(NH_4)_2SO_4$ prior to being fed into the bioreactor to achieve a specific ammonium level as shown in Table 10–4. For alkalinity and carbon source supplementation, the theoretical $NaHCO_3$ requirement for nitrification (7.1 g $CaCO_3$/(g NH_4^+–N)) was added into the wastewater.

Table 10–4 The test conditions of the PN process during different operational phases

Phases	Content	HRT/h	NH_4^+–N concentration/(mg/L)	DO/(mg/L)	Temperature/℃	pH
P_0	Set-up of the PN process	24	124–289	2	30	7.5–8.0
P_1 P_{11}	Capacity of the PN process	24	254–900	0.2–0.3	30	7.5–8.0
P_{12}		24–6.1	700	1.5	30	7.5–8.0
P_2	Feasibility of the PN process	6.1–24	254–1200	1.0–1.5	30	7.5–8.0
P_3 P_{31}	Stability of the PN process	8	650	1.0–1.5	30	7.5–8.0
P_{31}		8	1260	1.0–1.5	30	7.5–8.0
P_{31}		8	650	1.0–1.5	30	7.5–8.0

10.2.2.3 Reactor and operational conditions

An airlift bioreactor fabricated from Plexiglas was used for the establishment of the PN process. It had a total volume of 5.0 L and working volume of 4.0 L. The liquid temperature inside the reactor was controlled at (30 ± 1) ℃ in a room equipped with a thermostat. Air was supplied at the bottom of the reactor at a flow rate of 1.3 L/min, maintaining the DO concentration of 2 mg/L at the initial time. Later, the DO concentration was reduced to approximately 1 mg/L to conduct the subsequent experiments.

10.2.2.4 Experimental procedures

The experiments in this study were divided into four phases (P_0–P_3). The operational conditions in each phase are listed in Table 10–4.

10.2.2.5 Analytical procedures

The NH_4^+–N, NO_2^-–N and NO_3^-–N concentrations were determined spectrophotometrically. The values of the pH and DO concentration were measured using a pH meter (PHS–9V) and a DO meter (YSI-95), respectively. The denitrification activity, nitrite oxidizing activity, and Anammox activity of sludge was determined by batch test as described.

10.2.2.6 Feasibility of PN as a pretreatment for Anammox process

The PN process displayed good performance under steady-state conditions with a low effluent NO_3^--N concentration, as well as an NO_2^--N / NH_4^+-N of approximately 1.3, which was appropriate for an influent for the Anammox procedure. Then, for the substrate shocks, the inflow ammonium concentration was doubled to 1260 mg/L. A sudden high ammonium concentration led to nitrogen overload, which may exceed the reactor's capability and result in process instability. With the shock of substrate concentration, the effluent NH_4^+-N concentration became higher, and the NO_2^--N concentration was lower. After 1.5 HRT (12 h), the ammonium in the influent recovered to the steady-state level, and the shock test was completed. The effluent NO_2^--N / NH_4^+-N was below 0.5, and the effluent of the PN was not suitable for the subsequent Anammox process. The system recovered only after experiencing 3 HRT (24 h).

The feasibility of using PN to treat MSG wastewater was evaluated, and the combination controlling strategy was adapted to attain a stable PN process with a capacity for ammonium conversion and a nitrite yield of 1.83 mg N/ (L · d) and 87.0%, respectively. Additionally, the shock experiment validated that the PN reactor has a good resistance to overloading: even at a high concentration of substrate [1260 mg N/ (L · d)], the performance of PN is quickly restored after the cessation of the shock.

10.2.3 Rare earth wastewater

The production of rare earth produces a large amount of waste water, in which the concentration of ammonia nitrogen is between 8 and 20 g/L. If it is discharged directly, it will not only waste resources, but also pose a serious threat to the ecological environment. Therefore, effective measures should be taken to deal with it.

Anammox is the process of converting NH_4^+ and NO_2^--N into N_2 using Anammox bacteria under anoxic or anaerobic conditions. This method does not need additional carbon source, and has low sludge yield, low cost and no secondary pollution. It has been applied in the treatment of monosodium glutamate wastewater, aquaculture wastewater, pharmaceutical wastewater, etching wastewater and so on. In this study, the UASB was used as the reactor to treat ammonia-containing wastewater produced in the rare earth production process, combined with the Anammox process. The treatment effect of AnAOB on ammonia-containing wastewater was investigated by adjusting the temperature, HRT and the mass concentration of La^{3+} in raw water. In order to provide a new idea for the treatment of rare earth wastewater, the changes of microbial community structure in the reactor before and after treatment were analyzed by high-throughput technology.

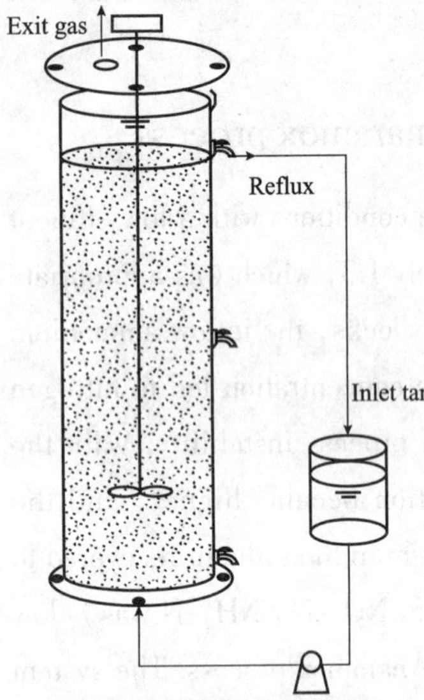

Figure 10–9 UASB Anammox reactor diagram

10.2.3.1 Set of test

The structure of the UASB Anammox reactor is shown in Figure 10–9.

The volume of the reaction zone is 12 L, and the top is sealed with a reserved vent hole. The water inlet and outlet are respectively located at the bottom and 15 cm away from the top. In the middle of the reactor is the suspended Anammox activated sludge, the bottom is provided with a sampling port, the outside is wrapped with a light shield to prevent the light from inhibiting the growth of bacteria, and the inside is installed with a stirrer. In order to strengthen the mud–water mixing in the reactor and prevent the concentration of the reactants at the inlet end from being too high, a magnetic circulating pump was used to circulate the inlet water. Install a heater into the tank to ensure constant internal temperature.

10.2.3.2 Wastewater of test

The rare earth wastewater containing NH_4^+–N used in the test was taken from a rare earth plant of Baotou Steel, sealed and stored with a pH of 7.7–8.4 and a hardness of 60–80 mg/L. The quality of wastewater is shown in Table 10–5. The water sample was filtered by a 0.45 μm filter membrane.

Table 10–5 The composition of rare earth wastewater containing ammonia nitrogen mg/L

COD	NH_4^+–N	NO_2^-–N	NO_3^-–N
60	10000	<5	<50

10.2.3.3 Test principle and method

Under anoxic or anaerobic conditions, NH_4^+–N can be converted into N_2 by AnAOB with NO_2^-–N as oxidant and hydroxylamine and hydrazine as intermediate products. The specific equation is

$$NH_4^+ + 1.32NO_2^- + 0.13H^+ + 0.066HCO_3^- \longrightarrow 0.066CH_2O_{2.5}N_{0.15} + 1.02N_2 + 0.26NO_3 + 2.03H_2O$$

The concentration of NH_4^+–N, N–(1–naphthalene) ethylenediamine and NO_3^-–N were determined by spectrophotometry. The concentration of NO_3^-–N was determined by ultraviolet spectrophotometry.

The temperature, pH and COD of wastewater were measured by portable dissolved oxygen meter. MLSS and MLVSS of sludge samples were determined by standard method. Microbial diversity and community structure in Anammox system were studied by 16SrRNA high-throughput sequencing technology. Samples S_1 and S_2 are granular sludge in the reactor before and after the introduction of rare earth wastewater, respectively. The continuous treatment method is used to treat the wastewater to ensure the anaerobic and light-shading environment and make the living conditions of anaerobic Anammox bacteria suitable. The addition of $NaHCO_3$ keeps the pH of the system stable, without external carbon source, and inhibits the competition between other strains and AnAOB. The mass concentration of NH_4^+–N in the wastewater is high, which is 10 g/L. Since the maximum treatment load of Anammox system is not known, the raw water is diluted 10 times, and the amount of NO_2^-–N in the water is controlled by the ratio of NH_4^+–N to NO_2^-–N according to the amount of substance 1 : 1.32. Adjust system temperature, hydraulic retention time and other parameters to monitor the concentration of NH_4^+–N and NO_2^-–N in the produced water. After the ammonia concentration of effluent was stabilized, the influent was added with different concentrations of rare earth lanthanum solution to analyze the effect of La^{3+} on the removal of ammonia nitrogen by Anammox process. The changes of bacterial community in the reactor were analyzed by high throughput sequencing technology.

10.2.4 Pharmaceutical wastewater

Wastewater containing antibiotics once emissions to the environment, most likely induced resistant bacteria in the receiving waters (antibiotic resistance bacteria, ARB) and resistance genes (antibiotic resistance genes, ARGs). High concentration of antibiotics will produce antibiotic selection pressure on microorganisms in the environment, making some microorganisms or bacteria resistant to antibiotics, which will lead to changes in community structure and even have adverse consequences on the ecosystem. In addition, even if antibiotics are removed from an area, ARGs and ARB in that environment may not disappear in the short term. Not only that, it has been reported that even if ARB is inactivated, the antibiotic resistance genes it carries can remain active long after being released into the environment until the next recipient bacterium is found.

10.2.4.1 Nitrogen removal effect of Anammox process on wastewater containing antibiotics

A large amount of antibiotic wastewater is produced in the process of drug research and production. The treatment of these pharmaceutical wastewater has always been the focus and difficulty of water pollution control. According to the different kinds of drugs produced and the different production processes, pharmaceutical wastewater can be divided into: antibiotics, synthetic medicine, Chinese patent medicine production of wastewater and washing water in the production of various pharmaceutical preparations. For example, antibiotic wastewater is mainly composed of nutrient solution, fermentation residues, extraction residues in the solvent extraction process, distillation residues dispersed after solvent recovery, adsorption waste discharged in the ion exchange process, bacterial infusion in water and insoluble antibiotics in the fermentation filtrate. Generally speaking, pharmaceutical wastewater has the main characteristics of large amount of wastewater, complex water quality, strong toxicity, high concentration of organic matter and suspended matter, high salt content, deep color, poor biodegradability, which belongs to the high-concentration industrial wastewater that is difficult to treat.

A two-stage combined process of anaerobic-aerobic-Anammox was used to treat mixed wastewater from a pharmaceutical company and an amylum plant in Inner Mongolia. The results show that the total removal rate of COD, the removal efficiency of NH_4^+-N, and TN are higher than 90% and 80%-95%, respectively, indicating that the combined process has achieved a good treatment effect. The integrated Anammox process was studied for the treatment of aureomycin wastewater after anaerobic and aerobic. The experimental results show that the removal rate of NH_4^+-N and TN in wastewater can reach 80%, and the maximum load of TN volume can reach 1.0 mg N/(L · d) when the integrated Anammox process is used to treat the secondary anaerobic dilution effluent. In the test of treating aerobic effluent, the removal rate of NH_4^+-N is higher than 90%, the maximum removal rate of TN is 80%, and the TN volume load is 0.6 mg N/(L · d). The results show that the treatment of high NH_4^+-N antibiotic wastewater by Anammox is technically and economically feasible. A study was conducted on the anaerobic effluent of natamycin wastewater treated by the nitrification and Anammox process. The test results showed that under the condition of long sludge age and low DO, the average removal rate of TN was 78%, and the highest was 92.5%. The final effluent concentration of NH_4^+-N was 8.38 mg/L and NO_2^--N concentration was 9.37 mg/L. The TN concentration was 38.12 mg/L, and good treatment effect was obtained.

10.2.4.2 Mechanism of action of antibiotics on Anammox bacteria

Anammox granules by bacteria and EPS. The main components of EPS are proteins and polysaccharides, as well as a small amount of lipids, nucleic acids and humic acids, which are synthesized and secreted by the bacteria in the granule. It can not only be used as a bonding agent to promote cell agglomeration and sludge granulation, but also is usually the first barrier for microbial cells, which directly contact and interact with substances in the water environment. Under the stimulation of exogenous toxic substances, Microbes will secrete more EPS to protect themselves and reduce their toxic effects. EPS plays an important role in the formation, maintenance and operation of Anammox granular sludge and is closely related to its two major components. Studies found that protein content was positively correlated with the hydrophobicity of bacterial cell surfaces. However, other research suggests that polysaccharides play a more crucial role in

Anammox granulation. Additionally, resistance to three broad-spectrum antibiotics, amoxicillin, flurprofen, and sulfamate dimethacil, was studied, revealing that during the long-term experiment, sludge activity was reduced by half in the first 3 days. Still, subsequently, Anammox bacteria gradually recovered, and sludge activity and heme levels increased, while the EPS level was higher than the initial level.

Anammox, on the other hand, has an extracellular polymer defense mechanism that helps increase its tolerance. It was demonstrated that the change of extracellular products was an obvious reaction of Anammox bacteria under antibiotic stress. The EPS matrix contains a wealth of functional groups, including carboxyl, phosphor, amine, and hydroxyl, which carry a negative charge at a neutral pH level. They can form organometallic complexes with polyvalent cations through electrostatic attraction or other interactions. The increase of EPS content in cells can enhance the force between the cells and the adjacent AnAOB, forming a cross-linking network to prevent the entry of heavy metals into the cells. OTC has electron donor groups and can form complexes with metal ions (such as Cu-OTC complexes), which can be further adsorbed on the surface of Anammox particles by electrostatic force to form EP-Cu-OTC complexes. They found that EPS yield was positively correlated with the specific growth rate of AnAOB. These behaviors can be considered as a defense strategy of AnAOB to reduce antibiotic intake.

10.3 Agricultural wastewater

10.3.1 Swine wastewater

The wastewater physicochemical characteristics make Anammox feasible in the swine wastewater treatment system. The influent was typical swine wastewater containing high NH_4^+-N (549.98 mg/L) with low C/N (0.85), whose physical and chemical properties were similar to landfill leachate. AnAOB are able to live in landfill leachate with NH_4^+-N/COD ranging from 0.96 to 2.00. In this treatment system, the influent COD was reduced from 549.98 mg/L to 288.67 mg/L after anaerobic digestion. NH_4^+-N was then removed very efficiently in the first-stage aerobic tank with red granules. AnAOB are forced to compete with denitrifying bacteria once the C/N of wastewater is more than 1. AnAOB can survive in laboratory reactors and wastewater treatment plants under a C/N of 0.5–2.8. However, their removal efficiency of NH_4^+-N (50%–97%) and TN (38%–93%) was quite different. In this study, the NH_4^+-N and total inorganic nitrogen removals of the wastewater treatment system were higher than the two swine wastewater treatment systems in Japan. One reason for this could be that the COD (288.67 mg/L) and C/N (0.6) in this study were more suitable for AnAOB to remove nitrogen, while low COD and C/N inhibit denitrifying bacteria. The COD (845–1285 mg/L) and C/N (1.3–2.8) in the other two swine wastewater treatment systems were too high, causing denitrifying bacteria to compete with AnAOB. In this study, the influent C/N was 0.85, and became 0.60 after anaerobic digestion, providing favorable conditions for AnAOB.

In this wastewater treatment process, pH was 7.46–7.99, a suitable range for AnAOB. Previous studies have shown that AnAOB live in the pH range 6.0–8.0, with optimal pH 7.2–7.6. Under optimal pH, inhibition of the Anammox enzyme activity by both pH and accumulation of free ammonia and free nitrous acid could be avoided in simulated wastewater. It was suggested that Anammox activity was highest when the pH was 7.85 in simulated wastewater. However, when the pH was maintained at 8.0, total inorganic nitrogen removal efficiency was the best (removal of NH_4^+-N was 98.4% and NO_2^--N was 99.5%). Suitable pH contributed to good nitrogen removal performance, and the pH in the studied system was in the suitable range for growth of AnAOB.

Many bacteria found in the red Anammox granule could provide a suitable living environment for Anammox bacteria (Figure 10–10). Red granules were observed mainly at the first-stage aerobic tank. The range of DO in the first-stage aerobic tank was 0.19–0.54 mg/L. Evaluations of DO tolerance of AnAOB have shown that influent DO must be lower than 2.5 mg/L. Anammox activity decreased 50% at DO of 2.3–2.8 mg/L. When DO was reduced from 2.3 to 2.8 mg/L to 1 mg/L, Anammox activity recovered. Used AnAOB to remove nitrogen in the SBR reactor of the WWTP, suggesting that controlling DO to below 0.3 mg/L can prevent excessive NO_2^--N accumulation and further oxidation to NO_3^--N. In several laboratory-scale reactors (Table 10–6), high nitrogen removal efficiencies of NH_4^+-N with 96% and TIN with 80% were recorded in a lab scale SBBR with swine wastewater, which could be attributed to its low DO (0.5 mg/L). According to previous studies, although NH_4^+-N removal efficiency reached 57%–70%, TIN removal efficiency was only 38%–55%, and some NH_4^+-N was converted to NO_x^--N. This could be attributed to their high DO values (3–4 mg/L and 5 mg/L). Some swine wastewater plants use the A/O process to treat swine wastewater. DO was controlled above 3.0 mg/L in the first-stage aerobic tank, providing sufficient DO for nitrifying bacteria to oxidize NH_4^+-N to NO_3^--N. Further, NO_3^--N was removed by denitrification. There was no report of Anammox in these swine wastewater treatment plants with controlled DO above 3 mg/L, and the main reason was that the DO (above 3 mg/L) placed AnAOB under stress. Therefore, strict control of DO is a key element of Anammox treatment of swine wastewater. In this study, DO was controlled at 0.19–0.54 mg/L in the first-stage aerobic tank, a condition that also produced NO_2^--N and prevented oxidation of NO_2^--N to NO_3^--N. Due to the large number of aerobic bacteria (including AOB), which consumed large amounts of DO during their growth, the DO of the first-stage aerobic tank (0.19–0.54 mg/L) also made it possible for AnAOB to occur.

Table 10-6 Comparison in performance of wastewater treatment using Anammox

Type of wastewater	Type of process	Influent parameter					pH	DO/ (mg/L)	Removal at steady state/ %			
		NH_4^+-N/ (mg/L)	NO_2^--N/ (mg/L)	NO_3^--N/ (mg/L)	COD or BOD/ (mg/L)	C/N or B/ N			NH_4^+-N	NO_2^--N	TN	COD
Swine wastewater	A/O; WWTP	404.8	29.95	49.36	288.67	0.60	7.46	0.19–0.54	100	69	94	59.7
Swine wastewater	WWTP	465–658.1	0–0.42	0.85–1.16	845–1285	1.3–2.8	8.1–8.8	0.2–0.51	68.7–85.3	—	48–80	—
Swine wastewater	Lab scale SBBR	418	—	—	437	1.05	7.8–8.0	0.2–2	50	—	48	87
Swine wastewater	Lab scale SBBR	519	—	—	288	0.75	7–8	0.5	96	—	80	76
Swine wastewater	Lab scale	76	111	—	141	0.7	7.4	5	70	80	55	25
Swine wastewater	Lab scale SHARON–ANAMMOX	213	323	—	—	—		3–4	57	76	38	—
Synthetic wastewater	Lab scale SBR	423	—	—	208–437	0.5–1	7.9	0.05–0.25	97	—	89	63.5
Synthetic wastewater	Lab scale SBR	250	—	—	250	1	6–7.35	0.17–0.35	94	—	82	92
Opto electronic industria	Lab scale SBR	567	—	—	100	0.18	7.8–8	0.1–5	95	—	93	79
Landfill–leachate	SNAD; WWTP	634	0	3	554	0.9	7.3–7.9	0.3	80	—	68	28

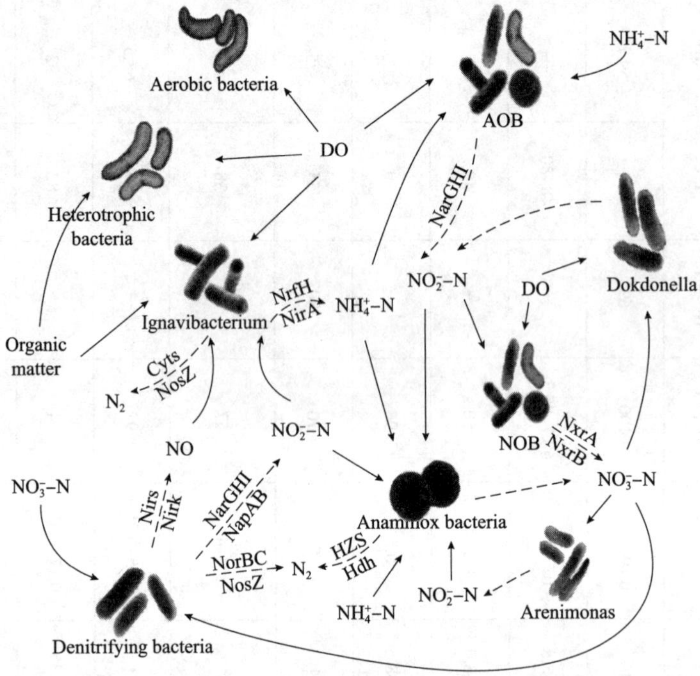

Figure 10–10 Collaboration network of nitrogen removal bacteria in the Anammox granules for a combined nitrification–Anammox–denitrification nitrogen removal pathway

10.3.2 Dairy wastewater

In order to prevent a large amount of biodegradable organic matter in the untreated dairy farm wastewater from having serious adverse effects on the microorganisms in the reactor, dilute biogas slurry of cow dung was used as PN/A influent at stage 3. Due to the drastic changes in the water environment, the operation performance of stage 3 gradually deteriorated in the first few days. On the 70th day, the removal rates of ammonia nitrogen and total nitrogen were 43% and 21%, with a high η value (50%). With the adaptation of microorganisms in the reactor, the water quality improved slightly, and the removal rates of ammonia and total nitrogen were 60% and 39% at day 76, and the η (33%) decreased. In order to further improve the effluent water quality, HRT was adjusted to 48h at stage 4, and the water quality was greatly improved. At the end of the reactor operation, the ammonia nitrogen, nitrous nitrogen and nitrous nitrogen were 43 mg/L, 17 mg/L and 32 mg/L, respectively. The removal rates of ammonia nitrogen and total nitrogen reached 82% and 62%, and the η value (23%) was further reduced. Getting closer and closer to the theoretical value.

The influent C/N ratio is the key factor to successfully start the PN/A process of dairy wastewater. The C/N ratio of diluted biogas slurry is 2.9, which is much higher than the appropriate C/N ratio of less than 0.5 reported in related literature. However, a large number of organic substances in diluted biogas slurry are difficult to degrade, and there are few degradable organic substances. According to relevant studies, the ratio of COD and NH_4^+-N removed by PN/A system is an important factor in PN/A process. During the third stage, the average ratio of C/N was 1.02, indicating that the diluted biogas slurry still contained a lot of biodegradable organic matter, which would affect the operation performance of Anammox, resulting in denitrifying bacteria and other heterotrophic microorganisms competing for inorganic nutrients and affecting the growth of AOB and Anammox bacteria. During the fourth stage, the average C/N ratio was 0.69, the biodegradable organic matter in biogas slurry of cow manure was reduced, and the domestication of AOB and AnAOB continued, and the PN/A integrated SBBR process was successfully started.

More interestingly, from stage three, the consumption of alkali in the reactor is greatly reduced, because the biogas slurry of cow manure contains a large number of refractory and easily degradable organic substances, as well as other complex components, which has a strong buffer capacity, which greatly reduces the consumption of alkali and further reduces the cost of the PN/A process. It is conducive to the future application of full-scale PN/A process to treat dairy farm wastewater.

10.3.3 Aquatic aquaculture wastewater

The inoculated sludge was a combination of Anammox sludge and denitrifying sludge, consisting of flocculated and granular sludge with a dark red color, taken from the anaerobic pond of a duck farming wastewater treatment system in Yunfu, Guangdong, China. The Anammox and denitrification performance was good, with a mixed liquor suspended solids of (3000 ± 50) mg/L. The wastewater quality in the mixed fish pond of an aquaculture base in Huadu, Guangdong Province was more in line with the requirements of SAD coupling process operation. The concentrations of NH_4^+-N, NO_2^--N, and COD in the actual wastewater were relatively low, which was not conducive to the enrichment growth of AnAOB and denitrifying bacteria. Therefore, the concentration of the synthetic wastewater was adjusted proportionally in the experiment to suit the enrichment of AnAOB and start-up of the SAD coupling process. The adjusted wastewater concentration exhibited a multiplicative relationship of the initial wastewater concentration. The HRT was adjusted to achieve the best removal effect in the subsequent treatment of actual industrial recirculating aquaculture systems wastewater. NH_4^+-N and NO_2^--N were supplemented in the form of $(NH_4)_2SO_4$ and $NaNO_2$, respectively, and sodium acetate was used as a readily biodegradable COD substrate, which was configured as required.

10.3.3.1 Experimental setup design

The main process unit consisted of a pilot-scale acrylic UASB reactor with an effective volume of 30 L, a diameter of 20 cm, and a height of 1.5 m. Its structure is shown in Figure 10-11. The reactor was equipped with a gas–liquid–solid three–phase separator inside.

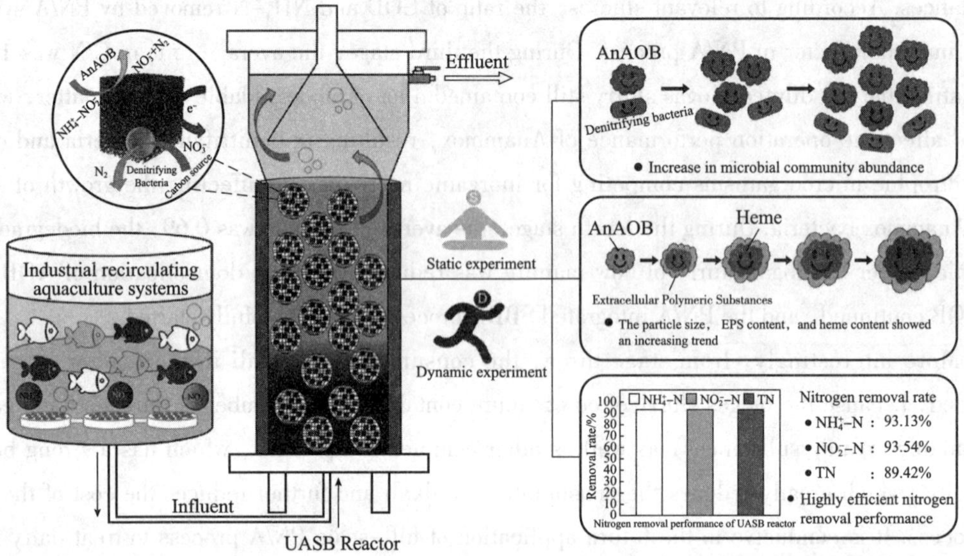

Figure 10–11 Reactor for the application of Anammox process to aquaculture wastewater

The filler was a porous suspension ball embedded with polyurethane black sponge. The outside of the reactor was covered with shade cloth. Throughout the experiment, a temperature controller was used to control the temperature, and a peristaltic pump was used to control the inlet and outlet water. The influent water was synthetic wastewater, and the DO concentration was reduced via N_2 aeration. The influent and effluent flow rates were kept constant, and HRT was set at 24 h.

10.3.3.2 Single–factor experiment of mixed sludge

The ideal optimal working parameters were first determined by reviewing several literature studies. These initial parameters are as follows: T = 30 ℃, pH = 8.0, solid–liquid ratio = 1∶1, DO = 0–0.3 mg/L, NH_4^+–N/NO_2^-–N = 1∶1.32, NH_4^+–N concentration = 80–90 mg/L, NO_2^-–N concentration = 90–100 mg/L, and COD concentration = 90–100 mg/L. Then, single-factor experiments were conducted with Anammox and denitrifying mixed sludge to investigate the effects of temperature, pH, solid–liquid ratio, and DO concentration on NH_4^+–N, NO_2^-–N, and NO_3^-–N in the synthetic wastewater. This led to the preliminary optimal working parameters of the SAD coupling process.

10.3.3.3 Static research test of the SAD coupling process

In this study, AnAOB sludge was mixed with denitrification sludge ata volume ratio of 4∶1 and then the mixture was inoculated in the UASB reactor. The volume of mixed sludge was 10 L; then, 20 L of synthetic wastewater was added, and the DO concentration of the synthetic wastewater was reduced to 0–0.3 mg/L via N_2 aeration. The reactor temperature was maintained at 30 ℃ using a temperature control switch; and the pH of the synthetic wastewater influent was 8. The wastewater was sampled and measured for NH_4^+-N, NO_2^--N, and TN regularly at 9: 00 am each day. Organic carbon source was added to enable the stable growth of the denitrifying bacteria. The UASB reactor alone was used for the static research test; chemicals and pH adjusters were added to the reactor. The reactor water volume was kept constant. This study aimed at achieving two main objectives via this approach. The first objective was to prevent sludge loss and enrich more AnAOB. The second objective was to alleviate the DO inhibition. The start-up process of the SAD coupling process under the static research test lasted for 85 days.

Researchers successfully enriched Anammox microorganisms with a purity of about 97.6% in a membrane bioreactor using Anammox granular sludge as the inoculum sludge. The doubling time was only 5 days. Furthermore, pre-experiments found that the basic removal of nitrogen-containing pollutants from the initial feed water took 4-5 days. Therefore, in this study, 5 days was used as a short cycle during the static research test. The static research test start-up process of the SAD coupling process was divided into five stages; and the specific duration of each stage was determined based on the actual level of nitrogen- containing pollutant removal. The short period (5 days) was used as the measurement unit, and the division point was when the NH_4^+-N, NO_2^--N, and TN removal rate reached 85% or more, respectively, at the end of any short cycle. This led to the establishment of the duration of each phase.

The reactor load was increased by gradually increasing the feed water substrate concentration. The nitrogen removal performance of the SAD coupling process was investigated during static operation. The study found that excellent nitrogen removal was achieved at influent COD of 150 mg/L, NH_4^+-N of 80 mg/L, and NO_2^--N of 160 mg/L, with almost all the NH_4^+-N, NO_2^--N, and NO_3^--N being removed. The results of small trial experiment revealed that the AnAOB activity was found to be slightly inhibited when the concentration of NO_2^--N exceeded 170 mg/L, thus the NO_2^--N concentration was set at 165 mg/L. The study found that the Anammox reactor was most effective at removing nitrogen when NO_2^--N / NH_4^+-N was equal to or lower than 1.32; therefore, the influent NO_2^--N / NH_4^+-N was set close to 1.32. Chemicals were added at the end of each short cycle, and the substrate concentration was increased at the end of each stage. The Anammox was determined to be effective when ΔNH_4^+-N/ΔNO_2^--N was close to 1 : 1.32, and brown-red Anammox granular sludge with large particle size was generated. Denitrification was determined to be effective when the NO_3^--N concentration was low and the COD reached a certain removal rate. The start-up of the SAD coupling process was considered successful when the first two conditions were met and the TN removal rate reached 89% or more.

10.3.3.4 SAD coupling process continuous flow test

After successful initiation of the SAD coupling process by the static research test, the continuous flow test was started. The nitrogen removal performance of the SAD coupling process under dynamic conditions was investigated by measuring NH_4^+-N, NO_2^--N, and

NO_3^--N concentrations in the influent and effluent. The substrate concentration was slowly increased when the removal rate of NH_4^+-N, NO_2^--N, and TN reached 85% or more. The operation of the coupling process was further stabilized. Unlike the static research test, the continuous flow test was controlled using a peristaltic pump with water inlet and outlet. The other conditions remained unchanged.

Questions

[1] What are the primary advantages of Anammox technology over traditional wastewater treatment methods in urban areas?

[2] Can Anammox technology be integrated with existing wastewater treatment plants in urban environments, and what are the considerations for such integration?

[3] How does Anammox contribute to the reduction of ammonia and nitrogen pollutants in urban wastewater, and what are the environmental benefits?

[4] What are the energy efficiency and cost-effectiveness of Anammox technology in the context of urban wastewater treatment?

[5] How can Anammox technology contribute to meeting strict environmental regulations for urban sewage discharge?

[6] What are the scalability and adaptability of Anammox technology for various urban wastewater treatment scenarios?

[7] What measures are taken to optimize the performance and reliability of Anammox systems in urban wastewater treatment?

[8] Can Anammox technology help reduce the carbon footprint of urban wastewater treatment facilities?

[9] What research and development efforts are underway to further enhance the efficiency and applicability of Anammox technology in urban wastewater treatment?

References

[1] Jetten M S, Wagner M, Fuerst J, et al. Microbiology and application of the anaerobic ammonium oxidation ('Anammox') process [J]. Current Opinion in biotechnology, 2001, 12 (3): 283-288.

[2] Chen C J, Huang X X, Lei C X, et al. Effect of organic matter strength on Anammox for modified greenhouse turtle breeding wastewater treatment [J]. Bioresource Technology, 2013, 148: 172-179.

[3] Liang Z, Liu J. Landfill leachate treatment with a novel process: Anaerobic ammonium oxidation (Anammox) combined with soil infiltration system [J]. Journal of Hazardous Materials, 2008, 151 (1): 202-212.

[4] Schmidt I, Sliekers O, Schmid M, et al. New concepts of microbial treatment processes for the nitrogen removal in wastewater [J]. FEMS Microbiology Reviews, 2003, 27 (4): 481-492.

[5] Tao Y, Gao Dw, Fu Y, et al. Impact of reactor configuration on Anammox process start up: MBR versus SBR [J]. Bioresource Technology, 2012, 104 (1): 73–80.

[6] Miao Y Y Y, Zhang J J H, Peng Y Z, et al. An improved startup strategy for mainstream Anammox process through inoculating ordinary nitrification sludge and a small amount of Anammox sludge [J]. Journal of Hazardous Materials, 2020, 384: 121325.

[7] Laureni M, Weissbrodt D G, Sziyak I, et al. Activity and growth of Anammox biomass on aerobically pretreated municipal wastewater [J]. Water Research, 2015, 80: 325–336.

[8] Xu X X, Ma B, Lu W K, et al. Effective nitrogen removal in a granule-based partial denitrification/Anammox reactor treating low C/ N sewage [J]. Bioresource Technology, 2020, 297: 122467.

[9] Kartal B, Kuenen J G, Van Loosdrecht M C M. Sewage treatment with Anammox [J]. Science, 2010, 328: 702–703.

[10] Mohammad-pajooh E, Weichgrebe D, Cuff G. Municipal landfill leachate characteristics and feasibility of retrofitting existing treatment systems with deammonification–a full scale survey [J]. Journal of Environmental Management, 2017, 187: 354–64.

[11] Ye J, Liu J, Ye M, et al. Towards advanced nitrogen removal and optimal energy recovery from leachate: a critical review of Anammox-based processes [J]. Critical Reviews in Environmental Science and Technology, 2020, 50 (6): 612–53.

[12] Tokutomi T, Shibayama C, Soda S, et al. A novel control method for nitritation: the domination of ammonia-oxidizing bacteria by high concentrations of inorganic carbon in an airlift-fluidized bed reactor [J]. Water Research, 2010, 44 (14): 4195–203.

[13] van Niftrik L, Jetten MSM. Anaerobic ammonium-oxidizing bacteria: unique microorganisms with exceptional properties [J]. Microbiology and Molecular Biology Reviews, 2012, 76 (3): 585–596.

[14] Yan Z, Li A, Shim H, et al. Effect of ozone pretreatment on biogranulation with partial nitritation – Anammox two stages for nitrogen removal from mature landfill leachate [J]. Journal of Environmental Management, 2022, 317: 115470.

Chapter 11 Sulfate-reducing ammonium oxidation (sulfammox)

Emphasis: Mechanism of sulfammox in presence of organic carbon; Mechanism of sulfammox without organic carbon; Environmental factors and operational conditions affecting sulfammox.

Nut: Effects of the NH_4^+/SO_4^{2-} ratio and COD addition on the sulammox process

11.1 Introduction

It is generally accepted that anaerobic ammonia-oxidizing bacteria (AAOB) oxidize ammonia (NH_4^+) to N_2 with NO_2^- as an electron acceptor. In fact, however, AAOB have a more comprehensive metabolism than initially assumed and other phenomena of "anaerobic" NH_4^+ oxidation have been discovered. In addition to NO_2^-, there may be other electron acceptors, including sulfate (SO_4^{2-}), for NH_4^+ oxidation under "anaerobic" conditions. This process is known as sulfate reducing ammonium oxidation (SRAO) or sulfammox.

In comparison with the conventional Anammox, sulfammox is easier to control as nitritation becomes unnecessary (SO_4^{2-} instead of NO_2^- serves as the electron acceptor). Besides, as a reducing process of SO_4^{2-}, it is also free of secondary pollution caused by sulfide (S^{2-}), which is toxic and harmful to human health and aquatic ecosystems. Sulfammox can also prevent interference with the conventional Anammox process caused by inhibition of S^{2-} or hydrogen sulfide (H_2S). Moreover, elemental sulfur (S^0) is formed and its recovery provides a valuable by-product. The recovered S^0 could be used as electron donor for autotrophic denitrification. Moreover, recovering S^0 from wastewater is also essential to reuse it as fertilizer or to re-enter production lines in other industries. The combination of the technology based on the Anammox process with the technology based on the sulfammox process would enable a balanced approach to the problem of specific industrial wastewater with high content of nitrogen compounds and SO_4^{2-} through their co-treatment in combined processes.

The sulfammox process has briefly been addressed in reviews on Anammox in marine environments and in the state of Anammox research in China. There are still a few publications on this process and finding them is not straightforward, as sulfammox also appears as SRAO or sulfate-dependent Anammox. Based on the Web of Science database, using the keywords "sulfate", "Anammox" and "wastewater", a number of publications and their citations appearing in 1999–2020 years are presented in Figure 11–1.

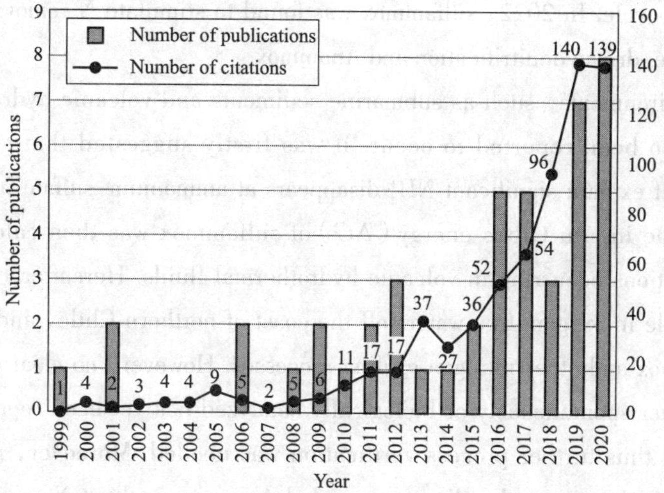

Figure 11–1 Number of publications based on keywords "sulfate", "Anammox" and "wastewater" and their citations in the Web of Science database in 1999–2020

11.2 Anthology of sulfammox studies

The negative ΔG values of the reaction Eq. (11–1) inferred that lithotrophic Sulfammox is metabolically feasible:

$$8NH_4^+ + 3SO_4^{2-} = 4N_2 + 3HS^- + 12H_2O + 5H^+,\ G_0 = -22\ kJ/mol \qquad (11\text{–}1)$$

In 2001, the researcher used a granular activated carbon anaerobic fluidized bed reactor (GACFBR) to treat vinasse and observed that a novel anaerobic process might occur in the system that can remove both NH_4^+ and SO_4^{2-}. Subsequently, in 2008, some researcher observed sulfammox in the absence of a carbon source, suggesting that NH_4^+ was first converted into NO_2^- with SO_4^{2-} as an electron acceptor, followed by the reaction of NO_2^- with the remaining NH_4^+ to produce N_2. In parallel, a new species of "Anammoxoglobus sulfate" was identified in their reactor. Yang et al. revealed that during the stabilization phase of Anammox, the NH_4^+/NO_2^- consumption ratio in the reactor was approximately 1 : 1.15 after 160 days, which was much higher than the theoretical 1 : 1.32 stoichiometry of NH_4^+ / NO_2^-. They assumed that SO_4^{2-} was the electron acceptor for NH_4^+ oxidation and NO_2^- was the only intermediate product in the simultaneous NH_4^+ and SO_4^{2-} removal process. Later, it was reported that sulfammox could occur independently and was positively affected by hydrazine and hydroxylamine. Recently, some researcher demonstrated sulfammox contributed to N elimination and the interaction between S and N cycles was more sophisticated than previously speculated. Furthermore, a previous report showed that sulfammox was most typically related to the coordinated metabolism of a consortium of bacteria that use both N and S, whereby Anammox bacteria predominate in the removal of N, and SRB and sulfur-oxidizing bacteria (SOB) participate in the S cycle. In 2022, sulfammox was found to stimulate N removal in the coupling process of thiosulfate-driven denitrification and Anammox.

In natural environments, such as submarine sediments and volcanic hydrothermal liquids, sulfammox has also been reported to occur. It was firstly suggested that sulfammox action [Eq. (11-1)] might explain significant NH_4^+ disappears at ammonium-sulfate interface in subsea floor sediments. The in situ Gibbs energy (ΔG) of sulfammox was then calculated as one of many possible reactions occurring in volcanic hydrothermal fluids. Hereafter, it was discovered a cryptic sulfur cycle in oxygen-free waters off the coast of northern Chile, and presumed that it may link to Anammox and other nitrogen cycling processes. However, no clear direct evidence of such processes either spontaneously or biologically catalyzed has yet been reported from natural environments, and thus further process verification was needed. Moreover, marine sediments accumulated from Mexican coastal sediments revealed Anammox without NO_2^-, indicating that the presence of high levels of Fe and Mn in this sediment, together with sulfammox, would facilitate NH_4^+ oxidation to N_2. The sulfammox process of up to 5–15 μg N_2/(g · d) represents a significant marine sink for fixed N. Their results also indicated that besides free sulfide, the sulfammox process also generates S^0 and sphalerite, a process that occurred in parallel with ammonium oxidation coupled to Fe (Ⅲ) reduction process. These reports suggest that sulfammox truly exists in nature, however, whether this process is spontaneous or biocatalytic in the natural environment remains unknown; therefore, further validation is required.

11.3 Mechanism of sulfammox

It has been shown that sulfammox can be achieved in conditions with organic carbon sources or without organic carbon sources. Therefore, the current explanations of the sulfammox mechanism are mainly divided into two cases, i.e., the presence and absence of organic carbon sources.

11.3.1 The presence of organic carbon sources in the influent

The mechanism of the sulfammox reaction under the condition with an organic carbon source was proposed as the biochemical reaction follows Eq. (11-2):

$$2NH_4^+ + SO_4^{2-} = N_2 + S^0 + 4H_2O,\ G_0 = -47.8\ \text{kJ/mol} \qquad (11\text{-}2)$$

The standard free energy of 47.8 kJ/mol favored the process. Sulfammox was assumed to be a stepwise reaction deriving from following three reactions:

$$4NH_4^+ + 3SO_4^{2-} = 4NO_2^- + 3S^{2-} + 4H_2O + 8H^+ \quad (11\text{-}3)$$

$$3S^{2-} + 8H^+ + 2NO_2^- \longrightarrow 3S^0 + N_2 + 4H_2O \quad (11\text{-}4)$$

$$2NH_4^+ + 2NO_2^- \longrightarrow 2N_2 + 4H_2O \quad (11\text{-}5)$$

As shown in Figure 11-2 (a), firstly, NH_4^+ and SO_4^{2-} react to produce S^{2-} and NO_2^-, then S^{2-} and part of NO_2^- continue to undergo sulfur autotrophic denitrification (SAD) to produce S^0 and N_2, and finally the remaining NO_2^- and NH_4^+ occur Anammox. Since the effect of the organic carbon on sulfammox is indirect, it is not reflected in the reaction equation. From a thermodynamic point of view, the reaction can occur. Later researchers also detected S^{2-} when initiating sulfammox in the presence of an organic carbon source, so the mechanism proposed as Eq. (11-2) is now generally accepted in the presence of organic carbon sources.

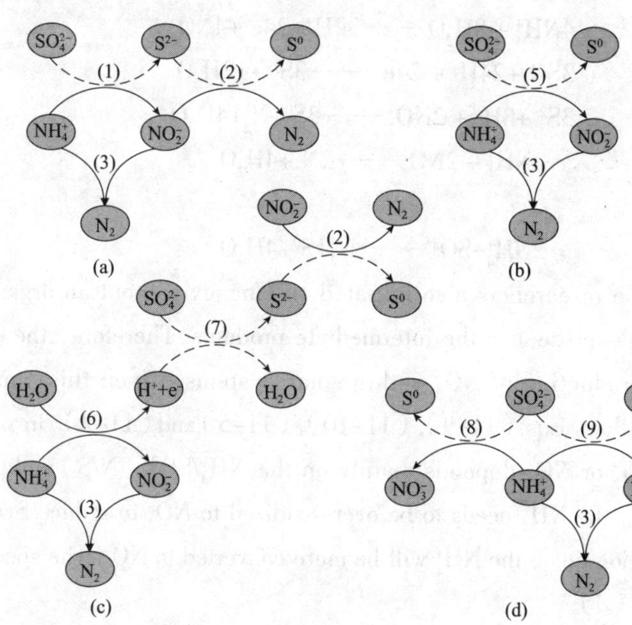

Figure 11-2 The mechanism of sulfammox proposed by different researchers

---- represents the sulfur conversion process, while — indicates the nitrogen conversion process, and the numbers indicate the corresponding reaction equations.

11.3.2 No organic carbon sources in the influent

There are three main sulfammox mechanisms in the conditions without organic carbon sources, which were proposed by several researchers. As shown in Figure 11-2 (b), it is investigated the sulfammox process mediated via Anammox organisms under autotrophic condition, and proposed a two-step metabolic pathway. In first step as shown in Eq. (11-6), the intermediate products were considered as nitrite, and elemental sulfur instead of sulfide assumed in Eq. (11-3).

$$NH_4^+ + SO_4^{2-} \longrightarrow NO_2^- + S + 2H_2O,\ G_0 = 314\ kJ/mol \tag{11-6}$$

The nitrite was then consumed by sequential Anammox reaction, Eq. (11–5). The summary equation proposed was identical to Eq. (11–2). In parallel, a new species of "Anammoxoglobus sulfate" was identified in their reactor, which could complete the first step of sulfammox [Eq. (11–6)] in autotrophic condition.

Some researchers then suggested that since the molecular structure of NH_4^+ is similar to that of methane, the reaction process of NH_4^+ and SO_4^{2-} may be similar to that of SO_4^{2-} reduction by anaerobic methane oxidation, and the reaction equations for NO_2^- and S^{2-} production were introduced as Eqs. (11–7), (11–8). That is, NH_4^+ first reacts with H_2O to produce NO_2^-, H^+, and e^-, then SO_4^{2-} reacts with H^+ and e^- to produce S^{2-} and H_2O, and next, S^{2-} reacts with NO_2^- to generate N_2 and S^0 [Eq. (11–2)], and finally the remaining NH_4^+ reacts with NO_2^- to generate N_2 [Eq. (11–3)], the specific process is shown in Figure 11–2 (c).

$$4NH_4^+ + 8H_2O \longrightarrow 32H^+ + 24e^- + 4NO_2^- \tag{11-7}$$

$$3SO_4^- + 24H^+ + 24e^- \longrightarrow 3S^{2-} + 12H_2O \tag{11-8}$$

$$3S^{2-} + 8H^+ + 2NO_2^- \longrightarrow 3S^0 + N_2 + 4H_2O \tag{11-4}$$

$$2NH_4^+ + 2NO_2^- \longrightarrow 2N_2 + 4H_2O \tag{11-5}$$

Total equation:

$$2NH_4^+ + SO_4^{2-} \longrightarrow N_2 + S^0 + 4H_2O \tag{11-2}$$

In addition, some researchers also initiated sulfammox without an organic carbon source and found that NO_3^- was present in the intermediate products. Therefore, the equations (11–6) cannot explain the production of NO_3^- within some systems. Given this, another sulfammox reaction mechanism follows Eqs. (11–9), (11–10), (11–5) and (11–2), in which the oxidation product of NH_4^+ is NO_2^- or NO_3^- depends mainly on the NH_4^+/SO_4^{2-} (N/S) molar ratio. When the N/S molar ratio is low, the NH_4^+ needs to be over-oxidized to NO_3^- to reduce SO_4^{2-}, while with the shortening of the reaction time, the NH_4^+ will be more converted to NO_2^-. The specific mechanism is shown in Figure 11–2 (d).

$$3NH_4^+ + 4SO_4^{2-} + 2HCO_3^- \longrightarrow 3NO_3^- + 4S^0 + 2H_2O + 2CO_3^- \tag{11-9}$$

$$NH_4^+ + SO_4^{2-} \longrightarrow NO_2^- + S^0 + 2H_2O \tag{11-10}$$

$$2NH_4^+ + 2NO_2^- \longrightarrow 2N_2 + 4H_2O \tag{11-5}$$

Total equation:

$$2NH_4^+ + SO_4^{2-} \longrightarrow N_2 + S^0 + 4H_2O \tag{11-2}$$

Recently, the sulfammox phenomena was found to occur only with high amounts of inoculant biomass under normal anaerobic condition ($0.2\ mg/L \leqslant DO < 0.5\ mg/L$), in other experimental conditions, sulfammox phenomena did not occur by using the same Anammox consortia. The authors proposed that Anammox bacteria do not possess the ability to oxidize ammonium with sulfate as electron acceptor and the assumed sulfammox could, in fact, be a combination of aerobic ammonium oxidation, Anammox and heterotrophic sulfate reduction processes (Figure 11–3).

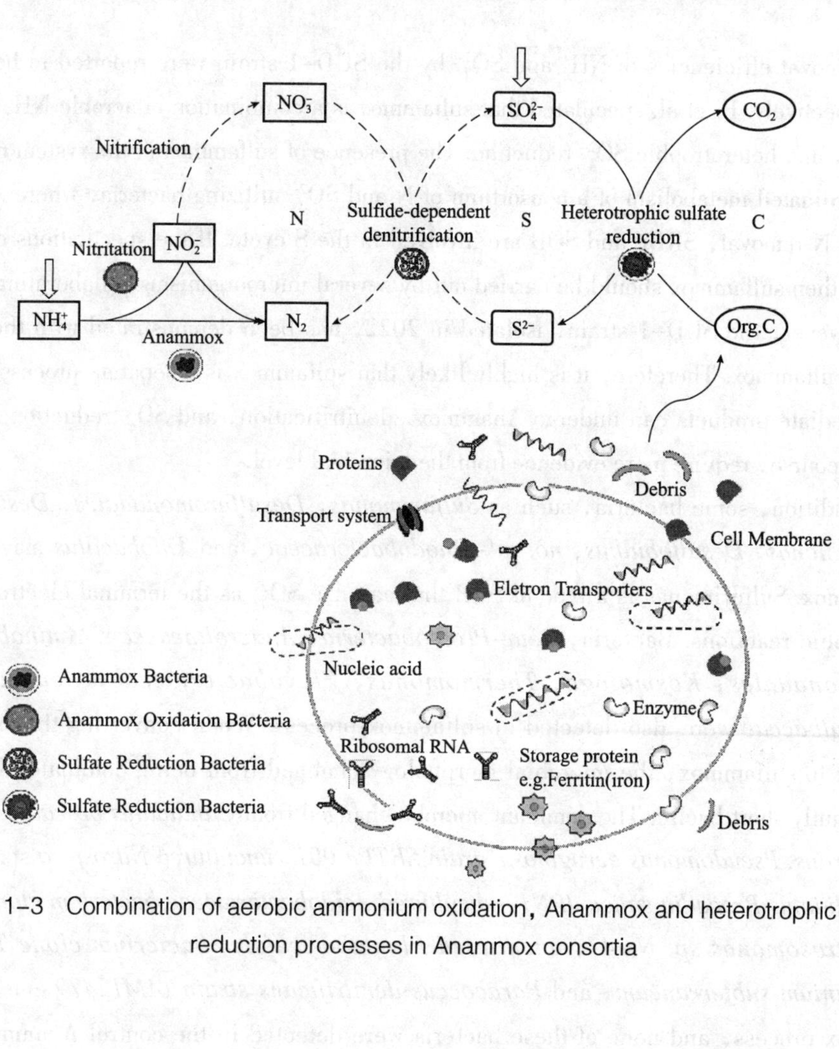

Figure 11-3 Combination of aerobic ammonium oxidation, Anammox and heterotrophic sulfate reduction processes in Anammox consortia

11.4 Characteristics of microbes in sulfammox

Sulfammox is a multistep process that involves various microbes and involves the utilization of different intermediate compounds. Recently, many bacteria associated with sulfammox have been identified and some pure strains have been isolated. A new species of the genus *Brocadia*, named *Anammoxoglobus* sulfate, were considered to be the functional community in the sulfammox and could offer a great biotechnological potential for NH_4^+ and SO_4^{2-} removal. A strain of ASR bacterium, belonging to *Bacillus* benzoevorans, in a stable operating sulfammox system was isolated and was demonstrated to have the ability to oxidize NH_4^+ via SO_4^{2-}. Uncultured bacteria *Candidate* division OD1, SHA-109, *Bacillus*, and *Candidatus Anammoxoglobus* were all shown to be important in sulfammox. In a report published in 2022, *Bacillus cereus* isolated from industrial wastewater, referred to as the SUD-1 strain, was demonstrated to have good sulfammox ability, with the optimum pH and growth temperature of 7 and (30 ± 2) ℃. In the sealed anaerobic reactor, the highest removal efficiencies of NH_4^+ and SO_4^{2-} by the SUD-1 strain were reported to be 67% and 80%, respectively. Bi et al. speculated that sulfammox is a combination of aerobic NH_4^+ oxidation, Anammox, and heterotrophic SO_4^{2-} reduction. The presence of sulfammox in the system may be due to the coordinated metabolism of a consortium of N and SO_4^{2-} utilizing bacteria, where Anammox-dominated N removal, SRB, and SOB are involved in the S cycle. If the speculations of them are accurate, then sulfammox should be carried out by several microorganisms collaborating with each other, however, the SUD-1 strain, isolated in 2022, has been demonstrated with the ability to carry out sulfammox. Therefore, it is highly likely that sulfammox is a separate process, only that its intermediate products can undergo Anammox, denitrification, and SO_4^{2-} reduction processes, which, of course, require more evidence from the microbial level.

In addition, some bacteria, such as *Sulfurimonas*, *Desulfuromonadales*, *Desulfovibrio*, *Desulfuromonas*, *Desulfobulbus*, *norank-Rhodobacteraceae*, and *Thiobacillus* may be related to sulfammox. Sulfurimonas is a type of SRB that can use SO_4^{2-} as the terminal electron acceptor for anaerobic reactions. Bacteria, *Beta-Proteobacteria*, *Anaerolineaceae*, *Aminobacterium*, *Xanthomonadales*, *Kosmotoga*, *Thermomonas*, *Flavobacteriaceae*, *Gemmatimonas*, *Desulfobulbaceae* were also detected in sulfammox process. When converting the system from Anammox to sulfammox, the microbial morphology changed from being dominated by cocci to predominantly short bacilli. The dominant microbe changed from *Candidatus brocadia* to *Bacillus benzoevorans*. *Pseudomonas aeruginosa strain SBTPe-001*, *uncultured* *Nitrospira* sp. clone 2A-14, *Candidatus Brocadia sinica JPN1*, *uncultured Acidobacteriaceae bacterium clone EfT107_A12*, *Nitrosomonas sp. Nm47*, *uncultured Planctomycetales bacterium clone T13J-B80*, *Desulfacinium subterraneum*, and *Paracoccus denitrificans strain IAM12479* emerged in the sulfammox process, and none of these bacteria were detected in the control Anammox system. Besides, bacteria belonging to *Planctomycetales* and *Verrucomicrobia* may be involved in the sulfammox process. For instance, uncultured bacteria clone P4 and uncultured planctomycete clone Amx-PAn30 appeared in the moving bed biofilm reactor (MBBR), while the *Verrucomicrobiales* bacterium clone De 2102 and uncultured bacterium clone ATB-KS-1929 were observed in the UASB. After 187 days of system operation, Pla 4_lineage, SM1A02, and *Planctomycetaceae* belonging to the genus Planctomycete exhibited higher abundance than their initial abundance, thus demonstrating their potential sulfammox ability. Sulfammox intermediates undergo other reactions that result in the coexistence of Anammox bacteria, sulfammox bacteria, SRB, and SOB in the system, as observed in *Candidatus Kuenenia*, *Candidatus Brocadia*, *Bacillus benzoevorans*, and *Thioplaca*.

The present findings show the high research value of microbes in the treatment of wastewater containing both NH_4^+ and SO_4^{2-}. Nevertheless, a detailed analysis of the physiological characteristics and structure of the microbes has not been performed extensively, which is primarily due to the low growth rate and specificity of the metabolism of the bacteria.

11.5 Environmental factors and operational conditions affecting sulfammox

11.5.1 Process medium and feeding options

Most of the sulfammox studies have been carried out with synthetic wastewater or growth media, but there have also been a few studies using real wastewater. Different compounds have been used as the SO_4^{2-} source in the medium (Table 11-1). The inoculum biomass originated from various sources (Table 11-2), including long-term operated Anammox reactors and anaerobic digesters. The cultivation experiments have been carried out with three feeding options:

(1) conventional Anammox was run at the beginning, and then NO_2^- was replaced with SO_4^{2-} as a new electron acceptor;

(2) SO_4^{2-} was used since the beginning without any addition of NO_2^-;

(3) SO_4^{2-} and NO_2^- were simultaneously used as electron acceptors during the whole study period.

Table 11-1 Environmental and operational conditions in sulfammox studies.

Source of SO_4^{2-}	COD addition	pH	Temperature/℃
$MgSO_4$/$FeSO_4$	No	7.8	35
$(NH_4)_2SO_4$	No	8.1-8.3	35
$(NH_4)_2SO_4$	No	8.1-8.6	30
K_2SO_4	No	8.5	30
Na_2SO_4	No	7.5-8.5	35
Na_2SO_4	No	7.5	30
$(NH_4)_2SO_4$	No	8-8.2	35
n.a.	Yes	7-8.5	14-15
n.a.	Yes	6.9-8.1	36
K_2SO_4	Yes	8.4	20
K_2SO_4	Yes	8.1 : 1	36
n.a.	Yes	7.8-8.3	36

n.a.: not available.

11.5.2 NH_4^+ / SO_4^{2-} ratio

Free ammonia (FA) and ionic ammonia are the two different forms of ammonium present in wastewater. Bacterial activity can be suppressed by the penetration of FA into the cell membrane. FA has been shown to interfere with fermentation and hinder protein degradation. Similarly, in SO_4^{2-} reduction systems, FA can alter the activity of SRB, which consequently affects the SO_4^{2-} reduction process. Additionally, microbial communities can be biologically affected by FA. As sulfammox involves a reaction between NH_4^+ and SO_4^{2-}, changes in NH_4^+ affect its performance. Furthermore, variations in the NH_4^+ concentration in the influent can directly cause changes in the N/S ratio during sulfammox. Therefore, the N/S ratio is considered to be one of the most important parameters in sulfammox. The N/S ratio typically indirectly affects the performance of sulfammox by affecting processes such as SO_4^{2-} reduction, denitrification, and Anammox. The N/S ratio changes the rates of NH_4^+ oxidation and SO_4^{2-} reduction, in addition to the end product of sulfammox. At an influent N/S ratio of 4, the microbes can convert NH_4^+ to NO_2^-, then the remaining NH_4^+ and NO_2^- occurred Anammox, whereas at an influent N/S ratio of 2, nitrification can change NH_4^+ to NO_3^-. An appropriate N/S ratio allows the NO_2^- reduction pathway to shift from Anammox to heterotrophic denitrification (HD) and SAD. In some reported experimental systems, the molar ratio of N/S consumption varied over a wide range from 2.0 to 20.0. Nevertheless, other works have indicated that the N/S ratio should be 0.7 to 2. At N/S ratio of 1 : (0.4 ± 0.03), NH_4^+ removal rates of (0.02 ± 0.01) kg N/(m^3 · d) can be achieved. Meanwhile, under N/S ratios of 1.481–2.963, the highest total nitrogen (TN) removal rate of 0.5 kg/(m^3 · d) was reported. The other researcher illustrated the multistep reaction of sulfammox can be explained based on N/S conversion ratios of 2.1–12.9. Additionally, the system was optimized at an N/S ratio of 3.0, with NH_4^+ and SO_4^{2-} removal efficiencies of 94.4% and 74.2%, respectively, by gradually increasing the N/S ratio (1, 1.5, 2, 2.5, and 3) under SO_4^{2-} concentration of around 100 mg/L. Increasing the influent N/S ratio primarily affects the activity of Anammox and sulfammox. Different N/S ratios contribute to the complexity of the sulfammox mechanism. When the N/S ratio was ≤1.5, sulfammox was used as the main pathway for N removal. At N/S ratios >1.5, Anammox became the primary pathway for N removal.

11.5.3 COD addition

Even though COD is not required for the sulfammox process, the experiments were performed either without COD addition or with COD addition. It is generally believed that sulfammox bacteria can survive under organic carbon sources; nevertheless, these carbon sources provide suitable growth conditions for both SRB and some heterotrophic microbes. Consequently, some bacteria compete with functional bacteria in the system for the carbon sources, thus inhibiting sulfammox bacteria and causing a long start-up time. Therefore, most current sulfammox studies have focused on inorganic carbon sources. When a carbon source is present in wastewater, sulfammox occurs simultaneously with HD. It was also suggested that nitrification, denitrification, Anammox, and sulfammox may occur simultaneously to remove NH_4^+, SO_4^{2-}, and COD from the wastewater. Additionally, the presence of an organic carbon source affects sulfammox by affecting the activity of Anammox bacteria; this is because Anammox may occur as an intermediate process. Anammox is strictly anaerobic, and as the concentration of organic matter increases, heterotrophic bacteria compete with Anammox for the substrate, thus deteriorating NH_4^+ removal. However, when the ratio of organic carbon to N is relatively low, Anammox can completely oxidize small-molecule organic acids to CO_2, and use CO_2 to synthesize the cell's own nutrients to partially metabolize the organic carbon. In general, the presence of a carbon source is critical to sulfammox performance and can result in distinct pathways and reactions.

11.5.4 Temperature and pH

The optimal temperature range for the sulfammox process is 25–35 ℃. The process temperature in different publications normally ranged from 30 to 36 ℃. In some case, despite a low temperature, the NH_4^+ and overall SO_4^{2-} removal efficiencies remained at a high level, i.e. 98.5% and 52.8%, respectively. Due to the combination of anaerobic-aerobic, continuous and batch processes adopted in this process, the Anammox and sulfammox coupled to remove nitrogen.

As an important environmental condition for biological processes, the effect of pH on sulfammox is typically shown via two scenarios. First, during the autolysis phase, the pH can disturb the growth and reproduction of bacteria; second, pH can indirectly affect the effective substrate of Anammox bacteria and thus the microbial activity. Most studies pertaining to sulfammox are based on pH 7.0–8.5. Sulfammox is suitable for alkaline environments, and increasing the pH of the reaction system can render sulfammox functional bacteria dominant and shorten the start-up time of the system, whereas pH > 8.5 will impair sulfammox functional bacteria. Different studies suggest that to achieve ideal performance, considerable research should be conducted based on different pH conditions to determine the optimal pH range and the effects of other variables on this relationship.

11.5.5 Spontaneity and oxidation-reduction potential

It has been described the spontaneity of the sulfammox reaction due to the ΔG^{θ} of the sulfammox being −45.35 kJ/mol. The reaction is obviously more difficult to proceed than conventional Anammox, which has ΔG^{θ} = −357 kJ/mol.

As the SO_4^{2-} dependent AAOB are obligate anaerobic bacteria, high substrate concentrations and a low oxidation-reduction potential (ORP)(< −100 mV) can intensify the sulfammox process. The calculated values of redox potential for the half reactions of reduction of N_2 to NH_4^+ and SO_4^{2-} to S^0 at pH = 8 was in the narrow range from −330 to −360 mV. Thus, SO_4^{2-} reduction and NH_4^+ oxidation could coexist together under anaerobic conditions.

11.5.6 Other factors

In addition to the influent N/S ratio, operating temperature, pH, and organic carbon source, other factors such as N_2H_4, NH_2OH, NO_2^-, NO_3^-, H_2S, and oxidation-reduction potential (ORP) also affect sulfammox. Sulfides inhibit microbial activity and denitrifying proteins, which may restrict sulfammox by affecting denitrification. HCO_3^- concentrations above 1000 mg/L were shown to inhibit sulfammox, whereas the addition of N_2H_4 and NH_2OH reinforced the effectiveness of sulfammox. Additionally, higher substrate concentrations and lower ORPs facilitated sulfammox. NO_3^- and NO_2^- can cause N conversion through denitrification and Anammox within the sulfammox system, and SO_4^{2-} removal through SAD, thus increasing the removal efficacy of NH_4^+ and SO_4^{2-} while inhibiting sulfammox. In addition to these factors, several other substances may affect sulfammox by affecting the S cycle. For example, some natural organic substances, such as humus and humic can serve as electron acceptors in the microbial S reduction process, thus partially oxidizing dissolved sulfides to SO_4^{2-}.

11.6 Applicable reactors and reported efficiencies

Until now, the sulfammox process has been studied in different reactors in terms of the flow conditions and biomass retention method. The applicable reactors are used to implement sulfammox, such as the GACFBR, UASB, MBBR, expanded granular sludge bed (EGSB), RBC, and a few self-designed reactors (Figure 11-4). The performance of these reactors and their ability to achieve sulfammox are reviewed in Table 11-2.

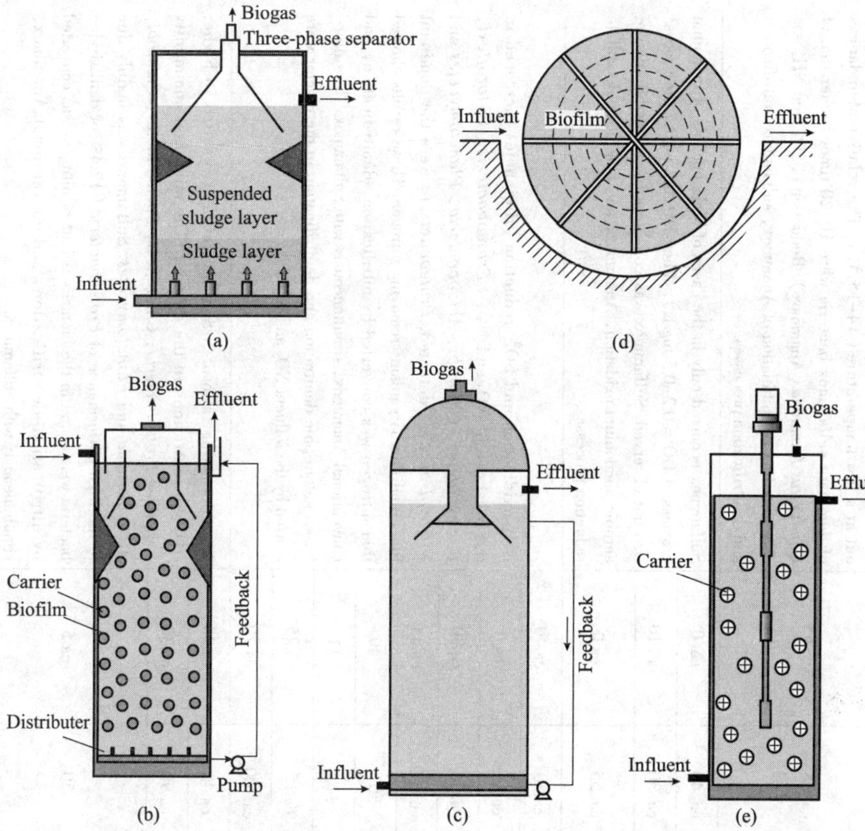

Figure 11-4 Schematic diagram of sulfammox in GACFBR (a), UASB (b), EGSB (c), RBC (d), and MBBR (e)

11.7 Conclusions

A systematic review of existing literature on sulfammox is presented herein. Sulfammox is mediated by microbes and is widely present in environments. The influent N/S ratio, presence or absence of organic carbon sources, pH, temperature, N_2H_4, NH_2OH, NO_2^-, NO_3^-, and H_2S could affect the effectiveness of sulfammox in wastewater treatment. Sulfammox is predominantly used in GACFBR, UASB, EGSB, RBC, and MBBR for wastewater treatment. In the future, the potential impacting factors need to be tested in bioreactors carrying out sulfammox process. Meanwhile, the operating parameters of sulfammox should be further optimized to determine the optimal parameters to achieve the desired treatment performance. The scale of the system requires to be expanded based on experiments, and more actual wastewater should be used to match the practical applications more closely. In addition, the characterization of key functional microbial communities and the analysis of microbial interrelationships should be undertaken in future research.

Table 11-2 Sulfammox based reactors and efficiency of NH_4^+, SO_4^{2-} removal

Reactocr	Onigin of biomass	Influent NH_4^+ / (mg/L)	Influent SO_4^{2-} / (mg/L)	NH_4^+ removal efficiency/%	SO_4^{2-} removal efficiency/%	Highlights of the study
Combining system: Upflow anaerobic sludge blanket (UASB), anoxic/oxic reactor (A/O), Anammox and Sulfammox reactor (ANAOR),	Landfill leachate	610–700	1870–1920	ca. 98	ca. 53	Landfill leachate was used as a substrate. The tests were carried out at a low temperature (14–15 ℃). The relative abundances of dominant sulfammox bacteria were 10–20 times higher that of *Candidatus Kuenenia* (Anammox). Reduction of SO_4^{2-} and NH_4^+ was considered as a combination of Anammox, sulfammox, nitrification and denitrification processes.
Continuous flow stirred tank reactor (CFSTR)	Long-term operation Anammox up-flow reactor	110	0–110	ca. 40	ca. 0	Sulfammox occurred only in the cases of high amounts of inoculum biomass at DO = 0.2–0.5 mg/L). When DO<0.2 mg/L, the process was not observed. Sulfammox was considered as a combination of aerobic ammonium oxidation, Anammox, and heterotrophic sulfate reduction processes.
		60	90	ca. 30	ca. 10	
		60	90	ca. 55	ca.0	
Self-designed circulating flowreactor (SDCF)	n.a.	120	183	ca.30	ca.40	NH_4^+ oxidization and SO_4^{2-} reduction efficiencies increased in the presence of NO_2^- and NO_3^-. *Proteobacteria*, *Chloroflexi*, *Bacteroidetes*, *Chlorobi*, *Acidobacteria*, *Planctomycetes* and *Nitrospirae* were detected. *Proteobacteria* were the dominant functional microorganisms removing nitrogen. These results showed that nitrogen was converted by nitrification, denitrification, and conventional Anammox, simultaneously with Sulfammox. The sulfur-based autotrophic denitration and denitrification in the reactor were caused by the influent NO_2^- and NO_3^-.
		160	216	ca.55	ca.0	
		110	116	ca. 75	ca.30	
		80	100	ca. 100	ca.45	
		120	183	30	40	
		160	216	11	11	
		160	216	ca.15	ca.25	
		90	133	ca.100	ca.70	
Self-designed circulating flowreactor (SDCF)	Mixed sludge, which consisted anaerobic granular sludge from a municipal wastewater plant and denitrification sludge from a continuous stirred-tank reactor	50	90	ca.40	ca.30	The increasing ratio of N/S in the influent resulted in higher NO_2^- concentrations in the effluent. The microbial community comprised *Proteobacteria*, *Chloroflexi*, *Bacteroidetes*, *Chlorobi*, *Acidobacteria* and *Planctomycetes*. Sulfammox was mainly due to the high performance of *Proteobacteria* (12.4% of denitrifying bacteria were found in the biomass). Part of nitrogen was converted by nitrification-denitrification, and conventional Anammox, simultaneously with Sulfammox.
		120	170	ca.90	ca.30	
		180	360	ca.20	ca.5	

Continued

Reactocr	Onigin of biomass	Influent NH_4^+ / (mg/L)	Influent SO_4^{2-} / (mg/L)	NH_4^+ removal efficiency/%	SO_4^{2-} removal efficiency/%	Highlights of the study
Expanded Granular Sludge Bed (EGSB)	Anaerobic hydrolysis acidification reactor	166–666	3600	40–58	64–71	The removal efficiency of SO_4^{2-} gradually improved as the influent NH_4^+ concentrations increased from 166–666 mg N/L to 1000–2000 mg N/L. At the same time, 71% NH_4^+ was removed. After increasing the NH_4^+ concentration to > 3000 mg N/L, the SO_4^{2-} reduction efficiency was reduced to 28%. SRB and denitrifying bacteria were mainly responsible for SO_4^{2-} and nitrogen removal.
		1000–2000		40–70	66–82	
		>3000		10–25	28	
Anaerobic sequencing batch reactor (ASBR)	Activated sludge from the aerobic tank of digested liquor	97	261	ca.88	ca.19	The presence of *Planctomycetes* revealed that Anammox was a highly involved pathway in NH_4^+ removal, even without NO_2^- in the feed. Other autotrophic denitrifying bacteria, related to species the *Paracoccus* Denitrificans, were also present. These bacteria utilize S^0 as an electron donor and produce SO_4^{2-}, and competitively use NO_2^- with Anammox.
Moving bed biofilm reactor (MBBR)	Well-established attached Anammox biofilm withdrawn from a lab-scale conventional Anammox reactor treating reject water	69	ca.70	ca.30	ca.10	Sulfammox tests were performed in MBBR at 20 ℃ and UASBR at 36 ℃. Very similar results of NH_4^+ and overall SO_4^{2-} removal were obtained in both reactors. The Sulfammox process took place as one reaction of the multiple complex interactions between N-compounds, S-compounds, and organics (primarily humic matter) resulting in a significantly higher removal ratio of NH_4^+ than the Sulfammox stoichiometry predicts. It was postulated that the phylum *Verrucomicrobia* could also be involved in sulfammox.
Upflow anaerobic sludge blanket reactor (UASBR)	Anaerobic sludge from a yeast factory wastewater treating facility (Salutaguse, Estonia)	69	ca.70	ca.25	ca.10	
Upflow anaerobic sludge blanket reactor (UASBR)	Reject water from anaerobic digestion of municipal wastewater sludge	221	193	ca.30	ca.20	Sulfammox and Anammox tests were carried out at 36 ℃ and 20 ℃, respectively. NO_2^- was proved to be a more efficient electron acceptor than SO_4^{2-}. The reduction of SO_4^{2-} and NH_4^+ was considered as a combination of sulfammox and denitrification processes.
Expanded bed reactor (EBR)	Lab-scale reactor treating NH_4^+ and SO_4 simultaneously for more than two years	229	163	ca.44	40	*Bacillus Benzoevorans* was isolated. Its optimum pH and temperature were 8.5 and 30 ℃, respectively. The reduction of SO_4^{2-} and NH_4^+ was considered as sulfammox only.

Continued

Reactocr	Onigin of biomass	Influent NH_4^+ / (mg/L)	Influent SO_4^{2-} / (mg/L)	NH_4^+ removal efficiency/%	SO_4^{2-} removal efficiency/%	Highlights of the study
Upflow anaerobic sludge blanket reactor (UASBR)	Nitrifying sludge in a municipal wastewater treatment plant	60	240	40	30	Sulfammox was successfully performed by changing NO_2^- into SO_4^{2-} as an electron acceptor. The reduction of SO_4^{2-} and NH_4^+ was considered as sulfammox only.
Expanded bed reactor (EBR)	Anaerobic digester in a municipal wastewater treatment plant	84–270	450–740	ca.40	ca.10	Sulfate-dependent anaerobic ammonium oxidation occurs with acclimated anaerobic digested sludge in the absence of organic matter. Anaerobic ammonium oxidation with sulfate does not tend to occur spontaneously due to its low ΔG, value. The experiment demonstrated that high substrate concentrations and low ORP may be favorable for sulfammox. The reduction of SO_4^{2-} and NH_4^+ was as a sulfammox only.
		30–90	80–200	ca.55	ca.43	
Non-woven rotating biological contactor (NWRBC)	Long-term operation Anammox up-flow reactor	ca.198	ca.528	ca.100	ca.70	Bacteria belonging to *Planctomycetales*, especially the new species *Anammoxoglobus Sulfate*, were identified as the functional community. The reduction of SO_4^{2-} and NH_4^+ was considered as a sulfammox only.
Anaerobic attached-growth bioreactor (AAGB)	Anaerobic activated sludge collected from an anaerobic continuous stirred tank reactor	50	57	ca.43	ca.59	Low removal of NH_4^+ was obtained with high removal of SO_4^{2-}, implying the existence of competition between SRB) and microorganisms responsible for using SO_4^{2-} and NH_4^+. Low COD, high SO_4^{2-} and high NH_4^+ loadings at pH = 7.8 could promoted sulfammox. The reduction of SO_4^{2-} and NH_4^+ was considered as a sulfammox only.
Granular activated carbon fluidized-Bed (GACFB)	Diluted vinasse originating from an ethanol distillery plant processing beet sugar molasses	<10	1000	50	80	The first report on the sulfammox process. The "anomalous" NH_4^+ removal was obtained in a granular activated carbon (GAC) anaerobic fluidized-bed reactor. The reactor treated vinasse from an ethanol distillery of sugar beet molasses. About 50% of the influent nitrogen load was removed from the liquid phase appearing as N_2 in the gas phase. Simultaneously, only 20% of the SO_4^{2-} initially present in the influent appears as S_2^- in the effluent or H_2S in the biogas, indicating that 80% of the sulfur was removed in sulfammox. The reduction of SO_4^{2-} and NH_4^+ was considered as a combination of sulfammox and denitrification processes.

Questions

[1] What is the main principle for Sulfammox process?

[2] What are the characteristics of Sulfammox bacteria?

[3] What are the natural factors that could affect the growth of Sulfammox bacteria?

[4] How the organics affects the Feammox process? Whether the mechanism of Sulfammox is different under inorganic or organic conditions?

[5] What kind of wastewater could be treated via Sulfammox-based technology?

References

[1] In't Zandt M H, de Jong AEE, Slomp C P, et al. The hunt for the most-wanted chemolithoautotrophic spookmicrobes [J]. FEMS Microbiol. Ecol. 2018, 94 : fly064.

[2] Bi Z, Wanyan D, Li X, et al. Biological conversion pathways of sulfate reduction ammonium oxidation in Anammox consortia. Front [J]. Environ. Sci. Eng., 2020, 14 : 38.

[3] Zhang D, Cui L, Madani R M A, et al. Effect of nitrite and nitrate on sulfate reducing ammonium oxidation [J]. Water Sci. Technol., 2019, 80(4): 634-643.

[4] Xu L Z, Zhang Q, Fu J J, et al. Deciphering the microbial community and functional genes response of Anammox sludge to sulfide stress [J]. Bioresour. Technol., 2020, 302 : 122885.

[5] Wisniewski K, di Biase A, Munz G, et al. Kinetic characterization of hydrogen sulfide inhibition of suspended Anammox biomass from a membrane bioreactor [J]. Biochem. Eng. J. 2019, 143, 48-57.

[6] Rios-Del Toro EE, Cervantes FJ, Anaerobic ammonium oxidation in marine environments : contribution to biogeochemical cycles and biotechnological developments for wastewater treatment [J]. Rev. Environ. Sci. Biotechnol., 2019, 18 : 11-27.

[7] Ucar D, Yilmaz T, Di Capua F, et al. Comparison of biogenic and chemical sulfur as electron donors for autotrophic denitrification in sulfur- fed membrane bioreactor (SMBR) [J]. Bioresour Technol., 2020, 299 : 122574.

[8] Fdz-Polanco F, Fdz-Polanco M, Fdz-Polanco N, et al. Simultaneous organic nitrogen and sulfate removal in an anaerobic GAC fluidised bed reactor [J]. Water Sci. Technol., 2001, 144(2): 15-22.

[9] Fdz-Polanco F, Fdz-Polanco M, Fdz-Polanco N, et al. New process for simultaneous removal of nitrogen and sulphur under anaerobic conditions [J]. Water Res. 2001, 35 : 1111-1114.

[10] Rikmann E, Zekker I, Tomingas M, et al. Sulfate-reducing anaerobic ammonium oxidation as a potential treatment method for high nitrogen-content wastewater [J]. Biodegradation, 2012, 23(4): 509-524.

[11] Wu L, Yan Z, Li J, et al. Low temperature advanced nitrogen and sulfate removal from landfill leachate by nitrite-Anammox and sulfate Anammox [J]. Environ. Pollut., 2020, 259, 113763.

[12] Rios-Del Toro E E, Cervantes F J, Coupling between Anammox and autotrophic denitrification for simultaneous removal of ammonium and sulfide by enriched marine sediments [J]. Biodegradation, 2016, 27(2-3): 107-118.

[13] Rios-Del Toro EE, Valenzuela, EI, Lopez-Lozano NE, et al. Anaerobic ammonium oxidation linked to sulfate and ferric iron reduction fuels nitrogen loss in marine sediments [J]. Biodegradation, 2018, 29(5): 429-442.

[14] Zhang F, Ren S, Liang H, et al. Efficient nitrogen removal and partial elemental sulfur recovery by combined nitrite-Anammox and sulfate-Anammox: a novel strategy for treating mature landfill leachate [J]. J. Clean. Prod., 2023, 399: 136553.

[15] Mohammed Madani R, Liang J, Cui L, et al. Novel simultaneous removal of ammonium and sulfate by isolated Bacillus cereus strain from sewage treatment plant [J]. Water Air Soil Pollut., 2022, 233(6): 185.

[16] Grubba D, Yin Z, Majtacz J, et al. Incorporation of the sulfur cycle in sustainable nitrogen removal systems - a review [J]. J. Clean. Prod., 2022, 372: 133495.

[17] Prachakittikul P, Wantawin C, Noophan PL, et al. Anammox-like performances for nitrogen removal from ammonium-sulfate-rich wastewater in an anaerobic sequencing batch reactor [J]. Environ. Sci. Health A. Tox. Hazard. Subst. Environ. Eng., 2016, 51(3): 220-228.

[18] Yang S, Zhu Y, Liu H, et al. Effects of N/S and ammonia concentrations on the process of sulfate reduction anaerobic ammonium oxidation [J]. Trans. CSAE., 2021, 37(16): 199-204.

[19] Zhang D, Cui L, Wang H, et al. Study of sulfate-reducing ammonium oxidation process and its microbial community composition [J]. Water Sci. Technol., 2019, 79: 137-144

[20] Cheng Y F, Zhang Z Z, Ma W J, et al. Response of the mainstream Anammox process to the biodegradable carbon sources in the granule-based systems: the difference in self-stratification of the microbial community [J]. Sci. Total Environ., 2022, 851(1): 158-191.

Chapter 12 Fe(Ⅲ) reduction coupled to anaerobic ammonium oxidation (Feammox)

Emphasis: Feammox microbial functions; The artificial and natural factors that affect the growth of Feammox microorganisms

Nut: Environmental factors that affect the Feammox process

12.1 Introduction

Nitrogen is a key element for living organisms, partially because it is crucial for the biosynthesis of cellular components, including proteins and nucleic acids, among others. Ammonium pollution in water systems is currently a severe problem, and this issue originated from influent ammonium, as well as conversion via denitrification and dissimilatory nitrate reduction to ammonium (DNRA). The widespread use and discharge of ammonium contribute to eutrophication, which causes detrimental changes to the water quality and structure of the ecosystem. Biological removal processes are the most commonly used methods in reality because they have low energy and chemical demands. Various novel technologies have been established to revolutionize the industrial nitrogen removal process. Anaerobic ammonium oxidation (Anammox) effectively solves the challenges in handling low-C/N sewage in practical applications at the appropriate high temperature conditions. Partial nitrification combined with Anammox (PN/A) and partial denitrification combined with Anammox (PD/A) have emerged as promising strategies for wastewater treatment plants.

Iron is a transition metal that can exist in a wide range of redox states, varying from −2 to +6. Numerous investigations have probed the use of iron substances for mitigating nitrogen pollution. Various iron morphologies have been applied to degrade nitrogen-containing inorganic and organic pollutants. Additionally, iron substances can increase the rate of nitrogen transformation in certain reactions, including Anammox and iron-dependent denitrification. Iron-containing active sites are also found in nitrogen elimination enzymes and cytochromes, which enrich microbial communities or influence the spatial structure of a coacervate.

Since insoluble iron oxides are ubiquitous electron acceptors in the environment, the anaerobic ammonium oxidation reaction coupled to Fe(Ⅲ) reduction (known as the Feammox process) has been developed and corroborated using 15N–NH_4^+ isotope labeling techniques. There has been a nearly 20-fold increase in the number of articles published over the last decade listing "Feammox" as a keyword, and this rise in publications has been accompanied by a gradually deepening understanding of the process (Figure 12–1).

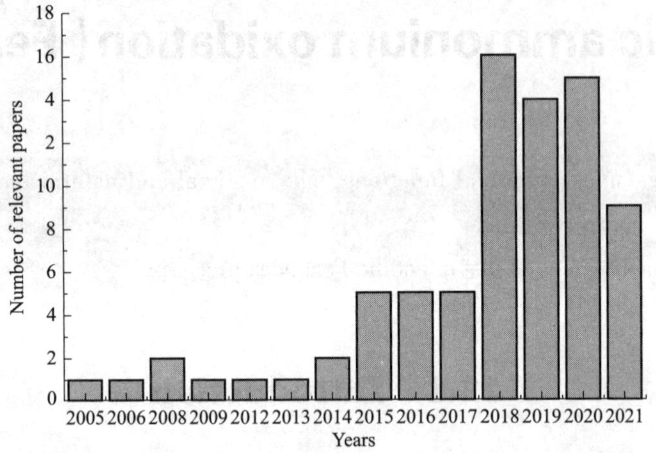

Figure 12–1 The number of articles discussing Feammox

These data are based on the number of articles mentioning "Feammox" and comprising a relevant description in the web of science database in April 2021.

This chapter highlights recent revelations regarding the Feammox process and summarize the inherent mechanism of the Feammox process and the relevant microorganisms, moreover, evaluate the influences of various environmental factors and potential approaches to promote community growth and discuss future prospects of implementing Feammox combined with other nitrogen removal methods for practical applications in low-C/N wastewater treatment plants.

12.2 A collection of existing investigations regarding the Feammox process

12.2.1 Anthology of Feammox studies

Feammox reactions are potential pathways for the conversion of ammonium (NH_4^+) to dinitrogen gas (N_2), nitrite (NO_2^-), or nitrate (NO_3^-) as the terminal products [see Eqs. (12–1), (12–2), and (12–3), respectively] based on experiments using the isotopic tracer technique (Figure 12–2).

$$3Fe(OH)_3 + 5H^+ + NH_4^+ \rightarrow 3Fe^{2+} + 9H_2O + 0.5N_2 \quad \Delta_r G_m = -245 \text{ kJ/mol} \qquad (12\text{–}1)$$

$$6Fe(OH)_3 + 10H^+ + NH_4^+ \rightarrow 6Fe^{2+} + 16H_2O + NO_2^- \quad \Delta_r G_m = -164 \text{ kJ/mol} \qquad (12\text{–}2)$$

$$8Fe(OH)_3 + 14H^+ + NH_4^+ \rightarrow 8Fe^{2+} + 21H_2O + NO_3^- \quad \Delta_r G_m = -207 \text{ kJ/mol} \qquad (12\text{–}3)$$

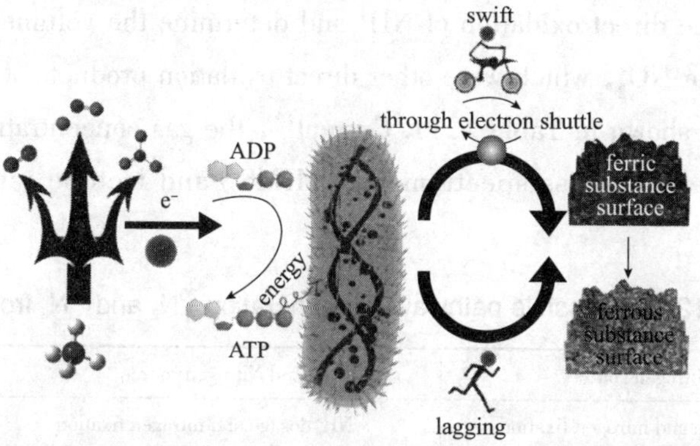

Figure 12-2 Schematic diagram of the Feammox process

blue spheres represent nitrogen, red spheres represent oxygen, white spheres represent hydrogen and the green spheres represent the phosphate group in the ATP and ADP

In these reactions, the Fe(Ⅲ)(hydr)oxides act as electron acceptors and convert to Fe(Ⅱ) under anaerobic conditions. Such biochemical reactions were first detected in timber soil without nitrogen or oxygen following the addition of $^{15}NH_4^+$ and the identification of $^{30}N_2$. Initial experiments suggested that NO_2^- was the endproduct of the nitrogen oxidation reaction. However, isotope tracer results indicated that NH_4^+ was oxidized directly to primarily generate N_2. Homogenous reactions were also investigated by employing the acetylene inhibition method. The Feammox reaction directly produced N_2, which made up 67.0%–93.2% of the entire Feammox reaction product mixture. Meanwhile, some NH_4^+ residue was oxidized to NO_3^- and NO_2^-, although the reaction is the most energy-efficient when it yields N_2.

12.2.2 Unveiling of Feammox in the environment and rate measurements

Short-range ordered (SRO) Fe minerals are more susceptible to redox effects. These Fe minerals often serve as the dominant terminal electron acceptors, and they are reduced during periods of oxygen depletion in the soil. This process is much more likely to occur in intermittent aerobic regions or in the deep soil that contains high levels of ferric iron and NH_4^+. Use of the $^{15}NH_4^+$ isotope tracer technique has revealed evidence of the Feammox reaction in heterogeneous soil and substrate mud habitats, such as in tropical rainforests, paddy fields, wetlands, riparian zones, iron mine drainage, and even ocean sediments. Studies conducted in manufactured sewage treatment plants ascertained that the process is related to anaerobic sludge digestion and activated sludge systems.

To date, $^{15}NH_4^+$ isotope tracing has been applied extensively to measure the rates of the reactions involved in the Feammox process. Feammox can generate $^{30}N_2$ or $^{29}N_2$, which might be a result of the Feammox process alone or the combination Feammox with Anammox, denitrification, or chemodenitrification processes. Acetylene is commonly used in such detection efforts to suppress the Anammox and denitrification reactions. Furthermore, it can help clarify how much of the N_2

was generated from the direct oxidation of NH_4^+ and determine the volume of N_2 originating from the reduction of NO_3^- or NO_2^-, which were other direct oxidation products of NH_4^+. Several potential reaction pathways are shown in Table 12–1. Currently, the gas concentration is often determined using membrane interface mass spectrometry (MIMS) and isotope ratio mass spectrometry (IRMS).

Table 12–1 Possible pathways for generating $^{30}N_2$ and $^{29}N_2$ from $^{15}NH_4^+$

Product	1st Nitrogen source	2nd Nitrogen source	Transformation process
$^{30}N_2$	$^{15}NH_4^+$ dosage and nitrogen fixation	$^{15}NH_4^+$ dosage and nitrogen fixation	Feammox
	$^{15}NH_4^+$ dosage and nitrogen fixation	$^{15}NO_2^-/^{15}NO_3^-$ engendered by the Feammox	Anammox
	$^{15}NO_2^-/^{15}NO_3^-$ engendered by the Feammox	$^{15}NO_2^-/^{15}NO_3^-$ engendered by the Feammox	Denitrification
	$^{15}NO_2^-/^{15}NO_3^-$ engendered by the Feammox	$^{15}NO_2^-/^{15}NO_3^-$ engendered by the Feammox	Denitrification and chemodenitrification
$^{29}N_2$	$^{15}NH_4^+$ dosage and nitrogen fixation	$^{15}NH_4^+$ aboriginal and nitrogen fixation	Feammox
	$^{15}NH_4^+$ dosage and nitrogen fixation	$^{15}NO_2^-/^{15}NO_3^-$ engendered by the Feammox and aboriginal	Anammox
	$^{15}NH_4^+$ aboriginal and nitrogen fixation	$^{15}NO_2^-/^{15}NO_3^-$ engendered by the Feammox	Anammox
	$^{15}NO_2^-/^{15}NO_3^-$ engendered by the Feammox	$^{15}NO_2^-/^{15}NO_3^-$ engendered by the Feammox and aboriginal	Denitrification
	$^{15}NO_2^-/^{15}NO_3^-$ engendered by the Feammox	$^{15}NO_2^-/^{15}NO_3^-$ engendered by the Feammox and aboriginal	Denitrification and chemodenitrification
	$^{15}NO_2^-/^{15}NO_3^-$ engendered by the Feammox	Amino compounds aboriginal	Codenitrification

Results from ^{15}N isotope tracing and those compiled in Table 12–1 illustrate that Feammox participates in the production of $^{30}N_2$ through various overall reaction processes. The rate of the Feammox reaction is considered to fall within the range of 0.04–0.59 mg N/ (kg · d) based on the detection of $^{30}N_2$; which is on the same order of magnitude as the Anammox rates determined in previous investigations. The nitrogen lost during the Feammox process varies and the reaction rate is significantly impacted by the nutrient state (Table 12–2). It is believed that Feammox accounts for a significant proportion of nitrogen removal in the soil, and its impact on nitrogen fertilizer cannot be ignored. In reality, since the production of $^{29}N_2$ is also relevant to the Feammox process, such reactions may have an even greater effect on nitrogen loss in the natural environment.

Table 12–2 Feammox reaction rates determined in various sampling locations and their contributions to the total nitrogen loss

Feammox colony source	Depth of the Feammox colony sample in the spot / cm	Feammox course rate/ [mg N/(kg · d)]	The loss of N via Feammox counts to the N dosage	References
Paddy soil	0–10	0.17–0.59	3.9%–31%	(Ding et al., 2014)
Uncultivated paddy soil	0–10	0.04	—	

Continued

Feammox colony source	Depth of the Feammox colony sample in the spot / cm	Feammox course rate/ [mg N/(kg · d)]	The loss of N via Feammox counts to the N dosage	References
Unploughed land	0–10	0.04	—	(Yi, 2019)
Paddy soil added fertilizer	0–20	0.205–0.319	8.0%–12.5%	—
Unfertilized paddy soil	0–20	0.047	—	—
Wetland substrate mud	0–5	0.24–0.36	0.04%–4.9% of total inorganic nitrogen	(Guan et al., 2018; Li et al., 2015)
Estuary	0–10	0.20–0.37	—	(Ding et al., 2017)
Tropical timber soil	0–10	0.32	—	(Yang et al., 2012)
Lake sediment	—	0.23–0.43	5.0%–9.2%	(Yao et al., 2019a)

12.2.3 Feammox microbial functions

The major challenge in studying Feammox lies in the identification and characterization of the microorganisms that carry out the Feammox process. Although the Feammox process is driven by microorganisms, state of the art research has thoroughly probed only one type of functional bacteria, i.e., *Acidimicrobiaceae* sp. A6 (ATCC, PTA-122488). These bacteria were isolated from rich cultures of a riparian wetland zone in New Jersey, USA, and have demonstrated NH_4^+ oxidation to NO_2^-. In the pioneering investigation, it was determined that this strain could use ferrihydrite as an Fe(Ⅲ) source and CO_2 as a carbon source. In terms of the enriched with the growth of the A6 culture via a mechanism of functional annotation from the metagenomic library. Among these oxygenase-related genes, there were two uncharacterized genes exhibiting 92.0%–93.0% similarity to particulate methane monooxygenases. This genomics analysis suggested that the A6 may also play a key role in the relevant microbial community.

The discovery of the A6 bacteria occurred long before the development of continuous-flow membrane reactors, which perform the NO_2^- conversion and achieve N_2O accumulation within an apparatus. The A6 thrive in a Feammox enrichment culture and can exhibit stable ammonium nitrogen transformation for more than one year, whereas pure A6 only remain active for one to two months. This distinction indicated that important cofactors or auxiliary growth factors produced by other microorganisms in the colony were missing in pure incubations, meaning that maintaining and fostering A6 required a longer time under those conditions. These results also demonstrated that more thorough explorations were necessary to morphological structure, microscopic examinations of the A6 colonies showed that they are grampositive and the cells are rod-shaped with lengths of 1.5–3 μm and widths of approximately 0.5 μm (Figure 12-3).

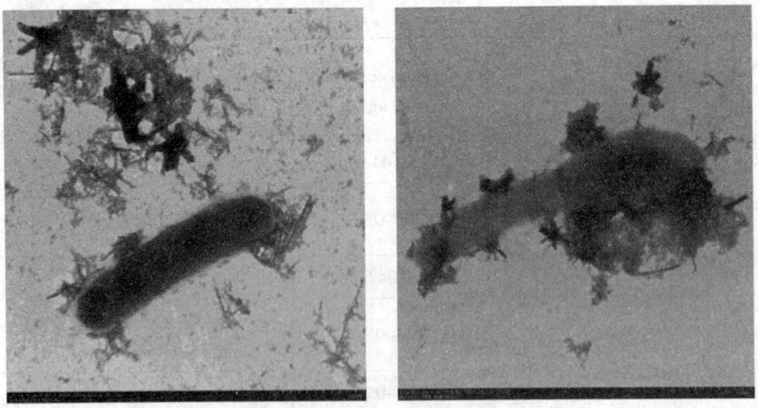

Figure 12–3 SEM images of *Acidimicrobiaceae* sp. A6

Further analysis using an Illumina HiSeq platform revealed a series of oxygenase-related genes (GenBank accession numbers: MG011983–MG012003) that were understand A6 and its relationship to Feammox. Although only a segment of the A6 ribosome's base sequence has been elucidated, the primer that can be used to distinguish the species has been identified. The primers encompassing the acm342f (5'- GCA ATG GGG GAA ACC CTG AC-3') - 439r (5'-ACC GTC AAT TTC GTC CCT GC-3') and primer set 33F (5'- GGC GGC GTG CTT AAC ACA T-3') - 232R (5'-GAG CCC GTC CCA GAG TGA TA-3') were used to quantify A6 via quantitative realtime PCR (qPCR) technology. The A6 has been isolated, but researchers have not yet established which functional genes and enzymes adversely affect the progress of the Feammox reaction.

Because of the limited information regarding pure A6, researchers have explored other types of iron-reducing bacteria (IRB) within the Feammox bacteria cluster. It has been proposed that *Geobacter* spp. and *Shewanella* spp. may play important roles during the reaction based on the positive correlations among (i) the abundance of these taxa, (ii) the rates of Fe (Ⅲ) reduction, and (iii) the rates of the $^{30}N_2$ production. Other bacteria, including Acidimicrobiaceae, are also believed to serve as latent driving species owing to their increasing relative abundances in bacterial experiments. Another investigation determined that *Geobacter*, *Anaeromyxobacter*, *Desulfosporosinus*, *Dechloromonas*, and *Geothrix* could participate in the Feammox process. The unclassified *Pelobacteraceae* and *Desulfovibrio* all increased in abundance following the addition of electron shuttles in the Feammox system, indicating that they might serve important roles in the Feammox process. Several functional bacteria have been discovered in soil from natural farmlands and identified on the basis of RNA-based stable isotope probing (RNA-SIP) technology, which relies on the peculiarity of DNA's semiconservative replication and high-throughput sequencing. *Geobacter*, *GOUTA*19, *Nitrososphaeraceae*, and *Pseudomonas* exhibit correlations with Feammox activity. In addition to the aforementioned microbial phylogenetic families, which may participate in the Feammox process, research has shown that *Exiguobacterium* spp. could also potentially promote the Feammox process.

It has also been determined that other microbial combination clusters can support the Feammox reaction. The co-culture known as HJ-4 consists of Anaerospora hongkongensis (85%) and Comamonadaceae (15%); in this combination system, A. hongkongensis is crucial for sulfur and iron recycling processes, and Comamonadaceae is responsible for the oxidation of NH_4^+ to NO_2^-. The investigation revealed that the sulfur, iron, and nitrogen cycles can be explored simultaneously in water and terrestrial ecosystems using co-culture methods.

Although the importance of IRB in terms of Feammox has been established, the extracellular respiration processes relating to Feammox remain unclear because of the various potential electron transfer pathways (Figure 12-4). Further investigations are needed to elucidate the Feammox mechanism, which could provide critical insights into extracellular respiration using NH_4^+ as the electron donor.

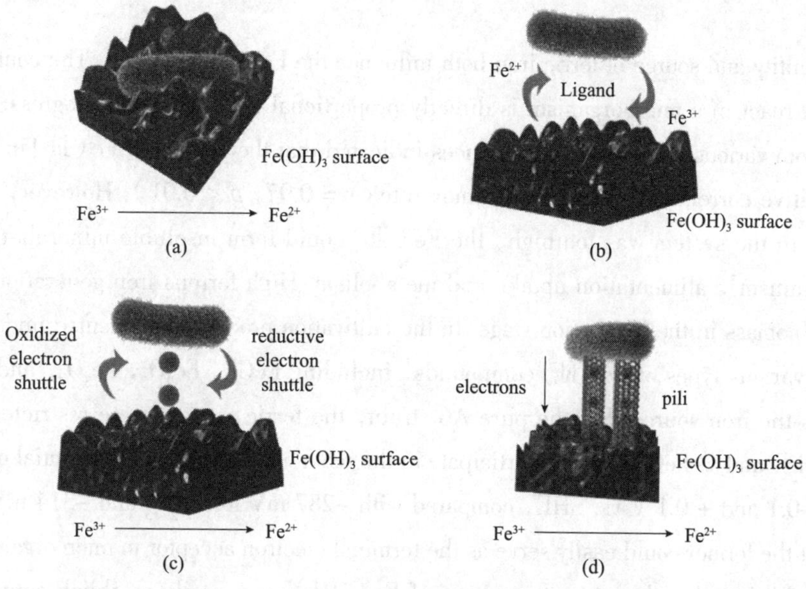

Figure 12-4 Potential pathways by which iron-reducing bacteria may transfer electrons to ferric iron surfaces

(a) direct contact, (b) the ligand, (c) the electron shuttle, or (d) the nanowire/e-pili

To date, several species of functional microorganism, including IRB and the A6 strain, have been detected in various Feammox habitats. However, the bioinformatics (e.g., metagenomics and metatranscriptomics) must be analyzed thoroughly to examine the functional genes and microbes in terms of the researching conditions and the complexity of their purification in the medley system. A deeper understanding of the Feammox colonies would also help verify the end-product of the Feammox process to facilitate Feammox applications. Furthermore, it would provide more crucial information about the NH_4^+ transformation process. Such future investigations will elucidate the details of nitrogen transformations in the nature environment and facilitate the implementation of the Feammox process for wastewater treatment (combined with other techniques).

12.3 The artificial and natural factors that affect the growth of Feammox microorganisms

Competition is fierce in the microbial world, and resources can be scarce. For microorganisms related to Feammox, suitable environmental factors can reduce the competition in harsh environments. Researchers have aimed to understand how the cultivation conditions are related to the relevant reaction rates, nitrogen removal percentages, and cultivation durations. As a result, diverse contributing factors have been explored in heterogenous experiments.

12.3.1 The ferric iron

The quantity and source of ferric iron both influence the Feammox process. The content of ferric iron that can react in a microorganism is directly proportional to the reaction progress. Analysis of paddy soil from various chronological sequences indicated that the content of Fe (Ⅲ) in the soil had a strong positive correlation with the Feammox rate ($r = 0.97$, $p < 0.01$). However, if the ferric iron content in the system was too high, the Fe (Ⅲ) could form insoluble minerals that blocked the microorganism's alimentation uptake and metabolism. High ferrous iron generation might also damage the biomass in the incubation stage. In the cultivation process implemented under laboratory conditions, various types of Fe (Ⅲ) compounds, including $FeCl_3$, Fe_2O_3, Fe_3O_4, and $Fe(OH)_3$ can serve as the iron source. For the pure A6 strain, the ferric sources were restricted, and only $Fe_2O_3 \cdot 0.5H_2O$ and α-FeOOH could participate in the reaction. The reduction potential of $Fe(OH)_3$ is between −0.1 and + 0.1 V vs. SHE, compared with −287 mV for Fe_2O_3 and −314 mV for Fe_3O_4, meaning that the former could easily serve as the terminal electron acceptor in microorganism reaction pathways. Additionally, the amorphous form of $Fe(OH)_3$ can easily establish contact with the relevant bacteria. The dissociation (K_{sp}) of ferric compounds is an additional aspect that governs various iron reduction reactions. The K_{sp} of $Fe(OH)_3$ is greater than that of Fe_2O_3 and magnetite, making $Fe(OH)_3$ more likely to participate in microbial iron reduction.

12.3.2 Soil pH and redox potential

The form and phase of iron in the soil are highly dependent on the soil redox potential (Eh) and pH; therefore, there parameters further influence the Feammox process and the distribution of the microorganism's alimentation uptake and metabolism. High ferrous iron generation might also damage the biomass in the incubation stage. In the cultivation process implemented under laboratory conditions, various types of Fe (Ⅲ) compounds, including $FeCl_3$, Fe_2O_3, Fe_3O_4, and Fe (OH)$_3$ can serve as corresponding microbial colony. The theoretic chemical equation suggests that the Feammox process requires acidic conditions (i.e., pH < 6.5) to satisfy the thermodynamic requirements. This is consistent with pH gradient experiments that evaluated the survival of Acidimicrobiaceae and revealed optimal growth at pH 4.5 (for A6) and 6.5. Further experiments exploring the activity of A6 indicated that the greatest quantity of ammonium was oxidized under conditions where pH = 4. However, Feammox reaction activity was also observed in the alkaline environments (i.e. pH between 8.32 and 8.46) of intertidal wetlands. This behavior was attributed to catalysis by functional Feammox microorganisms. The pH could also impact the end–product of NH_4^+ oxidation. For example, the concentration of protons in the system can affect the Gibbs free energy of the reactions involved. The much lower Gibbs free energy associated with N_2 production can partially explain why N_2 has been identified as the main product in various studies. However, the production of NO_2^- and NO_3^- has also been detected under neutral or alkaline conditions. Alternatively, some potential mechanistic insights are related to the fact that a higher pH can result in a low redox potential. Specifically, a higher concentration of OH^- can significantly reduce the concentration of free–form Fe (Ⅱ), leading to the production of NO_2^- and NO_3^- via Feammox. The pH can also influence the end–product of the reaction by modifying the microbial species. Various Feammox microorganisms have distinct suitable living conditions. For example, A6 favors a weakly–acidic environment and can oxidize NH_4^+ to NO_2^- instead of N_2. Therefore, in such conditions, the major end–product of the Feammox process is NO_2^-. In the soil sediments of the intertidal wetland, the Eh had a negative correlation with the reaction rate ($r = -0.87$, $p < 0.01$), indicating that the Feammox process favored lower Eh values within the examined scope. In addition, high concentrations of organic matter in low–Eh soil promote the release of structural Fe from clay minerals, leading to the formation of Fe (Ⅲ) oxides, which can participate in the Feammox reactions.

12.3.3 Dissolved oxygen

The dissolved oxygen content can also influence the ammonium oxidation pathway. The proportion of A6 and other putative Feammox bacteria in soil samplings have been shown to increase (from 5.0% to 68.0%) under incubation conditions containing low levels of dissolved oxygen (DO <0.02 mg/L). In contrast, higher levels of DO (0.8–1.0 mg/L) somewhat inhibit the proliferation of A6 (which only increases from 5.0% to 47.0%), suggesting that Feammox is a microaerophilic process.

12.3.4 Temperature

Various sources of the Feammox colony have distinct preferred temperatures, both in situ and ex situ. The ambient temperature plays a critical role in the seasonal distribution of Feammox by influencing the membrane transport protein affinity and controlling the oxygen consumption rate, thus creating a favorable anoxic atmosphere for the bacterial colony. Research investigations have shown that the Feammox reaction rate is much faster in the summer [0.05–0.19 mg N/(kg · d)] than in the winter [0.02–0.09 mg N/(kg · d)]. Seasonal shifts have also been observed for IRB, particularly the seasonally-variant abundance of Geobacteraceae in the tested locale. The optimal temperature for the Feammox cluster originating from Anammox sludge is 30 ℃. In contrast, addition of ferrous iron can accelerate the marine Feammox rate at only 15 ℃. For enriched soil, Feammox activity proceeds above 15 ℃, with an optimal incubation temperature of 20 ℃. Incubations in enrichment cultures with 79.0% Feammox bacteria removed approximately 8.0% more NH_4^+ at 20 ℃ than at 35 ℃, indicating that Feammox represents a promising strategy for municipal wastewater treatment that would reduce expenses and could be applied to various sources of polluted water without requiring heating.

12.3.5 Nitrite and nitrate

Feammox activity is susceptible to the NO_2^- concentration in the isolation strain incubation period. It was observed that when the NO_2^- content exceeded the threshold of 1.0 mmol/L, the reaction was completely inhibited, and the high nitrite concentration reduced the amount of active Acidimicrobiaceae in the incubation structure. However, the NO_2^- could stimulate the combined system comprising Feammox and other nitrogen transformation processes. Feeding 5 mg/L NO_2^- into the co-cultivation environment increased the ammonium conversion rate in an anaerobic sequencing batch reactor (ASBR) system, where the rate rose 53.8%, and the NO_2^- dosage was completely degraded in the reactor.

Adding NO_3^- into the Feammox bacterial enrichment system at regular intervals can promote nitrogen removal, and Fe(Ⅲ) may participate in catalytic activity because an Fe(Ⅲ)/Fe(Ⅱ) cycle was shown to emerge in the system. During this process, the Fe(Ⅲ) was reduced to Fe(Ⅱ), and the resulting Fe(Ⅱ) was oxidized by the NO_3^- under the action of iron-oxidizing bacteria (IOB). Then, the system continued the Feammox reaction with ammonium, which was confirmed using X-ray diffractometry (XRD), which verified the oxidation of the ferrous iron; the same conclusion was drawn in another study.

12.3.6 Carbon sources and electron shuttles

The impact of the carbon source on the Feammox reaction is still a topic of debate. Under natural conditions, organic matter (OM) can serve as the carbon source as well as the electron shuttles to boost the rate of this process. Total organic carbon has been reported to enhance electron transfer in the Feammox process by functioning as a recyclable electron mediator. However, ferric iron released from clay minerals can be stimulated by high concentrations of OM in sediment. The degradation of OM (algae or macrophyte) releases protons, which reduce the pH of the system. Feammox can also be induced by protons released following OM degradation.

However, the carbon source used by A6 was identified via 13C labeling experiments. The NH_4^+ oxidation was negligible without the addition of inorganic carbon, which indicated that the A6 bacteria require inorganic carbon to grow. Laboratory enrichment cultures are all incubated with $NaHCO_3$, which suggest that the process is autotrophic, and the Feammox bacteria can thrive in an oligotrophic environment.

Humic substances are quinone-rich and can behave as electron transit agents linking IRB with ferric iron. The quinone compound, 9,10-anthraquinone-2,6-disulfonate (AQDS) is often considered a model because it is quinone-rich. Notably, AQDS can assist the Feammox reaction in the acidic laterite of paddy soil and in experimental culture conditions. The $^{30}N_2$ concentration increased 319.0% after adding an appropriate dosage of AQDS into paddy soil samples. In addition, AQDS increased the total nitrogen removal rate by 28.5% in these experiments. The addition of AQDS can also ① accelerate the Feammox process, ② help A6 cells in the enrichment culture adhere to the anode of a microbial electrolysis cell (MEC) system, and ③ support iron cycling in the soil.

In general, quinones can somewhat simulate the redox capacity of humus. Biochar, which is a more complicated organic carbon source than quinones, can also facilitate electron transfer during the Feammox process through quinones and benzoquinones. The $^{30}N_2$ concentration increased 17.0% and 94.0% after adding AQDS to the ex situ and in situ ferrihydrite samples of paddy soil, respectively. The solid phase of biochar promotes the Feammox process at various pyrolysis temperatures. Comparing the biochar treated at 600 ℃ with the pyrolysis conducted at 300 ℃, the rate of $^{30}N_2$ production rose 29.2%–48.4% and the Feammox reaction rate increased 28.8% or 39.8% depending on the ferric iron supplement. AQDS has also demonstrated synergy with biochar when both were added into the system, and this combination boosted the Feammox reaction rate in the soil. The potentially Feammox-mediated nitrogen loss [0.13–0.48 mg N/(L · d)] increased by 336.0% and 161.0% in media enriched by AQDS and biochar, respectively, relative to the control. Both AQDS and the biochar that underwent pyrolysis at 600 ℃ were able to stimulate genetic copies of the A6 bacteria. In contrast, the biochar that experienced the 300 ℃ pretreatment substantially decreased the content of A6 bacteria (by approximately 40.3%), indicating a potential link between the Feammox activity of IRB and nitrogen transforming bacteria, although this reactivity requires further verification. The electron transfer capacities of the humic substances that drive the loss of nitrogen were also verified in ocean sediments, and this process was termed nature organic matter (NOM)-dependent Anammox. The results revealed that the humic substances could improve the efficiency of $15N_2$ production by 1.36 times according to the tracer analysis. The redox functional groups of NOM can serve as mediators to accelerate the Anammox process and then transfer the electrons to the final acceptor, i.e., Fe(Ⅲ) minerals in the anoxic ocean environment.

Alternatively, graphene nanoribbons can be employed as electron transfer mediators to expediate the Feammox reaction in mangrove forest environments. Graphene nanoribbons act as electrical bridges to facilitate electron transfer to the ferric iron, and they have been shown to enhance the iron reduction rate by 43.0%.

In addition to the aforementioned electron shuttles, there are additional mediators in the iron reduction phase that have not been probed in Feammox-related research, although they may demonstrate similar activity to those mentioned above. For example, potential electron transfer mediators include the riboflavin (RB) series, as well as phenazines and melanin. These chemical compounds are widely generated in IRB; however, these types of electron shuttles have not been probed in terms of how they may impact the Feammox reaction rate. Further investigations are required to evaluate the potential effects that these substances may have on the reaction.

12.3.7 In situ soil nutrient characteristics

The soil nutrient characteristics can have a remarkable impact on the Feammox reaction progress and rate, as well as the ferric iron reduction. The estimated Feammox rates increased from 0.047 to between 0.205 and 0.319 μg N/(g·d) following the addition of chemical and organic fertilizers in paddy soils. The application of organic or ammonium fertilizers induce the release of structural Fe from clay minerals and stimulate the formation of amorphous Fe(Ⅲ) oxides because the fertilizers reduce the soil pH. Long-term fertilizing can significantly modify the fundamental physical and chemical properties of the soil; specifically, it can elevate the total nitrogen and NH_4^+ contents in the soil to different degrees, which increases the abundance of functional Feammox bacteria. Additional dosing with phosphate can facilitate the growth of microorganisms and augment the relative functional gene abundance during the nitrogen removal process. The introduction of N and P also increase the relative abundance of certain species of IRB, such as *Pseudomonas*, *Geobacter*, and *Thiobacillus*, as shown in an investigation related to the Feammox rate acceleration. The higher Feammox reaction rates observed in farmland soils relative to river sediments were partially attributed to the root exudates and rhizodeposits of vegetation cultured in farmland and riparian soils. The enhanced nutrient bio-accessibility of the environment supports the activities of soil microbial communities and further influences the Feammox rate.

Several environmental factors are known to influence the Feammox process described above. Appropriate parameters can maximize the activity and enrichment of the functional colonies, which reduces the duration of their acclimation phase. Further optimization of the cultivation conditions could also make the Feammox reaction more competitive in real sewage treatment. However, the best parameters for different types of Feammox bacteria require further exploration, and molecular ecology-based research regarding Feammox promotion mechanisms is also necessary. Additionally, investigations combining pure bacteria culturing methods with molecular ecology technology may help elucidate detailed Feammox mechanisms and improve our understanding of Feammox promotion processes in natural systems.

12.4 Prospective: use of Feammox for practical large-scale wastewater treatment

Despite the fact that the Feammox reaction was only just recently discovered, this process has garnered significant attention as a new technology for removing nitrogen from sewage. Research efforts regarding this topic have comprised two main approaches. The first involves dosing ferric iron into the existing system to promote the Feammox reaction and ultimately achieve nitrogen removal in a reactor. The second approach focuses on controlling Feammox enrichment to bypass recalcitrant substances and probe whether the Feammox system can use the anode of the bioelectrochemical system (BES) to replace iron as the electron acceptor. In the trial application of Feammox, the inoculum sources included sludge from a wastewater treatment plant, cultivated Anammox sludge, wetland sediment, and the Feammox culturing colony. The colonies were all incubated with low-C/N culture conditions and various hydraulic retention times (HRT). Table. 12-3 summarizes the results from these Feammox applications for ammonium conversion in the laboratory and in autotrophic environments.

In the incubation system, the Feammox process is always concurrent with the Anammox process and nitrate dependent Fe oxidation (NDFO). For example, sludge cultivating experiments demonstrated that there would be some species related to Anammox bacteria retained in the sludge when using NO_2^- rather than ferric iron as the electron acceptor. Following the addition of ferric iron into the Anammox reactor, the excess NH_4^+ can be converted along with the NO_2^- (conversion ratio, $NO_2^- : NH_4^+ = 1.17$). The applicability of the Feammox reaction in the Anammox process was also confirmed in a previous report. The NO_2^- produced during the Feammox process can be further converted via Anammox, wherein the Anammox sludge is used to cultivate the Feammox functional bacteria. In fact, Fe (Ⅱ) and NO_3^- can also serve as matrix substrates for the NDFO process. By introducing NO_3^- regularly during the experiment, the nitrogen removal of the digester (initial total nitrogen = 1036.7 mg/L) reached 90.1% after a 98-day operation. It was determined that the NDFO process can augment the nitrogen removal efficiency and help to recycle the ferric iron. After 62 days of operation, NH_4^+ removal reached 67.6%, NO_3^--N removal reached 58.8%, and the total nitrogen removal was 63.2%.

Table 12–3 Summary of research results from engineered systems or real sewage using the Feammox technique.

Reactor and experimental scene	Substrate	Temperature	pH	Operation conditions	Nitrogen transform effects	Reference
Continuous flow membrane reactor	60 mg/L NH_4^+	25 ℃	4–5	150 days HRT: 2 days	AR: 64.5%	(Huang and Jaffé, 2015)
Anaerobic digestion reactor	$Fe(OH)_3$ and 1375.1 mg/L N	37 ℃	—	40 days	TNR: 20.1%	(Yang et al., 2018)
Fixed–bed reactor	NH_4^+ and Fe(Ⅲ)–EDTA–Na	35 ℃	—	42 days	AR: 30.6%	(Sawayama, 2006)
Anaerobic vials	NH_4^+–N 50 mg/L	32 ℃	6.5–7.0	HRT: 2 days	AR: 59.7% (29.9 mg/L)	(Li et al., 2016)
Biofilm reactor	75 mg/L NH_4^+	25 ± 3 ℃	4.5–5.0	HRT: 10 days	AR: 33.8%	(Wu et al., 2017)
Anaerobic sequencing batch reactor	50 mg/L NH_4^+	15–30 ℃	7.4–7.6	88 days HRT: 1 day	AR: 52.7% (28.4 mg/L)	(Yao et al., 2019a)
Biofilm reactor	Sponge iron	28 ± 1 ℃	—	300 days HRT: 6 h	AR: 42.3% [rate: 32.2 mg N/(L · d)]	(Yao et al., 2020a)
Constructed wetlands	500 mg 2–line ferrihydrite /kg sediment	—	—	—	AR: (25.0 ± 7.3)%	(Shuai and Jaffé, 2019)
Anaerobic vials	5.36 mmol/L iron oxides with domestic and mining sewage	25 ℃	4.5	10 days	Municipal sewage AR: 40.1% [rate: 8.17 g/(m^3 · d)] and 78.2% [rate: 4.15 g/(m^3 · d)]	(Li, 2017; Li et al., 2017)
Immobilizing with magnetic chitosan hydrogel beads	573.2 mg/L $Fe(OH)_3$ and 60 mg/L NH_4^+–N	25 ℃	4.5	16 days	AR: 53.6%	(Liu, 2018; Liu et al., 2018)

AR: ammonium removal. TNR: Total nitrogen removal. HRT: hydraulic retention time.

In addition, the Anammox reaction ultimately achieves nitrogen removal with the same products as Feammox and NDFO. Multiple Feammox reactions are believed to occur in Anammox sludge. Under anaerobic cultivation conditions, where Fe(Ⅲ) was the only electron acceptor in the reactor for 160 days with Anammox sludge acting as the bacterial source, some Anammox bacteria were still active in the sludge. These bacterial groups might play a critical role in the Feammox process. The NH_4^+ conversion rate was greater than 80.0%, and the total nitrogen removal rate reached 71.8% with direct gas production in the reactor where Feammox was the dominant pathway of gaseous nitrogen loss.

However, it is still unclear how the contribution of a single process impacts the levels of elemental nitrogen or the nitrogen removal rate. To date, numerous investigations have partially clarified the various contributions in the mixed system. After 120 days of active sludge acclimation in the ASBR reactor, the Anammox process dominated in the reactor within the first 0–7 h, and it was accompanied by the simultaneously consumption of NO_2^- and NH_4^+. In the subsequent period, the Feammox reaction clearly played a cardinal role, achieving NH_4^+ conversion equal to 12.2 mg/L. Quantitative analysis of the Feammox process indicated that it accounted for approximately 57.7% of the reactivity, and the remainder was attributed to the Anammox process. Nonetheless, the true proportion ascribed to NDFO is not well understood.

Feammox can also play an important role in bioelectrochemical systems. To explore this relationship, Fe_2O_3 powder was added to the system to evaluate the NH_4^+ removal performance by a BES coupled with Feammox. The results confirmed that the combination treatment enhanced the NH_4^+ removal by 50.4% and 38.8% relative to a single BES reactor and a ferric iron reactor, respectively; the highest NH_4^+ removal rate reached (80.62 ± 0.26) g N/(m^3 · d). Additionally, the current increased gradually from 0.05 to 4.30 mA in the BES reactor supplied with Fe(Ⅲ) (i.e., Fe(Ⅲ)–BES). The Fe(Ⅱ)/Fe(Ⅲ) cycle promoted NH_4^+ oxidation because the abundant *Thiobacillus* denitrificans in Fe(Ⅲ) BES could couple with NDFO to perform denitrification. Considering the expenses associated with maintaining a sufficient amount of bioavailable iron for large–scale sewage treatments and sludge disposal, research efforts have shifted to investigating anodes as replacements for the electron acceptor. Previous explorations of the Feammox process already confirmed the affinity of A6 bacteria to the anode by comparing the biomass content of the surrounding soil in wetland systems. It was estimated that the total cell number on the anode surface reached 2.8×10^8 cells/m^2. In this case, the maximum current density reached 376.1 mA/m^2 with pure A6 cultures in a microbial electrolysis cell, which achieved an average NH_4^+ removal of (1.14 ± 0.2) mmol. When the applied voltage was 0.3 V (vs. Ag/AgCl) and AQDS was added, cells adhered to the surface of the anode in the pure strain cultivation. However, there was no biofilm formation, and the majority of the cells were in the liquid. The results using another set of experimental parameters showed that the pure A6 strain produced a current when operating in an MEC with an applied voltage of 0.7 V. This confirmed its ability to use the anode as an electron acceptor in the absence of Fe (Ⅲ). The maximum coulombic efficiency was 16.4%, the current density was 4.2 A/m^3, and the NH_4^+ removal reached 0.52 mmol in the enrichment culture of the MEC, suggesting that the A6 bacteria can function in MEC systems.

The enrichment culture of A6 bacteria demonstrated the capacity to convert halogenated organic compounds to short chain organic acids (acetates) and free ions with a fairly low content of organohalides via the Feammox process. The degradation rates for trichloroethylene (TCE) and tetrachloroethylene (PCE) ranged from 32.0% to 55.0% with a dosage of 1 mg/L. For other organohalides, a 0.12 mmol (50.0%) reduction in perfluorooctanoic acid (PFOA) was detected in the A6 enrichment culture, and a 0.08 mmol (33.0%) reduction was observed for the pure A6 strain. The homogenous experiments with perfluorooctane sulfonate (PFOS) revealed 0.045 mmol (23.0%) reduction of PFOS by the pure A6 strain and 0.093 mmol (47.0%) reduction by the A6 culture. The mechanism governing this phenomenon is still unknown and cannot be explained by existing theories, although it may be linked to the monooxygenase-related genes discovered in the A6 bacteria. Alternatively, it might be a result of the redox situation generated by ammonium and ferric iron, and complicated by the synergistic interaction between A6 and other microorganisms in the enrichment culture during incubation.

The Feammox process can also reduce other high-valent elements besides Fe. The possibility of combining the N cycle with heavy metal immobilization has been proposed because A6 can withstand a surrounding environment containing 10.8–12.5 μmol uranyl acetate. This suggests a positive correlation between the bacterial content and the rate of uranyl consumption in the incubation system. Other projects have also investigated the reduction of selenium (Se) oxyanions via the Feammox process, suggesting that this method was suitable for reducing toxic and mobile Se oxyanions to insoluble or adsorptive forms.

Although Feammox has not yet been applied in wastewater treatment plants, it is certainly a promising approach, and it could be combined with other mature nitrogen removal techniques to provide several unique advantages. Feammox can potentially replace traditional nitrification methods that require an aeration period to oxidize ammonium to NO_3^- for further denitrification. Meanwhile the Feammox process can achieve NH_4^+ oxidation in an anoxic atmosphere, thereby producing NO_2^- and NO_3^-, which can be used in later manipulations. Furthermore, coupling the Feammox process with a BES could reduce the expense significantly compared with traditional strategies. A diagram depicting such a replacement is shown in Figure 12-5.

The Feammox process also has several advantages over Anammox, including its activity over a wide pH range, resistance to low temperatures and potential heavy metal resistance. Feammox can also be combined with BES, which together can significantly reduce the required cultivation period; furthermore, it could be combined with cathodic reduction to expand its substrate scope relative to Anammox.

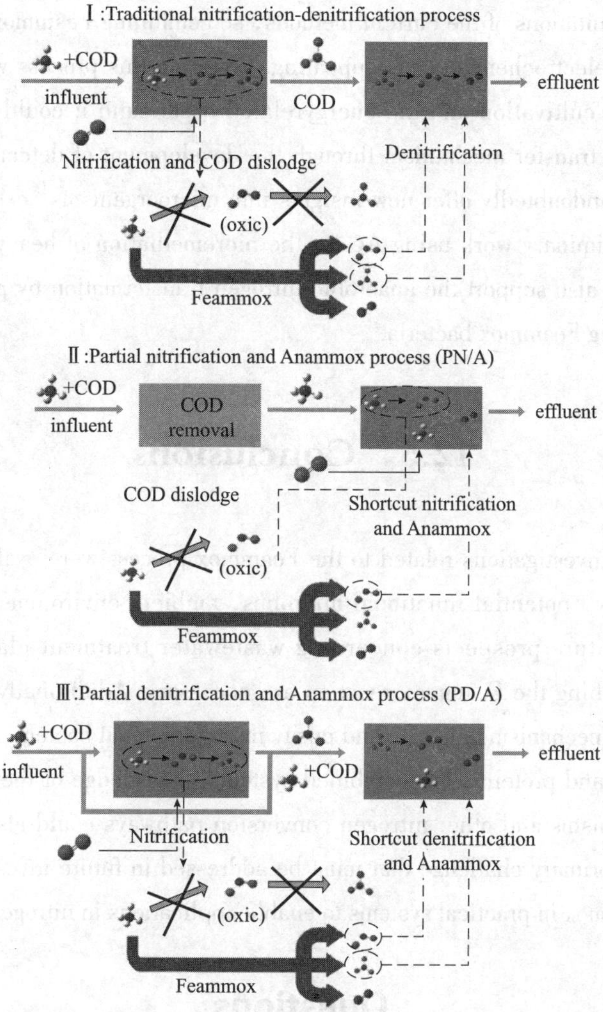

Figure 12–5 Potential future application of Feammox for nitrogen removal in a municipal waste water treatment plant

Furthermore, various conventional techniques and emerging methods are available to remove the Fe (Ⅱ) generated by the Feammox process, although they differ in their cost efficiencies, environmental benefits, and removal rates. For example, oxidation–precipitation filtration processes, the use of calcium carbonate–based minerals, separation through filtering media, and electro–coagulation could all significantly reduce the ferrous iron content in wastewater. Additionally, nanotechnology, such as halloysite nanotubes (HNTs), or membrane technology could also be employed to reduce the Fe (Ⅱ) content in low–C/N wastewater following the Feammox treatment process. Combining these newer technologies with conventional techniques may be the key for dealing with ferrous iron in wastewater.

To apply the Feammox reaction in practical systems aiming to achieve nitrogen removal, it would be beneficial to combine it with other nitrogen transformation processes. The coexistence of distinct nitrogen transformation pathways could enable more complete use of the product and reduce the existing limitations of the current methods, so combining Feammox with other emerging techniques (i.e., bioelectrochemistry) is appealing. Coupling this process with BES may reduce the time required for cultivation and the energyrelated costs, and it could also help clarify the extracellular electron transfer mechanism through the development of detection techniques. Such research efforts will undoubtedly offer new insights into microorganisms' extracellular respiration mechanism. The preliminary work using A6 for the bioremediation of heavy metal pollution and dehalogenation might also support the anaerobic nitrogen transformation by promoting recalcitrant pollutant removal using Feammox bacteria.

12.5 Conclusions

In this review, investigations related to the Feammox process were evaluated and compared. The reaction pathway, potential functional microbes, ambient environmental factors, existing applications, and future prospects concerning wastewater treatment plants were discussed. Continued efforts probing the Feammox reaction are necessary. Additionally, researchers should focus on the reaction mechanism to isolate and purify more functional bacteria and better understand the functional genes and proteins. For combined systems, knowledge of the interactions between Feammox microorganisms and other nitrogen conversion pathways could also be implemented in real applications. A primary challenge that must be addressed in future investigations involves the management of Feammox in practical systems to enable applications in nitrogen removal processes.

Questions

[1] What is the main principle for Feammox process?

[2] What are the characteristics of Feammox bacteria?

[3] What are the natural factors that could affect the growth of Feammox bacteria?

[4] How the pH and temperature affect the Feammox process?

[5] What kind of wastewater could be treated via Feammox-based technology?

References

[1] Ahmed M, Lin L S. Ferric reduction in organic matter oxidation and its applicability for anaerobic wastewater treatment: a review and future aspects [J]. Rev. Environ. Sci. Biotechnol., 2017, 16: 273-287.

[2] Coates J D, Cole K A, Chakraborty R, et al. Diversity and ubiquity of bacteria capable of utilizing humic substances as electron donors for anaerobic respiration [J]. Appl. Environ. Microbiol., 2002, 68: 2445-2452.

[3] Crane R A, Scott T B, Nanoscale zero-valent iron : future prospects for an emerging water treatment technology [J]. J. Hazard. Mater., 2012, 211-212 : 112-125.

[4] Esberg C, du Toit B, Olsson R, et al. Microbial responses to P addition in six South African forest soils [J]. Plant Soil, 2010, 329 : 209-225.

[5] Eusterhues K, Hädrich A, Neidhardt J, et al. Reduction of ferrihydrite with adsorbed and coprecipitated organic matter : microbial reduction by Geobacter bremensis vs. abiotic reduction by Na-dithionite. Biogeosciences, 2014, 11 : 4953-4966.

[6] Feng C, Liu Y, Li Q, et al. Quaternary ammonium compound in anolyte without functionalization accelerates the startup of bioelectrochemical systems using real wastewater [J]. Electrochim. Acta, 2016, 188 : 801-808.

[7] Huang S, Jaffé P R. Characterization of incubation experiments and development of an enrichment culture capable of ammonium oxidation under iron-reducing conditions [J]. Biogeosciences, 2015, 12 : 769-779.

[8] Kuypers M M, Marchant H K, Kartal, B. The microbial nitrogen-cycling network [J]. Nat. Rev. Microbiol., 2018, 16 : 263-276.

[9] Li H, Chen C, Wu Y, et al. Ammonium oxidation under iron-reducing conditions in three different wastewater treatment [J]. Water Purif. Technol., 2017, 36 (9): 14-22.

[10] Li X, Yuan Y, Yuan Y, et al. Effects of salinity on the denitrification efficiency and community structure of a combined partial nitritation- anaerobic ammonium oxidation process [J]. Bioresour. Technol., 2018, 249 : 550-556.

[11] Li H, Su JQ, Yang XR, et al. RNA stable isotope probing of potential Feammox population in paddy soil [J]. Environ. Sci. Technol., 2019, 53 : 4841-4849.

[12] Li J, Li J, Peng Y, et al. Insight into the impacts of organics on Anammox and their potential linking to system performance of sewage Partial Nitrification-Anammox (PN/A): a critical review [J]. Bioresour. Technol., 2020, 300 : 122655.

[13] Li N, Wan Y, Wang X. Nutrient conversion and recovery from wastewater using electroactive bacteria [J]. Sci. Total Environ., 2020, 706 : 135690.

[14] Ni SQ, Ni J, Yang N, et al. Effect of magnetic nanoparticles on the performance of activated sludge treatment system [J]. Bioresour. Technol., 2013, 143 : 555-561.

[15] Novak J T, Verma N, Muller C D. The role of iron and aluminium in digestion and odor formation. Water Sci. Technol., 2007, 56 : 59-65.

[16] Rahman M M, Shan J, Yang P, et al. Effects of long-term pig manure application on antibiotics, abundance of antibiotic resistance genes (ARGs), Anammox and denitrification rates in paddy soils [J]. Environ. Pollut., 2018, 240 : 368-377.

[17] Ren X, Chen Y, Guo L, et al. The influence of Fe^{2+}, Fe^{3+} and magnet powder (Fe_3O_4) on aerobic granulation and their mechanisms [J]. Ecotoxicol. Environ. Saf., 2018, 164 : 1-11.

[18] Rios-Del Toro E E, Valenzuela E I, López-Lozano N E, et al. Anaerobic ammonium oxidation linked to sulfate and ferric iron reduction fuels nitrogen loss in marine sediments [J]. Biodegradation, 2018, 29 : 429-442.

[19] Zhou L, Jiang Y, Wan Y, et al. Electron flow shifts from anode respiration to nitrate reduction during electroactive biofilm thickening [J]. Environ. Sci. Technol., 2020, 54 : 9593-9600.

Chapter 13 Anaerobic Methane Oxidation (AOM)

Emphasis: Classification of anaerobic methane oxidation processes; Functional microorganisms of anaerobic methane oxidation; Microbial metabolism of anaerobic methane oxidation; Physiological characteristics and ecological distribution of the anaerobic methanotrophic microorganisms.

Nut: Microbial metabolism of anaerobic methane oxidation.

13.1 Discovery and classification of anaerobic methane oxidation

13.1.1 Discovery of anaerobic methane oxidation

Methane (CH_4) is a greenhouse gas that is 20 to 30 times more potent than an equivalent amount of carbon dioxide and contributes up to a quarter of global warming, as methane levels increase from natural sources such as the decomposition of organic matter and anthropogenic sources such as the use of fossil fuels. The methane in the natural environment mainly comes from the anaerobic oxidation of organic matter, but most of the methane has been consumed by microbial aerobic and anaerobic oxidation processes before reaching the atmosphere.

The aerobic oxidation of methane is widespread in aerobic environments, where the methane is oxidized to carbon dioxide by aerobic methane-oxidizing bacteria. It has long been thought that methane oxidation is mediated by methane-oxidizing bacteria (mainly *α-Proteobacteria* and *γ-Proteobacteria*) under aerobic conditions, and bacteria obtain energy from this metabolic process. However, microorganisms capable of oxidizing methane under anaerobic conditions had never been discovered. Therefore, for most of the 20th century, many microbiologists insisted that methane was inert under anaerobic conditions. By 1976, it was shown that methane metabolism was closely associated with sulfate-reducing bacteria (SRB) in the anaerobic environment of the Cariaco Trench. In the same year, it was found that in marine sediments, methane concentration gradually decreased from the bottom sediment to the upper water, while sulfate concentration also gradually decreased from the upper water to the bottom sediment, indicating that sulfate might be the electron acceptor of methane in the anaerobic oxidation process. Since the 1980s, a series of studies have shown that methane and sulfate show generally similar distribution characteristics in marine sediments, suggesting that anaerobic methane oxidation (AOM) may occur under the combined action of anaerobic methane oxidation archaea and sulfate-reducing bacteria. By the early 2000s, strong evidence was provided for the hypothesis through stable isotope analysis, gene library construction, and fluorescence in situ hybridization (FISH), confirming that AOM was a biogeochemical process driven by microorganisms. The data showed that AOM consumed more than 90% of the methane produced on the ocean floor and about 50% of the methane produced in freshwater wetlands, limiting the amount of methane that enters the atmosphere. Therefore, AOM plays a very important role in the earth's carbon cycle and is crucial for maintaining global climate stability.

13.1.2 Classification of anaerobic methane oxidation processes

The AOM process can be coupled with a variety of electron acceptors. According to the electron acceptors, AOM can be generally divided into four categories: sulfate-dependent anaerobic methane oxidation (SAMO), denitrifying anaerobic methane oxidation (DAMO), metal-dependent anaerobic methane oxidation (metal-AOM), and other new types of anaerobic methane oxidation. In the ocean, 90% of the methane produced is consumed by the SAMO process. In the deep-sea environment with very low content of sulfate, iron/manganese oxides can replace sulfate as electron acceptor to complete methane oxidation. However, in intertidal ecosystems, the contribution rate of AOM process with nitrate/nitrite as electron acceptor to total AOM can reach 65.6% to 100%, which is an important methane sink in intertidal ecosystems. In recent years, studies have also reported the AOM processes using AQDS and Cr (Ⅵ) as new types of electron acceptors. Therefore, AOM is a redox reaction in which microorganisms take methane as the sole electron donor and sulfate, nitrate/nitrite, metal oxides, etc. as the electron acceptors under anaerobic conditions. Theoretically conjectural electron acceptors have been confirmed continually, and the diversity of electron acceptors makes the AOM process more common. Table 13-1 shows the AOM processes and corresponding Gibbs free energy of different electron acceptors.

Table 13-1 Reaction equations and standard Gibbs free energy for different AOM types

The types of AOM	Electron acceptor	Reaction equation	Gibbs free energy/(kJ/mol)
SAMO	SO_4^{2-}	$CH_4 + SO_4^{2-} \longrightarrow HCO_3^- + HS^- + H_2O$	-16.6
DAMO	NO_3^-	$5CH_4 + 8NO_3^- + 8H^+ \longrightarrow 5CO_2 + 4N_2 + 14H_2O$	-765.0
	NO_2^-	$3CH_4 + 8NO_2^- + 8H^+ \longrightarrow 3CO_2 + 4N_2 + 10H_2O$	-928.0

Continued

The types of AOM	Electron acceptor	Reaction equation	Gibbs free energy/(kJ/mol)
Metal–AOM	Mn^{4+}	$5CH_4+8MnO_4+19H^+ \longrightarrow 5HCO_3^-+8Mn^{2+}+17H_2O$	–991.7
	Fe^{3+}	$CH_4+8Fe^{3+}+3H_2O \longrightarrow HCO_3^-+8Fe^{2+}+9H^+$	–418.3
	Cr^{6+}	$3CH_4+4Cr_2O_7^{2-}+32H^+ \longrightarrow 8Cr^{3+}+CO_2+22H_2O$	–878.8
Other types of AOM	AQDS	$CH_4+4AQDS+3H_2O \longrightarrow HCO_3^-+H^++4AQH_2DS$	–41.0
	SeO_4^{2-}	$CH_4+SeO_4^{2-}+8H^+ \longrightarrow 4Se^0+3CO_2+10H_2O$	—
	$HASO_4^{2-}$	$CH_4+4HASO_4^{2-}+3H^+ \longrightarrow HCO_3^-+4H_2AsO_3^-+H_2O$	–299.6
	ClO_4^-	$CH_4+ClO_4^- \longrightarrow HCO_3^-+Cl^-+H^++H_2O$	–895.9

—: not available.

13.1.3 Functional microorganisms of anaerobic methane oxidation

So far, no pure culture of AOM microorganism has been obtained, so the research on AOM in natural environment are all based on molecular ecology methods. Based on the results of studies on 16S rRNA genes and functional genes, the microorganisms involved in AOM process include methane–oxidizing archaea with sulfate as electron acceptor, methane–oxidizing archaea with nitrate as electron acceptor, methane–oxidizing bacteria with nitrite as electron acceptor, and methane–oxidizing archaea with metal ions as electron acceptor.

Phylogenetic analysis shows that the anaerobic methanotrophic (ANME) archaea belong to the phylum *Euryarchaeota*, which consists of three main groups: ANME–1, ANME–2, and ANME–3. ANME–2 archaea are initially divided into three subgroups: ANME–2a, –2b, and –2c (ANME–2d clade is also found in later research). However, the sequence similarity among the three groups is only 75%–92%, and the phylogenetic distance is large. At the end of the 20th century, lipid analysis, 16S rRNA analysis, and FISH were used to preliminarily identify functional microorganisms involved in the SAMO process: ANME archaea oxidize methane and provide electrons, and SRB receive electrons and reduce sulfate. SRB usually coexist with ANME archaea, the main species are *Desulfosarcina*/*Desulfococcus* (DSS) and *Desulfobulbus* (DBB). ANME–2 and ANME–3 bind with SRB tightly, while ANME–1 archaea bind loosely with SRB. ANME–1 archaea are the earliest identified archaea of AOM. Further studies show that ANME–2 and ANME–3 are also involved in AOM, and ANME–2 closely relate to *Methanosarcinales*, while ANME–3 is closely related to *Methanococcoides*. The 16S rRNA gene sequences of these three groups of methane–oxidizing archaea are significantly different, suggesting that methane–oxidizing archaea evolve in multiple directions. The phylogenetic relationships of ANME–1, ANME–2, and ANME–3 are shown in Figure 13–1.

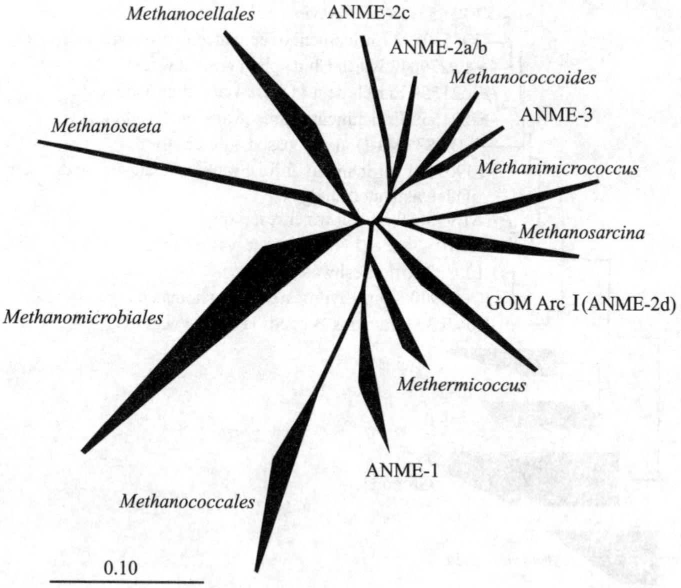

Figure 13-1 Phylogenetic overview of ANME archaea based on 16S rRNA gene sequences

Although the potential of methane as an electron donor for denitrification has long been recognized. But the first microbial evidence of the DAMO process has not been discovered until 2006 from the laboratory enrichment. DAMO microorganisms were successfully detected in the enriched culture by cloning sequencing and FISH. Moreover, sequencing results showed that the culture was mainly composed of a new branch of ANME archaea (GOM Arc I/ANME-2d, Figure 13-1) and bacteria belonging to NC10 phylum which had not yet obtained the pure culture. Due to the slow growth and symbiotic relationship of DAMO microorganisms, no pure culture has been obtained. In a mixed culture of DAMO microbes, NC10 bacteria are wrapped around ANME-2d archaea. However, further studies showed that when NO_3^- was used as an electron acceptor, ANME-2d archaea and NC10 bacteria catalyzed DAMO process together; while when NO_2^- acted as the electron acceptor, NC10 bacteria could catalyze DAMO process alone. Therefore, in the enrichment culture with nitrite as the sole electron acceptor, NC10 bacteria could exist alone in the absence of ANME-2d archaea, which provided conditions for the separate study of nitrite-type AOM. In subsequent studies, the ANME-2d archaeon was named as *Candidatus* Methanoperedens nitroreducens (*M. nitroreducens*). NC10 bacteria are Gram-negative, which are also unculturable to date and are far from related to other phyla bacteria (16S rRNA gene similarity is less than 85%). The 16S rRNA genes of NC10 bacteria are divided into four groups (A, B, C, D) (Figure 13-2), and almost all known NC10 bacteria performing nitrite-dependent AOM belong to group A. The denitrifying anaerobic methane oxidation bacterium was named as *Candidatus Methylomirabilis* oxyfera (*M. oxyfera*).

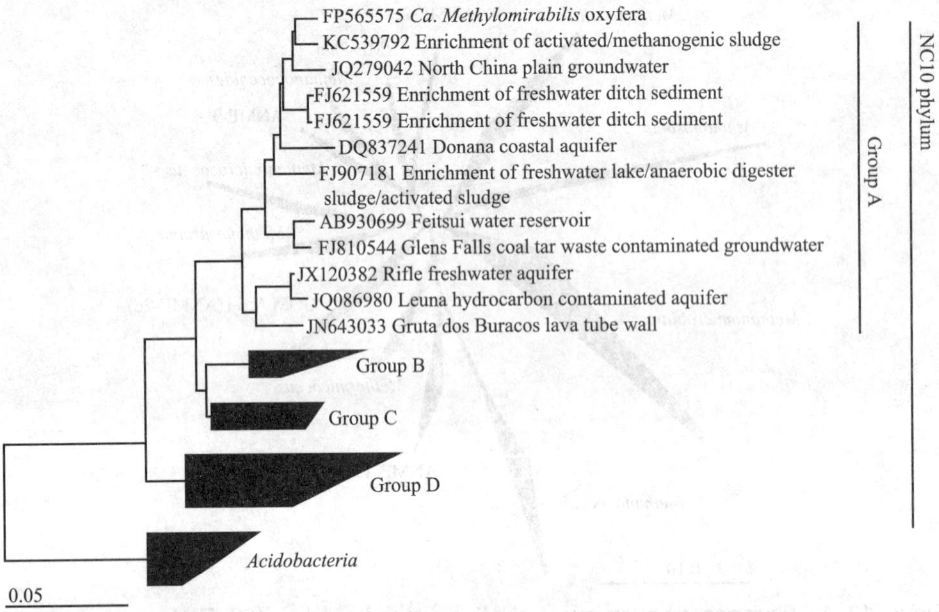

Figure 13-2 Phylogenetic overview of NC10 bacteria based on 16S rRNA gene sequences

Methane-oxidizing archaea using metal oxides as electron acceptors have not been well defined. Although the existence of metal-AOM in nature has been reported in many publications, the realization of metal-AOM using enriched cultures and pure cultures in laboratory is not widely reported. In the pore water of a mud volcano in eastern Taiwan, it has been found that almost all ANME subclasses, including ANME-1, ANME-2, and ANME-3, can participate in metal-AOM. It has also been shown that ANME-2a and ANME-2c in deep-sea sediments can oxidize methane in the presence of soluble ferric citrate and EDTA. The ANME-2d archaea obtained from the sediments of the Twente Canal in the Netherlands can reduce soluble Fe(Ⅲ), ferrihydrite, Fe(Ⅲ) and Mn(Ⅳ) minerals while anaerobically oxidizing methane. It has also shown that the anaerobic methanotrophic archaea related to *M. nitroreducens* are involved in AOM in the presence of iron-rich sediments. However, the above studies are based on mixed cultures. *Methanosarcina acetivorans* C2A, a pure culture isolated from the sediment containing ANME-1, can rely on the reduction of Fe(Ⅲ) for anaerobic methanotrophic growth, and can produce acetic acid in this process. However, it cannot completely oxidize methane to CO_2 under this Fe(Ⅲ)-containing condition. The anaerobic *Methanoperedens ferrireducens* was obtained by using reservoir sediment as the inoculum, iron hydroxide as the electron acceptor, and methane as the electron donor in a lab-scale bioreactor. ^{13}C labeling, ^{57}Fe labeling, metagenomic analysis, and metatranscriptomic analysis showed that the archaea could independently mediate AOM using Fe(Ⅲ) as a terminal electron acceptor, but the mechanisms had not been verified at the molecular level. Besides, the addition of methane and manganese oxides (birnessite) to a bioreactor inoculated with freshwater sediments has been studied. *Candidatus* Methanoperedens manganicus and *Candidatus* Methanoperedens (two new members of *Methanoperedenaceae*) are the main functional microbes in the enrichment culture, which are considered as Mn (Ⅳ)-dependent anaerobic methanotrophs through genomic and transcriptomic analysis, but the mechanism has also not been verified from the molecular level.

In addition, recent studies have identified other archaea involved in the AOM process, such as the thermophilic archaea *Candidatus* Ethanoperedens and the Asgard archaeon. With the further development of the research, more groups of anaerobic methane-oxidizing archaea may be identified by applying the 16S rRNA gene combined with functional gene markers.

13.2 Microbial metabolism of anaerobic methane oxidation

13.2.1 Sulfate-dependent Anaerobic Methane Oxidation (SAMO)

SAMO is the first kind of AOM process being discovered, methane is oxidized to CO_2, while sulfate is reduced to HS^- under anaerobic conditions.

At present, there are three hypotheses supporting the reaction mechanism of SAMO process, namely the reverse methanogenesis pathway, acetogenesis pathway, and methylogenesis pathway.

(1) Reverse methanogenesis pathway. The reverse methanogenesis hypothesis was first proposed in 1994 from in situ and laboratory observations of anoxic sediment samples from Cape Lookout Bight, North Carolina, suggesting that AOM was possible when sulfate was present and H_2 concentration was low enough. Since most of the enzymatic reactions involved in methanogenesis were reversible, a model of reverse methanogenesis pathway was proposed, and it was found that the genome of ANME archaea contained all functional genes in the methanogenesis pathway except the methylene-tetrahydromethanopterin reductase (Mer) gene, a key gene for methane formation. Methylene-tetrahydrofolate reductase (MetF) has a similar structure to Mer, and may be able to replace its function to promote the occurrence of AOM. Therefore, CH_3-CoM is generated from methane under the action of methyl-coenzyme M reductase (Mcr), then methyl is transferred to tetrahydromethanopterin to form CH_3-H_4MPT, which is catalyzed by MetF (similar to Mer) to form CH_2=H_4MPT. The CH_2=H_4MPT is finally converted into CO_2 after a series of enzymatic reactions such as dehydrogenation and hydrolysis. Methane is oxidized and releases electrons, and H_2O acts as an electron acceptor to produce H_2, which is in turn consumed by sulfate-reducing bacteria (SRB). Currently, through metagenomics and proteomics analysis, the reverse methanogenesis pathway is a generally recognized mechanism of SAMO process.

Among them, Mcr is the terminal enzyme in the process of methanogenesis, catalyzing the reduction of coenzyme B (H-S-CoB) and methyl-coenzyme M (CoM-S-CH_3) and the generation of methane and CoM-S-S-CoB. The gene of this enzyme is also present in ANME archaea, catalyzing the reaction of methane with CoM–S–S–CoB to produce coenzyme B (H–S–CoB) and methyl–coenzyme M (CoM–S–CH_3). After comparison, it is found that the phylogenetic relationship between *mcrA* gene (the functional gene of Mcr) and 16S rRNA gene of ANME archaea is very consistent. Therefore, the *mcrA* gene is also developed as a functional gene marker to detect the diversity and distribution of ANME archaea. A large number of subsequent studies have shown that *mcrA* gene contains five subgroups: a, b, c, d, and e. Through comprehensive phylogenetic analysis, it is found that the c and d subgroups correspond to ANME–2 archaea, the a and b subgroups correspond to ANME–1 archaea, and the e subgroup may be corresponding to ANME–3 or other methane–oxidizing archaea.

(2) Acetogenesis pathway. Since some microorganisms do not use H_2 as substrate in the process of methane oxidation but use acetate or methyl–containing compounds, the acetogenesis pathway is proposed as a supplement to the reverse methanogenesis pathway. This pathway contains two reactions. One is that ANME archaea use two molecules of CH_4 to produce H_2 and acetic acid, and SRB then consumes the acetic acid. The second is that ANME archaea use CO_2 to oxidize CH_4 to produce acetate, which is then consumed by SRB. These two reactions are more energetically favorable and consistent with lipid isotope analysis results from previous studies. But in most environments, low acetate concentration and fast reaction rate make detection difficult, thus it is hard to be further confirmed.

(3) Methylogenesis pathway. In 2008, methyl sulfide was found significantly inhibiting the methane oxidation rate, thus a methylogenesis pathway was proposed on this basis. It is saying that ANME archaea use methane and HS^- to generate CH_3SH, which is used by SRB to generate HCO_3^- and HS^-, and the HS^- can be recycled as the substrate of ANME archaea.

With further research, it has been found that ANME archaea can complete the SAMO process independently of SRB, and it is speculated that they reduce sulfate to zero sulfur (S^0) or other sulfides, but the specific metabolic pathway remains to be explored.

13.2.2 Denitrifying Anaerobic Methane Oxidation (DAMO)

DAMO process mainly includes nitrate–dependent anaerobic methane oxidation and nitrite–dependent anaerobic methane oxidation. The former oxidizes methane and reduces nitrate to nitrite through the reverse methanogenesis pathway under the catalysis of ANME–2d archaea, and the latter oxidizes methane and reduces nitrite to nitrogen through an intra–aerobic pathway under the action of *M. oxyfera* bacteria.

(1) The reverse methanogenesis pathway of ANME–2d archaea. Similar to the reverse methanogenesis pathway of SAMO process, Mcr is the key enzyme to activate methane. ANME–2d archaea oxidize methane along the reversed pathway of methanogenesis, releasing electrons for nitrate reduction. As shown in Figure 13–3, methyl–coenzyme M is generated from methane under the action of Mcr, which is then catalyzed by translocating methyl transferase (Mtr) and MetF (similar to Mer) to form $CH_2{=}H_4MPT$, which is finally converted into CO_2 through a series of enzymatic reactions. Among them, Mcr plays a role in activating methane. Mcr and its coenzyme F_{430} are important markers for the identification of reverse methanogenesis. Because ANME-2d archaea only contain the coding gene for nitrate reductase but not the enzymes required in the subsequent steps of denitrification, they could only carry out partial denitrification to reduce NO_3^- to NO_2^-, and need the cooperation of *M. oxyfera* bacteria to further convert the generated NO_2^- to nitrogen.

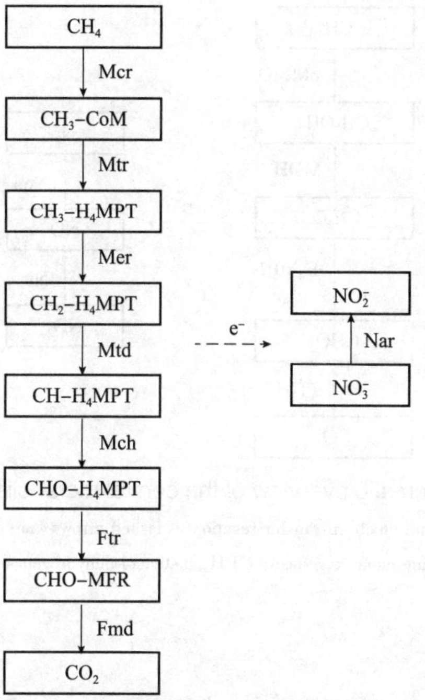

Figure 13-3 Schematic overview of the central metabolism of the ANME-2d archaea

Note: Biochemical (solid arrows) and electron transfer reactions (dashed arrows) are depicted schematically and do not indicate stoichiometries. Mcr, methyl-CoM reductase; Mtr, Na^+ translocating methyl transferase; Mer, methylene-tetrahydromethanopterin reductase; Mtd, methylene-tetrahydromethanopterin dehydrogenase; Mch, methenyltetrahydromethanopterin cyclohydrolase; Ftr, formyl transferase; Fmd, formylmethanofuran dehydrogenase; Nar, nitrate reductase.

(2) Intra-aerobic pathway of *M. oxyfera* bacteria. It was found in a subsequent study on DAMO bacterial enrichment that although the oxygen concentration in the system was consistently below the detection limit (no extracellular oxygen), metagenomic sequencing showed that *M. oxyfera* bacteria contained key enzymes of aerobic methane oxidation, particulate methane monooxygenase (pMMO) and terminal respiratory oxidase, which meant that O_2 was produced in the enrichment culture. Thus an intra-aerobic pathway was proposed for *M. oxyfera* bacteria. Isotopic tracer experiments showed that methane oxidation could only be activated in the presence of nitrite. In this pathway, NO_2^- is first reduced to NO by nitrite reductase, and then 2 molecules of NO are catalyzed to produce N_2 and O_2 under the action of an unknown NO dismutase. 75% of the

O_2 produced is used for CH_4 oxidation, and the remaining 25% is consumed by terminal respiratory oxidase, which is used for normal respiration. CH_4 is oxidized to formaldehyde (HCHO) under the action of various enzymes, and then enters cellular anabolism. Meanwhile, formaldehyde is finally oxidized to CO_2 and H_2O. This intra-aerobic pathway of DAMO bacteria has very important biological significance, indicating that there is the fourth biological oxygen generation pathway on Earth, bypassing the denitrifying intermediate N_2O to directly convert NO to N_2 and O_2, that is, NO dismutase replaces N_2O reductase and catalyzes NO to produce oxygen. However, the NO dismutase in this denitrification process has not yet been discovered (Figure 13-4).

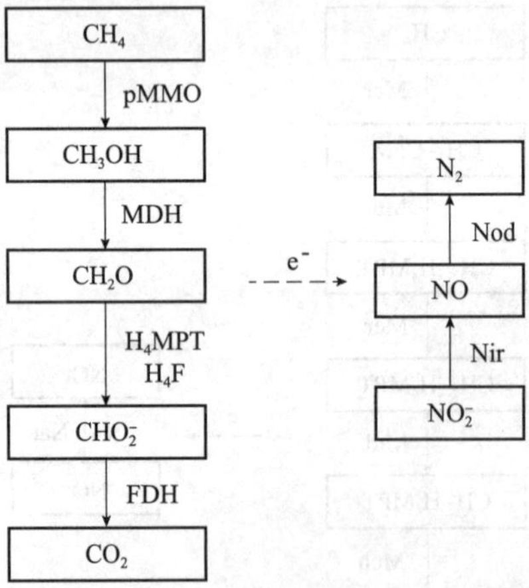

Figure 13-4 Schematic overview of the central metabolism of the bacterium

Note: Biochemical (solid arrows) and electron transfer reactions (dashed arrows) are depicted schematically and do not indicate stoichiometries. pMMO, particulate methane monooxygenase; MDH, methanol dehydrogenase; FDH, formate dehydrogenase; Nir, cd1 nitrite reductase; Nod, NO dismutase.

13.2.3 Metal-dependent Anaerobic Methane Oxidation (metal-AOM)

Iron/manganese-dependent AOM using Fe (Ⅲ) or Mn (Ⅳ) as electron acceptor coupled methane oxidation is the earliest discovered metal-AOM. Sediment samples from the Eel River Basin in California were collected, then sulfate, birnessite, and ferrihydrite were provided, respectively. It was found that the reduction rate of manganese was higher than that of iron, but both of them were lower than that of sulfate. From the perspective of Gibbs free energy, iron/manganese is more suitable than sulfate as the electron acceptor for AOM. However, maybe because iron/manganese oxides are solid which are difficult to be utilized by microorganisms, and the reaction involves the conversion among gas, liquid and solid phases, the metabolic process is more complex than other types of AOM. It has been suggested that methane oxidation in iron/manganese-dependent AOM is also conducted via the reverse methanogenesis (the same as SAMO and DAMO); Fe (Ⅲ) reduction is catalyzed via the multiheme c-type cytochromes (MHCs), and Mn (Ⅳ) reduction depends on the combined action of MHCs and the S layer. Therefore, in the material metabolism of AOM microorganisms, there are not only similar metabolic pathways, but also specific pathways due to the differences in microbial species and electron acceptors. In addition, it has been found that Cr (Ⅵ) can be reduced to Cr (Ⅲ) coupled with methane oxidation, which is a novel metal electron acceptor other than iron/manganese. At present, the electron acceptors of metal-AOM are expanding constantly. Studies have shown that vanadium which also belongs to the transition metals (the same with iron, manganese, and chromium), as well as metalloid arsenic and antimony can be used as electron acceptors for methane oxidation. However, the corresponding metabolic mechanisms need further verification.

Metal-AOM usually occurs in the deep-sea environment where sulfate is absent or present with extremely low concentration, and the microorganisms that drive the reaction have been poorly studied. Researchers have reached different conclusions on the in-situ observation of different environmental samples. The complexity and particularity of environmental samples may be important reason for the changes in microbial metabolism and community structure. In general, the functional microorganisms mediating methane oxidation in metal-AOM are similar to those in SAMO process, while the microorganisms mediating metal oxides reduction are diverse due to different habitats and electron acceptors.

13.2.4 Anaerobic methane oxidation with other new electron acceptors

In addition to the above three types of AOM, researchers have successively discovered AOM using biochar, selenate, chlorate, and electrode as electron acceptors, expanding the range of electron acceptors for AOM. However, since these processes have been discovered in laboratory, it is necessary to further explore the relevant evidence of their involvement in element cycling in natural habitats, to determine the ecological significance and application potential of these novel electron acceptors.

ANME-2 archaea may be the functional microorganisms driving methane oxidation in these new processes. It is found that ANME-2a and ANME-2c have the potential of coupling methane oxidation with AQDS reduction, ANME-2d can use biochar and humus as electron acceptors for anaerobic methane oxidation. The DAMO microbes (DAMO archaea and DAMO bacteria) are found to be able to couple methane oxidation with selenate reduction. These results suggest that ANME-2d archaea involved in nitrate-dependent AOM may have the ability to use a variety of electron acceptors. In addition, it has been found in laboratory enrichment culture that ANME-2d and its bacteria partner can facilitate the occurrence of metal-AOM and show unique catalytic potential for SAMO process. This may be due to the fact that ANME-2d archaea have a variety of MHCs, and thus show an advantage over other ANME groups in using sulfate and even insoluble metal oxides and biochar as electron acceptors. Therefore, the AOM processes with different types of electron acceptors show certain specificity, which require corresponding functional microorganisms to drive the reactions, but it also indicates that ANME archaea have the potential to coexist with a variety of different species of microorganisms.

13.3 Physiological characteristics and ecological distribution of the anaerobic methanotrophic microorganisms

13.3.1 Anaerobic methanotrophic archaea

ANME-1&2&3 archaea mainly distribute in cold spring areas of the ocean. In the freshwater system, they mainly distribute in extreme environments such as craters, ice sheets, hot springs, and so on. These microorganisms are strictly anaerobic and cannot survive in aerobic environment. The main distribution characteristics of ANME-1&2&3 are shown in Table 13-2.

Table 13-2 Overview of the environmental distribution of ANME archaea

ANMEs	Family	Regions discovered
ANME-1	Similar to *Methanosarcinales/Methanomicrobiales*	Volcanic craters, abyssal sea, etc.
ANME-2a/b/c	Belong to *Methanosarcinales*	Cold spring areas, the Black Sea, freshwater wetlands, landfills, etc.
ANME-2d	Belong to *Methanoperedens*	Wetlands, paddy fields, lakes, rivers, wastewater treatment plants, etc.
ANME-3	Similar to *Methanococcoides*	Extreme environments (ice sheet, hot springs, etc.), volcanic craters

SAMO microbes mainly distribute in environments where methane and sulfate coexist such as deep-sea sediments and mud volcanoes, as well as artificial environments such as landfills and sulfur-containing wastewater treatment plants. The functional microorganisms that conduct SAMO process vary greatly in different environments, and the complexity of environmental samples becomes one of the important factors restricting the exploration of the mechanism.

ANME-1&2&3 can mediate the SAMO process in marine or freshwater environments. ANME-1 archaea have a typical rod-shaped cell morphology and can fluoresce spontaneously under UV light, which is a remarkable feature of methanogenic archaea containing coenzyme F_{420}. ANME-1 and ANME-3 can exist either as single cells or as symbiotes with SRB, while ANME-2 archaea often exist as symbiotes with SRB. ANME-1&2&3 are evolutionary branches of methanogenic archaea, but so far, no pure culture of ANMEs has been obtained. ANMEs grow slowly and the doubling time is as long as 2-7 months. Therefore, the enrichment is difficult, which limits the corresponding study of physiological characteristics.

DAMO microbes distribute widely in natural habitats and can be found in wetlands, river sediments, intertidal zones, paddy soils, forest soils, wastewater treatment plants, and marine sediments, indicating that they can adapt to different habitats and contribute significantly to the global carbon/nitrogen cycle. In addition, the researchers explore the biodiversity and activity of DAMO archaea and DAMO bacteria in intertidal ecosystems, and it is found that the contribution of DAMO archaea may be more significant than that of DAMO bacteria. However, more studies are needed in the future to explore the differences in biodiversity and activity of DAMO archaea and DAMO bacteria in habitats, to evaluate the ecological significance of these two kinds of microorganisms.

ANME-2d archaea are also found in nature, including wetlands, paddy soils, lake and river sediments, wastewater treatment plants, and so on. ANME-2d belongs to the family of *Methanoperedens*. They are irregular coccus with a diameter of 1–3 μm. The optimum growth temperature is 22–35 ℃ and pH is 7.0–8.0. Because ANME-2d archaea can only reduce nitrate to nitrite, and cannot continue to reduce nitrite to nitrogen. As a result, ANME-2d archaea tend to coexist with *M. oxyfera* in the environment. But recent research has found that ANME-2d can counteract the toxic effects of high levels of NO_2^-. AOM activity is discovered in the lake sediment with low sulfur concentration and high iron concentration in Denmark, and demonstrate that ANME-2d archaea are responsible for SAMO and/or Fe-AOM. Therefore, ANME-2d archaea not only exist in environments with high nitrate concentration, but also in freshwater environments rich in other electron acceptors, further supporting the importance of ANME-2d for methane emission control in freshwater systems.

ANME-2d archaea also grow slowly and have long doubling time, so they are difficult to enrich and separate for pure cultures. In order to further study the distribution and abundance of ANME-2d in different habitats, it is necessary to improve the detection methods and their sensitivity. Therefore, the researchers have designed a series of primers and probes. Current studies have proved that ANME-2d can exist in various ecological environments through molecular biology and phylogenetic analysis, but there are few studies on the mechanism and related ecological significance. Due to the widespread existence of ANME-2d in freshwater environments, its influence on the carbon cycle in global freshwater ecosystem and greenhouse effect may be seriously overlooked.

13.3.2 Anaerobic methanotrophic bacteria

M. oxyfera, the functional microbe in nitrite-dependent anaerobic methane oxidation (N-DAMO), also spread widely in natural habitats (Table 13-3). In addition to the 16S rRNA gene, many current studies are also based on the functional gene of methane monooxygenase (*pmoA* gene), a key enzyme of aerobic methane oxidation in the N-DAMO process, but there are few studies on the NO dismutase (Nod) which directly convert NO to N_2 and O_2. At present, quantitative and phylogenetic analyses of the functional gene-*nod* have been carried out, but the spatial structure and biochemical reaction mechanism of the Nod enzyme need to be further explored.

Table 13-3 Overview of the environmental distribution of N-DAMO bacteria.

Environment	Location	Nutrient conditions
Freshwater ditch	Ooijpolder, Netherlands Twentekanaal, Netherlands	Eutrophic, high nitrite, high methane
Various wetlands	China	Oligotrophic-eutrophic, temperatures -25-80℃, pH 5-9
Minerotrophic peatland	Brunsummerheide, Netherlands	Oligotrophic, influence by agricultural groundwater
Freshwater lakes	Yunnan Plateau, China	Various levels of reactive N and C
Lake sediment	Lake Biwa, Japan	Mesotrophic
Lake sediment	Lake Constance, Germany	Oligotrophic
River sediment	Qiantang River, China	Eutrophic, polluted by domestic and industrial effluents, high reactive N and P contents
Agricultural soils	Nanjing City, China	Slightly acidic pH, moderate N loading
Paddy field soils	Japan	Fertilized soils
Paddy field soils	China	Fertilized soils, eutrophic, high N and C load
Wastewater treatment plants	Netherlands China	Long sludge retention times, low biological oxygen demand/N rations
Coal tar-contaminated aquifer	South Glens Falls, USA	High CH_4 content, moderate reactive N load, contamination with naphthalene
Contaminated aquifers	Cape Cod, USA Banisveld, Netherlands	Elevated N load, contaminated with leachate and wastewater
Coastal wetland	Mai Po, China	Brackish, eutrophic
Coastal sediments	Xiaogan Island, China	Marine
Estuary	Jiaojiang, East Sea, China	Eutrophic, moderate salinity, high pollution with polycyclic aromatic hydrocarbons
Sea sediments	South China Sea, China	—
Oceanic oxygen minimum zone	Mexico and Costa Rica	Anoxic core zone with detectable CH_4 and NO_2^-

—: not available.

High-throughput and metagenomic sequencing techniques have been used to investigate the community structure, abundance, and ecological value of the unculturable *M. oxyfera* bacteria. At present, the environmental investigations on *M. oxyfera* have been carried out in many different habitats, such as lakes, rivers, coastal and arid lands. By analyzing the effects of environmental factors on *M. oxyfera*, it is found that the diversity of *pmoA* gene increases with temperature. Carbon sources, nitrogen sources, pH, and oxygen concentration can affect *the M. oxyfera* community individually or collaboratively. Microorganisms such as nitrifying bacteria, Anammox bacteria, and methanogenic archaea in the environment can also develop synergistic or competitive interspecific relationships with *M. oxyfera* bacteria.

In addition to the potential contribution of controlling methane emission in natural ecosystems such as paddy fields and wetlands, attention should also be paid to the impact of various nitrogen sources, carbon sources, and antibiotics caused by human activities on the AOM process in wetland and other ecosystems. At the same time, with the progress of various human activities, the original balance of water, salt, and heat in the wetland ecosystem is broken, which caused the stress response of *M. oxyfera* community, the mechanism need to be further studied.

13.4 Enrichment of anaerobic methane-oxidizing microorganisms

Due to the limitation of cultivation techniques, ANME archaea have not been obtained in pure culture. However, through laboratory enrichment, researchers have made important progress in studying the AOM mechanism. ANME archaea cultivation is often more technically complex than other anaerobic microorganisms, and the following considerations need to be taken into account when culturing: ① ANMEs grow very slowly; ② an oxygen-free environment is needed; ③ the water solubility of CH_4 is poor.

The ANMEs enrichment has been conducted with continuous anaerobic bioreactor, which is designed to continuously provide nutrients and remove metabolites. ANME-2a archaea are obtained by continuous high-pressure bioreactor enrichment, proving that they can produce acetic acid through AOM-SR. In addition, a batch bioreactor was used to enrich ANME with the reservoir sediments, and after running for about 220 days, a new species of ANME archaea with coupled iron reduction was discovered. Polyurethane sponge is a porous material that can also be used as bioreactor filler to promote the adhesion and aggregation of biomass, and methane can effectively diffuse through porous sponges. One of the biggest advantages of using bioreactor for enrichment is that it can improve the solubility of CH_4 under high pressure, thus shortening the enrichment time.

When the culture pressure is not high than 0.25 MPa, conventional serum bottles are widely used for enrichment. It was found in serum bottles that manganese-dependent AOM was catalyzed by ANME-1/*Methanococcoides*/ANME-3 together with bacteria. Researchers incubated arsenic-contaminated wetland soils from 7 areas of China for 7-20 days in dark, and found that arsenate reduction caused by AOM accounted for 26%-49% of the total arsenic release in soils. Phylogenetic analysis showed that the enriched ANMEs were mainly ANME-1 and ANME-2, while ANME-3 was hardly detected. In the study of humus-dependent AOM, culture bottles were also used for cultivating the mangrove wetland sediments. It was found that humus could promote AOM and metal reduction, and could also act as an electron acceptor for AOM. By controlling different environmental conditions such as temperature, salinity, and alkalinity, multiple parallel experiments can be conducted in batch experiments. Batch experiments are relatively easy to control and operate, especially for slow-growing microorganisms such as ANMEs that need to maintain strict anaerobic conditions.

13.5 Application potential of anaerobic methane oxidation

In engineering applications, methane as the simplest hydrocarbon, is also a potential energy material. Methane recovery from anaerobic wastewater treatment and sludge digestion is an important way to realize closed-loop operation and energy self-sufficiency of sewage treatment plants, as well as measure to achieve carbon emission reduction goal. Microbial methane oxidation coupled with different electron acceptor reduction is the basis of pollutant removal using AOM process.

The advantage of AOM process is that the methane produced by anaerobic wastewater treatment and side-flow sludge digestion can provide electron donors for AOM microorganisms. Improving the efficiency of gas collection in the anaerobic section and digester can provide sufficient carbon sources for AOM, and can avoid secondary pollution caused by the addition of chemicals in traditional biological wastewater treatment due to insufficient carbon sources. At the same time, the utilization of methane by AOM can also help mitigate the greenhouse effect caused by methane emission, with both energy and economic benefits. The shortcomings of AOM process are mainly due to slow microbial growth and substrate utilization. However, coupled with other processes, the pollutant removal rate can be greatly improved. For example, DAMO-Anammox is used as a post-treatment unit for wastewater nitrogen removal. SAMO and metal-AOM can be used to treat sulfur/heavy metal-containing wastewater. There are also research works using AOM to degrade organic matter (methyl orange). Due to the lack of understanding of the functional microorganisms and reduction mechanism of AOM coupling with other electron acceptors, the application of AOM to pollution control is still in the exploration stage.

Due to the potential extracellular electron transport capacity of AOM microbes, the use of microbial fuel cells (MFC) to cultivate AOM microbes or co-culture with other electrogenic microbes can achieve electrical energy recovery, overcoming the bottleneck of methane transport and storage. However, affected by the reactor configuration, microbial species, and environmental factors, there are great differences in the electricity generation performance. The first DAMO-MFC system demonstrated only weak power generation (maximum voltage of 25 mV). In the follow-up study, engineering bacteria and *Geobacter* were used to form mixed culture, which greatly improved the electrical performance of the system and kept approaching the power density of non-gas substrate MFCs. In the future, more in-depth research is needed on the stability of electricity generation performance and conditions optimization for AOM microbes.

Questions

[1] How many types can AOM be generally divided according to the electron acceptors? Please describe them.

[2] Based on the current research results of 16S rRNA and functional genes, what are the main functional microorganisms involved in the AOM process?

[3] Please describe the reverse methanogenesis pathway of ANME archaea and intra-aerobic pathway of *M. oxyfera* bacteria.

[4] What are the considerations need to be taken into account when culturing the ANME archaea?

[5] Please analyze the application potential of AOM process.

References

[1] Reeburgh W S. Methane consumption in Cariaco Trench waters and sediments [J]. Earth and Planetary Science Letters, 1976, 28 (3): 337-344.

[2] Barnes R O, Goldberg E D. Methane production and consumption in anoxic marine-sediments [J]. Geology, 1976, 4 (5): 297-300.

[3] Welte C U, Rasigraf O, Vaksmaa A, et al. Nitrate-and nitrite-dependent anaerobic oxidation of methane [J]. Environmental Microbiology Reports, 2016, 8 (6): 941-955.

[4] Raghoebarsing A A, Pol A, Van De Pas-Schoonen K T, et al. A microbial consortium couples anaerobic methane oxidation to denitrification [J]. Nature, 2006, 440 (7086): 918-921.

[5] Hoehler T M, Alperin M J, Albert D B, et al. Field and laboratory studies of methane oxidation in an anoxic marine sediment: evidence for a methanogen-sulfate reducer consortium [J]. Global Biogeochemical Cycles, 1994, 8 (4): 451-463.

[6] Valentine D L, Reeburgh W S. New perspectives on anaerobic methane oxidation [J]. Environmental Microbiology, 2000, 2 (5): 477-484.

[7] Moran J J, Beal E J, Vrentas J M, et al. Methyl sulfides as intermediates in the anaerobic oxidation of methane [J]. Environmental Microbiology, 2008, 10 (1): 162-173.

[8] Haroon M F, Hu S, Shi Y, et al. Anaerobic oxidation of methane coupled to nitrate reduction in a novel archaeal lineage [J]. Nature, 2013, 500 (7464): 567-570.

[9] Beal E J, House C H, Orphan V J. Manganese- and iron-dependent marine methane oxidation [J]. Science, 2009, 325 (5937): 184-187.

[10] Fu L, Li S W, Ding Z W, et al. Iron reduction in the DAMO/*Shewanella oneidensis* MR-1 coculture system and the fate of Fe(Ⅱ) [J]. Water Research, 2016, 88: 808-815.

[11] Yang S, Lv Y, Liu X, et al. Genomic and enzymatic evidence of acetogenesis by anaerobic methanotrophic

archaea [J] . Nature Communication, 2020, 11 (1) : 3941.

[12] Cai C, Leu A O, Xie G J, et al. A methanotrophic archaeon couples anaerobic oxidation of methane to Fe (Ⅲ) reduction [J] . Isme Journal, 2018, 12 (8) : 1929–1939.

[13] Shi L D, Guo T, Lv P L, et al. Coupled anaerobic methane oxidation and reductive arsenic mobilization in wetland soils [J] . Nature Geoscience, 2020, 13 (12) : 799–805.

[14] Ding J, Lu Y Z, Fu L, et al. Decoupling of DAMO archaea from DAMO bacteria in a methane–driven microbial fuel cell [J] . Water Research, 2017, 110 : 112–119.

Chapter 14 Hydrogen-based denitrification

Emphasis: H_2-based denitrification, membrane biofilm reactor, control factors, reactor type, biofilm management.

Nut: Control factors

14.1 Introduction

Hydrogen (H_2), the most abundant element in the universe, has a wide range of applications, including serving as a critical raw material in chemical industries, a reducing agent, an energy carrier, and a fuel. Various technologies have been employed for hydrogen production, encompassing biologically biomass dark or photo fermentation (utilizing agricultural and municipal wastes, algae, etc.), renewable energy sources (such as solar and wind power), fossil fuels (coal, oil, natural gas), and nuclear energy. H_2 can act as an electron donor for microbial respiration, contingent upon the redox potential of the system. Theoretically, all oxidized pollutants can be reduced by H_2. H_2 can be utilized for the bio-reduction of oxidized pollutants in water and wastewater, including NO_3^--N, ClO_4^-, BrO_3^-, Cr (Ⅵ), SeO_4^{2-}, $H_2AsO_4^-$, U (Ⅵ), and others. Notably, nitrate is a significant pollutant that can lead to water body eutrophication, prompting increased attention and the study of various methods for its removal from water and wastewater.

The practical application of many gaseous substrates is limited by their low aqueous solubility, despite their efficiency and versatility. In activated sludge processes, for example, continuously bubbling air is frequently used to supply sufficient oxygen for microbial metabolisms to degrade pollutants. However, this aeration method is energy-intensive and can remove volatile organic compounds (VOCs) and greenhouse gases (e.g., methane and nitrous oxide) from the wastewater, rendering it unsuitable for applying H_2 as an expensive fuel. A more energy-efficient approach to supply dissolved gases to microorganisms is needed, such as membrane aeration. The membrane's lumen can be pressurized with a gas, which diffuses through the membrane wall to a biofilm attached the membrane's outer surface, referred to as the membrane biofilm reactor (MBfR) process.

H_2-MBfR is characterized by utilizing H_2 as the electron donor for the autotrophic denitrifying bacteria growing on the membrane, which has the economic advantages without the involvement of organic carbons to achieve efficient nitrogen removal over the conventional heterotrophic denitrification. During this process, H_2 can be delivered in a bubble-free form through diffusion across the nonporous wall of the membrane as a safety control.

H_2 is one of the most thermodynamically favorable electron donors for denitrification. Besides, the high diffusivity of H_2 in biofilms supports the high denitrification activity to remove NO_3^-–N in the attached-growth systems. The residual H_2 due to its low solubility in water (0.182% v H_2/v H_2O at 20 ℃ and 1 atm) can be easily removed by air stripping. However, the relatively high cost, low solubility, and safe handling and storage of H_2 limit its utilization in full-scale plants.

14.2 Fundamental of H_2-based denitrification

H_2 is a clean energy source. Unlike the common denitrification using organic substances as an electron donor, which results in the production of CO_2 and gaseous N_2O that contribute to global warming, the microbial metabolism of H_2 produces water and N_2 in the hydrogenotrophic denitrification [Eq. (14-1)]. However, practical application of this autotrophic denitrification has been hindered by the low delivery efficiency of H_2 to microorganism as well as the low metabolic activity of the denitrifying microorganisms due to the low solubility of H_2 in water. In addition, the challenges of low biomass retention under anoxic conditions and safe operation of H_2 need to be addressed for the full-scale implementation of hydrogenotrophic denitrification.

$$2NO_3^- + 5H_2 \longrightarrow N_2(g) + 4H_2O + 2OH^- \quad (14\text{-}1)$$

Producing H_2 in situ has emerged as a novel process to overcome the limitations associated with H_2 diffusion, which can provide a performance advantage to support the attached-growth systems, leveraging high diffusivity of H_2 through biofilms to promote nitrate reduction. The technologies including membrane, electrochemical and syntrophic metal corrosion have been developed for the in-situ production of H_2. Particularly, MBfR is most commonly applied because of its several benefits.

Figure 14-1 shows the processes in an H_2-based MBfR. In this system, H_2 is transported to the biofilm through the gas-preamble membrane, while nitrate or nitrite is transported from the water to the biofilm phase. The denitrifying bacteria that attach to the membrane utilize H_2 as an electron donor and reduce nitrate or nitrite to nitrogen in the biofilm. Several key processes govern the denitrification in the H_2-based MBfRs, including the H_2 and nitrate diffusion rate and the metabolism of denitrifying bacteria. Using gas-permeable membranes for hydrogenotrophic denitrification is an effective method for improving H_2 delivery efficiency and reducing explosion risks through micro-bubble or bubble-less aeration. This type of H_2 supply is referred to as in situ-like H_2 production, as the gas-permeable membranes serve as both H_2 diffusers and biofilm carriers, maximizing the availability of H_2 for denitrifiers.

The biofilms in MBfRs exhibit distinct characteristics compared to conventional biofilms, primarily due to the counter-diffusional behaviors of electron donors and acceptors as depicted in Figure 14-1 (b). In H_2-based MBfRs, the counter-diffusional biofilms involve the transport of nitrate from the bulk liquid into the biofilm, while H_2 is supplied and diffuse through the membrane into the biofilms. The co-existence of H_2 and NO_3^- allows autotrophic denitrification in a concentration-dependent manner in biofilm. This counter-diffusional profile also enables the formation of diverse microbial niches in the biofilm based on microscopic observations and mathematical modeling, which is beneficial for nutrient removal. Moreover, these biofilms demonstrate increased resilience to toxic shocks or inhibitory compounds, as the microorganisms in the inner regions of the biofilms can maintain relatively high activity with protection against external environmental stresses. However, there are several critical challenges in controlling biofilm thickness and aeration rate regarding counter-diffusion management to optimize MBfR performance.

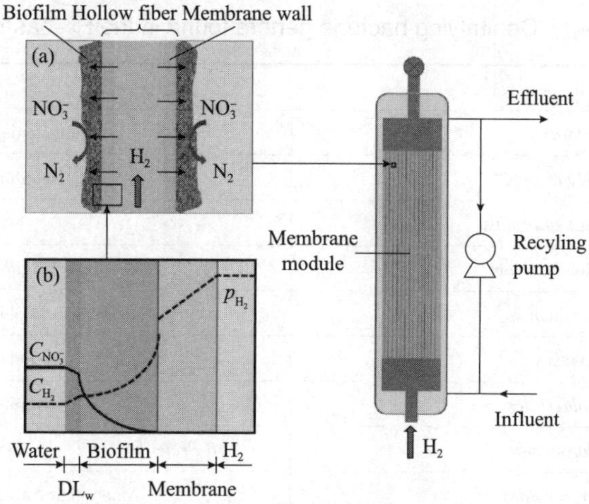

Figure 14-1 Principle of H_2-based MBfR (Wu, et al., 2018)

14.3 Microorganism involved in H_2-based MBfRs

Denitrifying bacteria are characterized by a diverse phylogenetic distribution across various genera. The dominant genera commonly found in H_2-based MBfRs are listed in Table 14-1. While many denitrifying bacteria are affiliated with the γ-subclass of the *Proteobacteria* phylum, they are also present in the α- and β-subclasses. Consequently, in H_2-based MBfRs, α-, β- or *γ-Proteobacteria* are frequently prevalent for denitrification. Notably, *Paracoccus denitrificans* (Table 14-1) has been extensively documented for its role in H_2-driven denitrification, alongside Flavobacteria and Sphingobacteria.

Additionally, pure cultures of *Ralstonia metallidurans* and *Rhodocyclus* sp. have been employed for hydrogenotrophic denitrification in biofilm reactors. *Pseudomonas* and *Acinetobacter* have been reported as the dominant genera in mixed cultures of hydrogenotrophic denitrifiers.

However, when multiple electron acceptors (NO_3^-–N, ClO_4^-, and SO_4^{2-}) are present, the genera with broader electron-accepting capabilities, particularly those in ε-*Proteobacteria* class, may become more prominent. Typical heterotrophic denitrifiers include *Pseudomonas* and *Azonexus*, while *Hydrogenophaga* is a genus of facultative denitrifying bacteria. Notably, autotrophic denitrifiers encompass genera such as *Thiobacillus*, *Paracoccus*, *Ferrobacillus*, and *Leptothrix*. Some well-known autohydrogenophs include *Paracoccus denitrificans*, *Alcaligenes eutrophus*, *Pseudomonas pseudotiara*, *Pseudomonas pseudoflava*, *Ochrobactrum anthropic*, *Pseudomonas stutzeri*, *Paracoccus pantotrophus*, and *Rhodopseudomonas sphaeroids*. Furthermore, *Aquiflexum balticum* and *Marinobacter hydrocarbonoclasticus*, known for their salt tolerance and high salinity selectivity, have been identified in H_2–based MBfRs for Ⅸ brines treatment.

Table 14–1 Denitrifying bacteria genera found in the H_2–based MBfRs.

Denitrifying bacteria genera	
Acidovorax sp.	Also able to degrade phenol
Acinetobacter sp.	Heterotrophic
Alcaligenes eutrophus	
Aquiflexum balticum	A member of *Sphingobacteria*, salt tolerant
Aquaspirillum sp.	An oligotrophic denitrifier
Azonexus	Heterotropic
Burkholderiales	Heterotrophic
Dechloromonas	*β-Proteobacteria*, able to reduce ClO_4^-, autotrophic
Dehalococcoides	Autotrophic, able to reduce TCE
Geothrix	Able to reduce iron–
Hydrogenophaga	Autotrophic
Hydrogenophilales	Autotrophic
Marinobacter hydrocarbonoclasticus	*γ–Proteobacteria*, high salinity selective, for Ⅸ brines
Methyloversatilis	Autotrophic
Paracoccus sp.	*Proteobacteria*, aerobic
Propionivibrio	A genus of *Rhodocyclaceae*
Pseudomonas sp.	Hydrogenotrophic, *γ–Proteobacter*
Rhodobacter	A genus of *Rhodocyclaceae*
Rhodocyclaceae	*α-*, *β-Proteobacteria*
Rhodocyclales	Autotrophic
Rhodocyclus	A genus of *Rhodocyclaceae*
Rubrivivax	Nonsulfur purple bacteria
Sphaerotilus	Sulfur–oxidizing
Sulfuricurvum	Sulfur–oxidizing, autotrophic
β-Proteobacteria HTCC379	

14.4 The key control factors

14.4.1 Membrane materials

The membranes play a crucial role in the functionality and cost-effectiveness of MBfRs for H_2-based applications. The permeability of membranes significantly impacts hydrogen diffusion and biofilm growth, making membrane design a critical factor in the development of H_2-based MBfRs. An ideal membrane for MBfRs should be able to withstand high hydrogen pressures, facilitate easy attachment of biofilm, and remain cost-effective. Microporous membranes allow the high gas transfer rates due to the significantly greater gas transfer rate compared to dense materials. However, to prevent bubbling and the loss of gas and biofilm consequently, microporous membranes must be operated at lower intramembrane gas pressures. These materials are also more prone to clogging and wetting, as conditioning films can render pores hydrophilic and fill them with water. The common microporous membrane materials utilized in MBfR studies include polyethylene, polyvinylchloride, polypropylene, and polysulfone.

The initial application of MBfRs utilized composite membranes, which comprised a thin, dense membrane layer coated on a second microporous membrane layer, facilitating rapid gas transfer and maintaining a high bubble point. However, they are generally more expensive than conventional membranes. The non-porous layer is designed to promote a high-pressure operation for hydrogen diffusion without forming bubbles.

Dense membranes, on the other hand, have the advantage of resistance to clogging or wetting problems Although the diffusional resistance of gas is relatively high in dense membranes, higher intramembrane pressures can be applied to overcome the diffusion limitations. Subsequently, the development of MBfRs sought to reduce membrane costs by employing lower-cost porous hollow fiber membranes and non-porous silicone tubes due to their efficient gas diffusion and cost-effectiveness. Silicone membrane, known for its relatively high permeability, has been employed, but it is difficult to fabricate such membrane in larger outer diameters. Furthermore, specific membranes have been developed for MBfR applications, such as a polymethyl pentene membrane with a thin membrane wall designed for high gas transfer rates and surface area-to-volume ratios. Another example is a polypropylene membrane with a thickness of 40-60 μm, offering economic advantages due to the manufacturing process. While dense membranes impose greater resistance to gas transfer than microporous materials, they have the benefits of preventing clogging or wetting problems, making them a good option for MBfR applications.

A comparative analysis of the three membranes was illustrated in Figure 14-2, to evaluate their operational pressures, removal fluxes, and acclimation times. The results indicate comparable performance in terms of hydrogen pressures and nitrate fluxes between the porous membrane and silicone tubes and the composite membrane. Despite this, the composite membrane exhibited a better performance in acclimation time. The silicone tubes, while enabling high efficiency in gas diffusion with low cost, require biomass immobilization due to poor bacterial attachment. Novel combined bacterial supports, such as silicone-coated, reinforced fiberglass membrane fiber and silicone-coated ferro-nickel fibrous slag, are developed to address this issue, but it still requires a minimum acclimation period of 60 days.

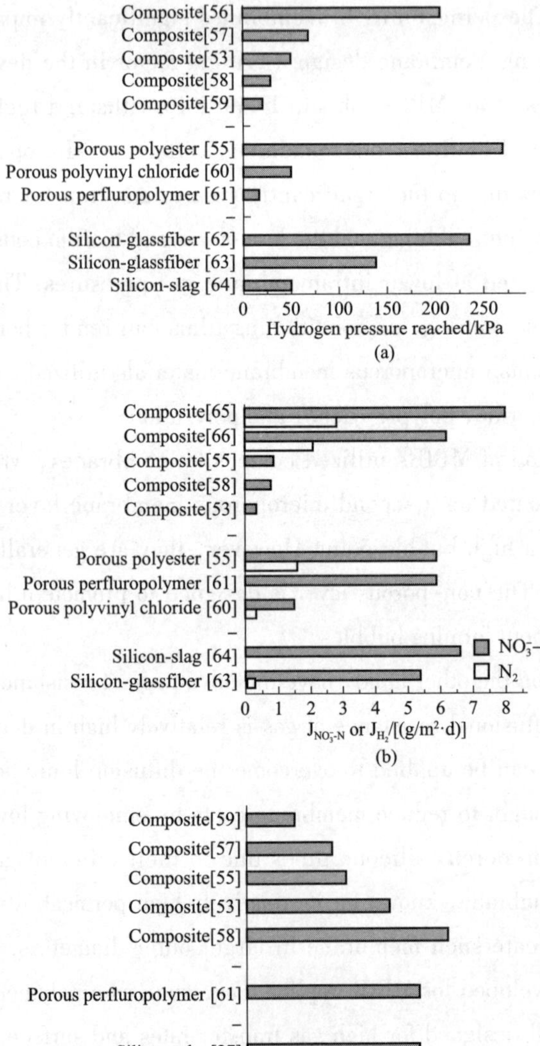

Figure 14-2 Influence of membrane materials on denitrification in H_2-based MBfRs

(a) hydrogen pressure; (b) denitrification rate; and (c) acclimation time

This comparative analysis sheds light on the performance and cost considerations of different membrane materials and structures in H_2-based MBfRs, providing valuable insights for future membrane design and selection in such applications.

14.4.2 Reactor types

Two primary reactor designs have been specifically engineered for H_2-based MBfRs to achieve a fully mixed operation. The tubular type MBfR is a commonly utilized configuration in MBfR, necessitating a recirculation rate ten times larger than the feed flow rate to ensure complete mixing. The recirculation rate significantly impacts system performance, as evidenced by increased denitrification efficiency with higher recirculation rates. Figure 14-3 (b) illustrates a continuous stirred tank reactor (CSTR) with hollow fiber MBfRs installed, where membrane modules are submerged in the reactor and a complete mixing on the membrane surface can be achieved by using a magnetic stirrer placed at the bottom of the reactor. This CSTR-type MBfR has been operated stably for over 300 days and meanwhile has achieved a high denitrification rate without the circulating process, demonstrating its feasibility of integrating into existing wastewater treatment systems. Figure 14-3 (c) shows an in-situ bench-scale MBfR in a simulated wetland, featuring seven bundles of polypropylene fibers sealed into H_2 ports to deliver H_2 with the other fiber ends sealed with glue. Each bundle comprised 12 fibers. A stir bar (325 rpm) was used for mixing in the bottom section.

14.4.3 Biofilm management

The rate of initial biofilm attachment is a critical factor influencing the start-up of MBfRs. The mechanical stability of biofilm upon the attachment plays a pivotal role in maintaining a healthy and evenly biofilm structure. Membrane surface properties influence the attaching process. Increased attachment rates are generally found to be correlated with heightened surface roughness and surface charge (i.e., positive or negative). Additionally, membrane hydrophobicity is a factor, although the evidence is less definitive. Some studies indicate a direct correlation between bacterial attachment and membrane hydrophobicity, while others suggest that a hydrophobic surface does not necessarily lead to maximum adhesion.

Notably, the membranes with rough surfaces, such as microporous membranes, favor microbial attachment. For instance, Suzuki, et al. (2000) observed reduced time for start-up and enhanced performance by abrading silicone membranes with sandpaper to increase the membrane roughness. Furthermore, the membranes woven into fabric or gathered into bundles tend to improve biomass retention. Although the membrane surface remains unaltered, the interstices between membranes, as well as the threads used in weaving, create a heterogeneous surface with many areas protected from water shear. For flat membranes, in which significant biofilm sloughing is prone to occur, attaching a non-woven scrim to the membrane surface encourages biofilm attachment and prevents sloughing within the depth of the scrim.

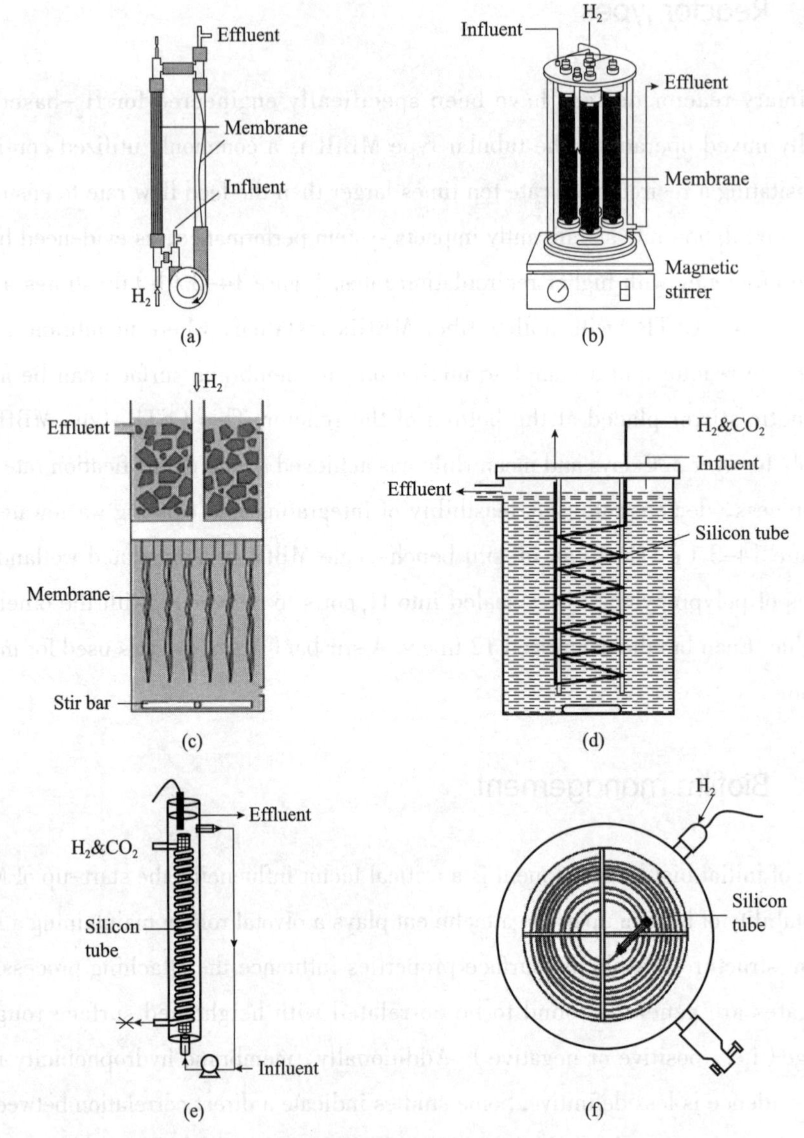

Figure 14-3 Different reactor types of H_2-based MBfRs

(a) a turbular reactor with hollow fibers; (b) a magnetic stirring reactor with hollow fibers; (c) in situ MBfR with hollow fibers; (d) silicon tube wound around a support; (e) a turbular reactor with a silicon tube wound; (f) a spiral wound silicon tube

The electrostatic properties of the membrane surface can be modified to enhance the biofilm's initial attachment and protect the biofilm from shear stress. As bacterial cells are always negatively charged on their surfaces, it has been reported that increasing the positive charge of a polyethylene membrane coupled with optimizing the membrane structure in brush format can significantly improve the initial adhesion of nitrifying bacteria that are known as poor biofilm-forming microorganism. Another approach for promoting attachment involves the seeding of bacteria. For example, the silicone membranes covered with fibrous slag and a mesh layer can be inoculated by nitrifying bacteria before their implementation in a reactor. This approach provides a high surface area and protects the slow-growing bacteria from shear stress. However, if the biofilm is too thick, the nitrifying activity might be negatively impacted.

14.4.4 H_2 pressure

The primary electron donor, H_2, has been shown to significantly impact denitrification rates. Lee and Rittmann (2002) identified hydrogen pressure as the key factor in controlling the denitrification efficiency. A complete nitrate removal can be achieved at high H_2 pressure levels. An increase of H_2 pressure from 3.4 to 65 kPa leads to linearly increase of H_2 flux from 0.4 to 1.3 g/(m^2 · d) through the membrane and thereby the increased nitrate reduction rate. Apart from nitrate, other electron acceptors such as O_2 and NO_2^- might be involved in the redox reaction catalyzed by various bacteria.

Furthermore, H_2 has been observed to influence microbial community structure and biofilm growth in MBfRs. When the availability of H_2 is limited, an increase in denitrifying bacteria abundance can be achieved under higher H_2 pressures. Conversely, by limiting the H_2 supply, biofilm growth decreases and fiber membrane performance diminishes. In addition, when H_2 is not sufficiently supplied, incomplete denitrification and nitrite accumulation might occur due to that nitrite reductase is more sensitive to hydrogen than nitrate reductase.

14.4.5 Nitrate loading

Nitrate reduction was found to be significantly influenced by nitrate loading. Under a constant hydrogen pressure, an increase in nitrate loading leads to higher effluent concentrations and reduced removal efficiency. Specifically, a sharp decrease in total nitrogen removal can reach up to 80% when the nitrate loading increases to 1.36 g N/(m^2 · d) for groundwater treatment.

Moreover, an increase in nitrate loading can cause an immediate increase in denitrification, producing strong base, which in turn inhibits the denitrification. Interestingly, nitrite accumulation can be observed in this process when a high nitrate loading rate of 192 g N/(m^2 · d) is applied, suggesting a complex relationship between nitrite accumulation and nitrate loadings and hydrogen pressures.

14.4.6 pH

As the conversion of nitrite to nitric oxide theoretically produces alkalinity, which can lead to the precipitation of many ions. It has been identified that the optimal pH for denitrification is 7.7–8.6, with maximum efficiency at 8.4. However, a low or high pH can retard the denitrification process and lead to the accumulation of intermediates such as NO_2^-–N, NO, and N_2O. It has been reported that denitrification is inhibited at pH > 7.8, while complete denitrification can be achieved at neutral pH. This suggests that an alkaline environment of high pH values is detrimental to the growth of denitrifying bacteria.

Adding inorganic acid or carbon dioxide has been suggested as a means to maintain an optimal pH for denitrification. However, inorganic precipitates tend to form under high pH conditions, which negatively impact biological denitrification processes. For H_2–based MBfRs, in addition, the buildup of mineral solids inside the biofilm and at the biofilm–membrane interface increases mass–transport resistance for H_2 diffusion and decreases the nitrate and nutrient transport rate. The calcification of fibers, induced by the deposition and accumulation of $CaCO_3$ on the surface of the biofilm, causes fiber breakage. The net change in alkalinity results from the balance between denitrification which increases the alkalinity and precipitation which reduce the alkalinity, which significantly depends on the presence of CO_2. For the systems with high carbonate concentration and high pH, precipitation can lead to a net negative change in alkalinity. Chemical control of foulants such as reacting EDTA with cations can prevent precipitation, but it should be cautious that EDTA is poorly degradable in conventional wastewater treatment processes. The long–term effects of mineral solid precipitation on H_2–based MBfR performance require further exploration.

14.5 Applications of H_2–based MBfRs

H_2–based MBfRs have been utilized for treating various water, including groundwater, domestic wastewater, aquaculture water, and ion exchange (IX) brine (Table 14–2). Notably, these systems have the advantage of simultaneous removal of nitrate and perchlorate from groundwater. Zhao, et al. (2014) studied two pilot–scale H_2–based MBfRs that treat groundwater with perchlorate levels of 200 μg/L and nitrate levels of 9 mg N/L, achieving notable removal of perchlorate to as low as 4 μg/L and complete elimination of nitrate. Furthermore, the oxygen of high concentrations in the groundwater did not adversely impact the MBfR's performance. In other complex field applications, H_2–based MBfRs also have the potential to achieve complete nitrate reduction without the accumulation of nitrite or nitrous oxide.

For the treatment of ion exchange (IX) brine, which is commonly used for removing perchlorate and nitrate from groundwater, high removal fluxes of nitrate and perchlorate were observed in H_2-based MBfR, comparable to those in MBfRs for treating freshwater. In aquaculture wastewater treatment, there are elevated levels of nitrite, nitrate, and phosphorus. The denitrification-aeration system (DAS) with H_2-based MBfR incorporated yields high-quality effluent that meets the standards for recirculation.

Bench-scale investigations of hydrogenotrophic denitrification for wastewater treatment with H_2-based MBfR have demonstrated a high degree of purification with effluent nitrate concentrations below 10 mg/L under controlled pH and membrane fouling conditions. Moreover, by coupling H_2-based MBfR with conventional wastewater treatment plants (WWTPs) for removing nitrogen, the total nitrogen concentrations of well below 2.0 mg/L in the effluent can be achieved, suggesting its potential as an alternative approach for conventional activated sludge processes.

Table 14-2 H_2-dependent autotrophic or mixotrophic denitrification.

Electron donor	Source	Reactor*	Influent	T/ ℃	Electron acceptor	N removal rate/ [kg N/(m^3 · d)]	N removal /%
H_2	aeration	HF-MBfR	SW[a]	25	NO_3^-	1.20	93.8
H_2	aeration	ASMBR	SW	25-28	NO_3^-	0.11	100
H_2	aeration	HF-MBfR	SE[b]	20	NO_3^-	0.35	97
H_2	aeration	HF-MBfR	SW	—[d]	NO_3^-	0.33	91.5
H_2	aeration	MDPBR	SE	30/15	NO_3^-	0.25/0.24	97.6/96.5
H_2	aeration	PHRR	SW	25	NO_3^-	1.06	94.8
H_2 (e^-)	electrodes	MEC	SW	30	NO_3^-	0.02	98
H_2	electrodes	MEC	SW	25	NO_3^-	0.04	97
H_2	electrodes	MEC	SW	30	NO_3^- / NO_2^-	0.88×10^{-4}	88/85
H_2	electrodes	BECW	SW	25-30	NO_3^-	0.17	78.9
H_2 (acetate)	electrodes	MFC	SW	22	NO_3^-	0.08	91.2
H_2 (acetate)	electrodes	MFC	SW	22	NO_3^-	0.01	86.7
H_2 (COD)	electrodes	MFC	PE[c]	-10-30	NO_3^-	—	76.2
H_2/sulfur	electrodes	MFC	SW	10	NO_3^-	1.21	96.6
H_2 (acetate)	electrodes	MFC	SW	30	NO_3^- / ClO_4^-	—	86.0
H_2 (sulfide)	electrodes	MFC	SW	20-30	NO_3^-	0.02	96.6
H_2 (sulfide)	electrodes	MFC	SW	30	NO_3^-	0.03	38.2

[a] Synthetic wastewater; [b] Secondary effluent from municipal wastewater treatment plant; [c] Primary effluent; [d] Data unavailable; * HF-MBfR: Hollow-fiber membrane biofilm reactor; ASMBR: Anaerobic submerged membrane bioreactor; MDPBR: Membrane diffusion packed-bed bioreactor; PHRR: Pressurized high-rate reactor; MEC: Microbial electrolysis cell; UPBR: Up-flow pecked-bed reactor; BECW: Bioelectrochemically-assisted constructed wetland; MFC: Microbial fuel cell.

Although there are many advantages in MBfR over traditional biofilm treatment processes, the concerns of inconsistent performance consistency and low cost-effectiveness need to be addressed for its scale-up. To maintain consistent performance, the MBfR should contain sufficient amounts of active biomass to grow on the membrane surface and at the same time avoid membrane fouling. This is challenging, as the biofilm needs to be maintained at an optimum thickness for strong metabolic activity, but properly controlling biofilm thickness is difficult. One potential approach is to use mechanical shear and gas sparging to balance biofilm growth and detachment. However, it is largely unknown how the detachment affects the microbial community structure which is crucial for stable reactor operation, especially in multispecies MBfRs.

Additionally, the cost-effectiveness of the MBfR is an important factor in its large-scale application, including the membrane capital costs and life expectancy. Membrane manufacturers, material scientists and environmental engineers should collaborate to develop appropriate membrane materials and modules for specific applications. The performance of the membranes should be tested under realistic water/wastewater conditions, and their cost-effectiveness and manufacturing feasibility should be considered.

Gas back-diffusion is another area that needs further study, as it also impacts reactor performance. The rate of back-diffusion, especially in a biofilm that produces gaseous species, should be determined. Moreover, for dead-end membranes, the operational strategy such as the frequency and duration of gas purges shall be determined.

Finally, many other aspects other than gas diffusion affect the choice of process design in the full-scale application of MBfR. For instance, the settling characteristics of MBfR solids, especially for hybrid applications, should be investigated together with the biofilms. When membrane-based post-filtration is applied, the potential membrane fouling caused by MBfR-treated effluents should be thoroughly investigated for optimal overall system performance.

Questions

[1] What is the mechanism of MBfR?

[2] Please list four main factors of MBfR and describe how to achieve a high denitrification performance.

[3] Please compare the advantage and disadvantages of H_2-based denitrification with other autotrophic denitrification.

[4] Please analyze the application potential of H_2-based denitrification.

References

[1] Ahn C H, Oh H, Ki D, et al. Bacterial biofilm-community selection during autohydrogenotrophic reduction of

nitrate and perchlorate in ion-exchange brine [J] . Applied Microbiology and Biotechnology, 2009, 81 (6) : 1169-1177.

[2] Celmer-Repin D, Hwang J H, Cicek N, et al. Autotrophic nitrogen-removing biofilms on porous and non-porous membranes [J] . Environmental Technology, 2010, 31 (12) : 1391-1401.

[3] Celmer D, Oleszkiewicz J, Cicek N, et al. Hydrogen limitation – a method for controlling the performance of membrane biofilm reactor for autotrophic denitrification of wastewater [J] . Water Science and Technology, 2006, 54 (9) : 165-172.

[4] Chang C C, Tseng S K, Chang C C, et al. Reductive dechlorination of 2-chlorophenol in a hydrogenotrophic, gas-permeable, silicone membrane bioreactor [J] . Bioresource Technology, 2003, 90 (3) : 323-328.

[5] Cheng J, Long M, Zhou C, et al. Long-term continuous test of H_2-induced denitrification catalyzed by palladium nanoparticles in a biofilm matrix [J] . Environmental Science & Technology, 2023, 57 (32) : 11948-11957.

[6] Chu L, Wang J. Comparison of polyurethane foam and biodegradable polymer as carriers in moving bed biofilm reactor for treating wastewater with a low C/N ratio [J] . Chemosphere, 2011, 83 (1) : 63-68.

[7] Chu L, Wang J. Denitrification performance and biofilm characteristics using biodegradable polymers PCL as carriers and carbon source [J] . Chemosphere, 2013, 91 (9) : 1310-1316.

[8] Chung J, Li X, Rittmann B E. Bio-reduction of arsenate using a hydrogen-based membrane biofilm reactor [J] . Chemosphere, 2006, 65 (1) : 24-34.

[9] Chung J, Nerenberg R, Rittmann B E. Bio-reduction of soluble chromate using a hydrogen-based membrane biofilm reactor [J] . Water Research, 2006, 40 (8) : 1634-1642.

[10] Chung J, Nerenberg R, Rittmann B E. Bioreduction of selenate using a hydrogen-based membrane biofilm reactor [J] . Environmental Science and Technology, 2006, 40 (5) : 1664-1671.

[11] Chung J, Rittmann B E. Bio-reductive dechlorination of 1, 1, 1-trichloroethane and chloroform using a hydrogen-based membrane biofilm reactor [J] . Biotechnology and Bioengineering, 2007, 97 (1) : 52-60.

[12] Dahab M F, Lee Y W, Bogardi I. A rule-based fuzzy-set approach to risk analysis of nitrate-contaminated groundwater [J] . Water Science and Technology, 1994, 30 (7 pt 7) : 45-52.

[13] Di Capua F, Pirozzi F, Lens P N L, et al. Electron donors for autotrophic denitrification [J] . Chemical Engineering Journal, 2019, 362 : 922-937.

[14] Drysdale G D, Kasan H C, Bux F. Assessment of denitrification by the ordinary heterotrophic organisms in an NDBEPR activated sludge system [J] . Water Science and Technology, 2001, 43 (1) : 147-154.

[15] Hwang J H, Cicek N, Oleszkiewicz J A. Achieving biofilm control in a membrane biofilm reactor removing total nitrogen [J] . Water Research, 2010, 44 (7) : 2283-2291.

[16] Hwang J H, Cicek N, Oleszkiewicz J A. Long-term operation of membrane biofilm reactors for nitrogen removal with autotrophic bacteria [J] . Water Science and Technology, 2009, 60 (9) : 2405-2412.

[17] Jiang M, Zheng J, Perez-Calleja P, et al. New insight into CO_2-mediated denitrification process in H_2-based membrane biofilm reactor : an experimental and modeling study [J] . Water Research, 2020, 184 : 116-177.

[18] Karataş S, Hasar H, Taşkan E, et al. Bio-reduction of tetrachloroethen using a H_2-based membrane biofilm reactor and community fingerprinting [J] . Water Research, 2014, 58 : 21-28.

[19] Lai C Y, Yang X, Tang Y, et al. Nitrate shaped the selenate-reducing microbial community in a hydrogen-based biofilm reactor [J]. Environmental Science and Technology, 2014, 48(6): 3395-3402.

[20] Lee K-C, Rittmann B E. Applying a novel autohydrogenotrophic hollow-fiber membrane biofilm reactor for denitrification of drinking water [J]. Water Research, 2002, 36(8): 2040-2052.

[21] Lee K C, Rittmann B E. Applying a novel autohydrogenotrophic hollow-fiber membrane biofilm reactor for denitrification of drinking water [J]. Water Research, 2002, 36(8): 2040-2052.

[22] Lee K C, Rittmann B E. A novel hollow-fibre membrane biofilm reactor for autohydrogenotrophic denitrification of drinking water [J]. Water Science and Technology, 2000, 41(4-5): 219-226.

[23] Li H, Sun R, Zhang X, et al. Characteristics of denitrification and microbial community in respect to various H_2 pressures and distances to the gas supply end in H_2-based MBfR [J]. Frontiers in Microbiology, 2022, 13: 1023402.

[24] Li P, Xing W, Zuo J, et al. Hydrogenotrophic denitrification for tertiary nitrogen removal from municipal wastewater using membrane diffusion packed-bed bioreactor [J]. Bioresource Technology, 2013, 144: 452-459.

[25] Liang S, Xiao K, Wu J, et al. Mechanism of membrane filterability amelioration via tuning mixed liquor property by pre-ozonation [J]. Journal of Membrane Science, 2014, 454: 111-118.

[26] Lu C, Gu P, He P, et al. Characteristics of hydrogenotrophic denitrification in a combined system of gas-permeable membrane and a biofilm reactor [J]. Journal of Hazardous Materials, 2009, 168(2-3): 1581-1589.

[27] Martin K J, Nerenberg R. The membrane biofilm reactor (MBfR) for water and wastewater treatment: Principles, applications, and recent developments [J]. Bioresource Technology, 2012, 122(0): 83-94.

[28] Martin K J, Picioreanu C, Nerenberg R. Multidimensional modeling of biofilm development and fluid dynamics in a hydrogen-based, membrane biofilm reactor (MBfR) [J]. Water Research, 2013, 47(13): 4739-4751.

[29] Nerenberg R. The membrane-biofilm reactor (MBfR) as a counter-diffusional biofilm process [J]. Current Opinion in Biotechnology, 2016, 38: 131-136.

[30] Ontiveros-Valencia A, Tang Y, Krajmalnik-Brown R, et al. Managing the interactions between sulfate- and perchlorate-reducing bacteria when using hydrogen-fed biofilms to treat a groundwater with a high perchlorate concentration [J]. Water Research, 2014, 55: 215-224.

[31] Park J Y, Yoo Y J. Biological nitrate removal in industrial wastewater treatment: which electron donor we can choose [J]. Applied Microbiology and Biotechnology, 2009, 82(3): 415-429.

[32] Pavissich J P, Aybar M, Martin K J, et al. A methodology to assess the effects of biofilm roughness on substrate fluxes using image analysis, substrate profiling, and mathematical modelling [J]. Water Science and Technology, 2014, 69(9): 1932-1941.

[33] Shen Z, Wang J. Biological denitrification using cross-linked starch/PCL blends as solid carbon source and biofilm carrier [J]. Bioresource Technology, 2011, 102(19): 8835-8838.

[34] Shen Z, Yin Y, Wang J. Biological denitrification using poly (butanediol succinate) as electron donor [J]. Applied Microbiology and Biotechnology, 2016, 100(13): 6047-6053.

[35] Shen Z, Zhou Y, Hu J, et al. Denitrification performance and microbial diversity in a packed-bed bioreactor

using biodegradable polymer as carbon source and biofilm support [J] . Journal of Hazardous Materials, 2013, 250-251 : 431-438.

[36] Suzuki Y, Hatano N, Ito S, et al. Performance of nitrogen removal and biofilm structure of porous gas permeable membrane reactor [J] . Water Science and Technology, 2000, 41 (4-5) : 211-217.

[37] Tian T, Yu H-Q. Denitrification with non-organic electron donor for treating low C/N ratio wastewaters [J] . Bioresource Technology, 2020, 299 : 122686.

[38] Van Ginkel S W, Kim B O, Yang Z, et al. Effect of NaCl on nitrate removal from ion-exchange spent brine in the membrane biofilm reactor (MBfR) [J] . Water Science and Technology, 2012, 65 (1) : 100-104.

[39] Van Ginkel S W, Lamendella R, Kovacik Jr W P, et al. Microbial community structure during nitrate and perchlorate reduction in ion-exchange brine using the hydrogen-based membrane biofilm reactor (MBfR) [J] . Bioresource Technology, 2010, 101 (10) : 3747-3750.

[40] Van Ginkel S W, Tang Y, Rittmann B E. Impact of precipitation on the treatment of real ion-exchange brine using the H2-based membrane biofilm reactor [J] . Water Science and Technology, 2011, 63 (7) : 1453-1458.

[41] Wan D, Liu H, Qu J, et al. Using the combined bioelectrochemical and sulfur autotrophic denitrification system for groundwater denitrification [J] . Bioresource Technology, 2009, 100 (1) : 142-148.

[42] Wu J, Huang X. Effect Of Mixed Liquor Properties On Fouling Propensity In Membrane Bioreactors [J] . Journal Of Membrane Science, 2009, 342 (1-2) : 88-96.

[43] Wu J, Le-Clech P, Stuetz R M, et al. Novel filtration mode for fouling limitation in membrane bioreactors [J] . Water Research, 2008, 42 (14) : 3677-3684.

[44] Wu J, Li H, Huang X. Indigenous somatic coliphage removal from a real municipal wastewater by a submerged membrane bioreactor [J] . Water Research, 2010, 44 (6) : 1853-1862.

[45] Wu J, Yin Y, Wang J. Hydrogen-based membrane biofilm reactors for nitrate removal from water and wastewater [J] . International Journal of Hydrogen Energy, 2018, 43 (1) : 1-15.

[46] Wu W, Yang L, Wang J. Denitrification performance and microbial diversity in a packed-bed bioreactor using PCL as carbon source and biofilm carrier[J] . Applied Microbiology and Biotechnology, 2013, 97 (6) : 2725-2733.

[47] Wu Y, Li Y, Ontiveros-Valencia A, et al. Enhancing denitrification using a novel in situ membrane biofilm reactor (isMBfR) [J] . Water Research, 2017, 119 : 234-241.

[48] Xing W, Wang Y, Hao T, et al. pH control and microbial community analysis with HCl or CO_2 addition in H_2-based autotrophic denitrification [J] . Water Research, 2020, 168 : 115200.

[49] Xu C, Wang X, An Y, et al. Potential electron donor for nanoiron supported hydrogenotrophic denitrification : H_2 gas, Fe (0) , ferrous oxides, Fe^{2+} (aq) , or Fe^{2+} (ad) ? [J] . Chemosphere, 2018, 202 : 644-650.

[50] Yang G, Wang J. Co-fermentation of sewage sludge with ryegrass for enhancing hydrogen production : Performance evaluation and kinetic analysis [J] . Bioresource Technology, 2017, 243 : 1027-1036.

[51] Yin Y, Wang J. Biohydrogen production using waste activated sludge disintegrated by gamma irradiation [J] . Applied Energy, 2015, 155 : 434-439.

[52] Zhang Y, Zhong F, Xia S, et al. Autohydrogenotrophic denitrification of drinking water using a polyvinyl

chloride hollow fiber membrane biofilm reactor [J] . Journal of Hazardous Materials, 2009, 170 (1) : 203–209.

[53] Zhao H P, Ontiveros–Valencia A, Tang Y, et al. Removal of multiple electron acceptors by pilot–scale, two–stage membrane biofilm reactors [J] . Water Research, 2014, 54 : 115–122.

[54] Zhao H P, Ilhan Z E, Ontiveros–Valencia A, et al. Effects of multiple electron acceptors on microbial interactions in a hydrogen–based biofilm [J] . Environmental Science and Technology, 2013, 47 (13) : 7396–7403.

[55] Zhao H P, Ontiveros–Valencia A, Tang Y, et al. Using a two–stage hydrogen–based membrane biofilm reactor (MBfR) to achieve complete perchlorate reduction in the presence of nitrate and sulfate [J] . Environmental Science and Technology, 2013, 47 (3) : 1565–1572.

[56] Zhao H P, Ontiveros–Valencia A, Tang Y, et al. Removal of multiple electron acceptors by pilot–scale, two–stage membrane biofilm reactors [J] . Water Research, 2014, 54 : 115–122.